Springer Series in Optical Sciences

Founding Editor
H. K. V. Lotsch, Nußloch, Baden-Württemberg, Germany

Volume 244

Springer Series in Optical Sciences is led by Editor-in-Chief William T. Rhodes, Florida Atlantic University, USA, and provides an expanding selection of research monographs in all major areas of optics:

- lasers and quantum optics
- ultrafast phenomena
- optical spectroscopy techniques
- optoelectronics
- information optics
- applied laser technology
- industrial applications and
- other topics of contemporary interest.

With this broad coverage of topics the series is useful to research scientists and engineers who need up-to-date reference books.

Reuven Gordon
Editor

Advances in Near-Field Optics

Springer

Editor
Reuven Gordon
Department of Electrical and Computer Engineering
University of Victoria
Victoria, BC, Canada

ISSN 0342-4111 ISSN 1556-1534 (electronic)
Springer Series in Optical Sciences
ISBN 978-3-031-34741-2 ISBN 978-3-031-34742-9 (eBook)
https://doi.org/10.1007/978-3-031-34742-9

This Springer imprint is published by the registered company Springer Nature Switzerland AG
The registered company address is: Gewerbestrasse 11, 6330 Cham, Switzerland

Paper in this product is recyclable.

Preface

Thirty years after the first Near-Field Optics (NFO) conference, NFO16 was held in Victoria, BC Canada, in 2022. As is customary, a one-day summer school preceded the four-day conference. We invited summer school lecturers that reflected the evolution of NFO, and more generally nanophotonics. The topics covered in that summer school are mainly compiled in this volume. Nearfield optics and nanophotonics makes use of conventional electromagnetics theory, for which there are many techniques particularly well suited to subwavelength structures (Chap. 1—Analytical Methods for Near-Field Optics and Plasmonics). The field has evolved towards a greater understanding of what is possible in terms of the ultimate limits (Chap. 2—Fundamental Limits to Near-Field Optical Response) and what is the quantum nature of extremely confined electromagnetic modes considering losses/gain and open cavities (Chap. 3—Quasinormal Mode Theories and Applications in Classical and Quantum Nanophotonics). Of course, there has been a long history of applying near-field techniques to imaging and spectroscopy that makes use of confinement at the nanoscale (Chap. 4—Probing the Optical Near-Field), and there has also been more recent interest on how active devices such as lasers (Chap. 5—On-Chip Nanoscale Light Sources) and modulators will be made from nanostructured metals, semiconductors and insulators in the years to come. I believe that NFO16 provided a glimpse of where NFO will impact the future. Now that we are seeing individual proteins (as described by plenary speaker Philip Kukura and others at the conference), will the future allow us to see protein structure and the amino acid sequence? Surpassing classical noise limits with quantum light may help us get there (as described by plenary speaker Warwick Bowen). And as the Information Age evolves beyond Moore's law, speeds approaching THz and exploiting quantum nature, it is almost certain that NFO will play a role in these future technologies as well. NFO is the natural convergence of nanotechnology and light, and as such, technological advances are also being seen, in the areas of medicine, energy and metasurface optical elements (as described by plenary speakers Naomi Halas, Teri Odom, and others in the field).This volume is far from comprehensive, but in some ways provides the most complete collection of tools to allow future scientists and engineers to make the most of NFO in providing new technologies to the world. As such, I hope it will be useful to this community and also that it will be improved upon in the years to come.

Victoria, BC, Canada
April 2023

Reuven Gordon

Acknowledgements

SH acknowledges funding by the Natural Sciences and Engineering Research Council of Canada (NSERC), the Canadian Foundation for Innovation (CFI), Queen's University, Canada, and the Alexander von Humboldt Foundation through a Humboldt Research Award.

RB thanks the French Research Agency (ANR) for funding support: ADVANSPEC, POPCORN, STRONG-NANO projects and Graduate School NANO-PHOT (École Universitaire de Recherche, contract ANR-18-EURE-0013). LD gratefully acknowledges the financial supports received from the Air Force Office of Scientific Research (AFOSR) awards n° FA8655-20-1-7039, project POEEMS and the "Plan Cancer" managed by the French ITMO Cancer n° 17CP077–00, project HEPPROS.

Contents

Contributors

Renaud Bachelot Light, Nanomaterials, Nanotechnologies (L2n) Laboratory, CNRS EMR 7004, University of Technology of Troyes (UTT), Troyes Cedex, France

Ludovic Douillard Université Paris-Saclay, CEA, CNRS, SPEC, Gif sur Yvette, France

Sebastian Franke Department of Physics, Engineering Physics and Astronomy, Queen's University, Kingston, ON, Canada
Technische Universität Berlin, Institut fur Theoretische Physik, Nichtlineare Optik und Quantenelektronik, Berlin, Germany

Reuven Gordon Department of Electrical and Computer Engineering, University of Victoria, Victoria, BC, Canada

Qing Gu Department of Electrical and Computer Engineering, North Carolina State University, Raleigh, NC, USA
Department of Physics, North Carolina State University, Raleigh, NC, USA

Stephen Hughes Department of Physics, Engineering Physics and Astronomy, Queen's University, Kingston, ON, Canada

Jacob Kokinda Department of Electrical and Computer Engineering, North Carolina State University, Raleigh, NC, USA

Xi Li Department of Electrical and Computer Engineering, North Carolina State University, Raleigh, NC, USA

Owen D. Miller Department of Applied Physics, Yale University, New Haven, CT, USA

Juanjuan Ren Department of Physics, Engineering Physics and Astronomy, Queen's University, Kingston, ON, Canada

Chapter 1
Analytical Methods for (Near-Field) Optics and Plasmonics

Reuven Gordon

Abstract An overview of methods to analyze optics with an emphasis on near-field and plasmonic applications. Methods include the transfer matrix method for multilayer structures, the effective index method, unconjugated orthogonality for lossy/gain systems, mode matching, quasistatic methods, perturbation theory, and complex coupled mode theory.

1.1 Maxwell's Equations

This section will introduce Maxwell's equations for electromagnetics in media. The goal is to end up with the transfer matrix approach. The transfer matrix approach can be used for all problems of layered media with translational symmetry in two directions, including waveguiding. It can also be generalized to systems with other symmetries (such as rotational), but that will not be discussed here.

Gauss's law for the electric displacement gives:

$$\nabla \cdot \mathbf{D} = \rho_f \tag{1.1}$$

where $\mathbf{D}$ is the displacement vector and ρ_f is the free charge. The boldface denotes a vector. In matter, we write $\mathbf{D} = \epsilon_0 \epsilon_r \mathbf{E}$ as a constitutive relation, which tells you the displacement of the charge in response to the applied electric field $\mathbf{E}$. $\epsilon_0 = 8.85 \times 10^{-12}$ F/m. In general, ϵ_r can be a tensor, which means it can have different responses along different directions, but also have off-diagonal components, which means that there will be a displacement of charge in response to field in an orthogonal direction.

R. Gordon (✉)
Department of Electrical and Computer Engineering, University of Victoria, Victoria, BC, Canada
e-mail: rgordon@uvic.ca

R. Gordon (ed.), *Advances in Near-Field Optics*, Springer Series in Optical Sciences 244, https://doi.org/10.1007/978-3-031-34742-9_1

This relation tells you that the total field coming out of a surface can be related to the free charge contained in the volume surrounded by that surface. This is free charge, not the bound charge of the atoms.

Obeying duality, which is the symmetry under the exchange of electric and magnetic quantities (considering that there is no magnetic charge), we have Gauss's law for the magnetic field:

$$\nabla \cdot \mathbf{B} = 0 \tag{1.2}$$

where $\mathbf{B}$ is the magnetic flux density. By analogy, $\mathbf{B} = \mu_0 \mu_r \mathbf{H}$, where $\mu_0 \epsilon_0 = c^{-2}$ where c is the speed of light in vacuum (3.0×10^8 m/s) and $\mathbf{H}$ is the magnetic field. In the visible and near-infrared region of the spectrum, μ_r is typically close to 1. There is an argument based on kinetic inductance about why this is the case, and there has been some discussion about how this can be contradicted with metamaterials [1], which is of interest for creating negative refractive index [2] and perfect lensing not subject to the usual Abbe limit of diffraction [3].

Faraday's law is given by:

$$\nabla \times \mathbf{E} = -\frac{\partial \mathbf{B}}{\partial t} \tag{1.3}$$

If we have a change in the magnetic flux density through an area will create a voltage in a loop going around that area.

Ampere's law is found by analogy:

$$\nabla \times \mathbf{H} = \frac{\partial \mathbf{D}}{\partial t} + \mathbf{J} \tag{1.4}$$

where we have introduced $\mathbf{J}$ as the current density. The constitutive relation (material response) for the current density is $\mathbf{J} = \sigma \mathbf{E}$, which is the linear relationship called Ohm's law (although the behavior can also be nonlinear in general).

We will assume a time dependence for all these quantities $\mathbf{E} = \mathbf{E_0} e^{-i\omega t}$, where we have adopted the convention of using a negative sign in front of the angular frequency ω. A positive sign is also possible, but the negative sign gives a positive phase for propagation in the forward direction and so this has a nice physical "meaning." We are using le système international (SI) units.

Do we really need four relations for harmonic dependence? Consider taking the divergence of Eq. (1.3). Since the divergence of a curl is zero, this naturally gives Eq. (1.2). Similarly, the divergence of Eq. (1.4) yields the continuity relation:

$$\nabla \cdot \mathbf{J} = -\frac{\partial \rho_f}{\partial t} \tag{1.5}$$

which says that the charge flowing into a surface is equal to the rate at which the charge is increased in the volume enclosed by that surface. So the essential dynamics are captured by Eqs. (1.3) and (1.4). It is not surprising then that the finite-difference time-domain method is one numerical approach to solve Maxwell's equations that relies only on these two equations on a Cartesian grid.

It is easy to show by enclosing a boundary that the continuity relations require the normal components of **D** and **B** to be continuous across the boundary and the tangential components of **E** and **H**. For cases where there is surface charge or surface current (which are idealized scenarios and arguably not physical), the normal component of the displacement vector steps by the surface charge and the tangential component of the magnetic field by the surface current.

Taking the curl of Eq. (1.3) and combining with Eq. (1.4) yields the Helmholtz wave equation (after using some identities):

$$\nabla^2 \mathbf{E} = \frac{\epsilon_r \mu_r}{c^2} \frac{\partial^2 E}{\partial t^2} \tag{1.6}$$

which has plane wave solution:

$$\mathbf{E} = \hat{x} E_0 \exp\left(-i\omega t + ikz\right). \tag{1.7}$$

with $\omega/k = \pm c/\sqrt{\epsilon_r \mu_r}$ as the phase velocity. The refractive index is $\pm\sqrt{\epsilon_r \mu_r}$, for which the negative root is taken when both μ_r and ϵ_r are negative. It is noted that for $\sqrt{\epsilon_r \mu_r} < 1$, the phase velocity is faster than the speed of light in vacuum: this does not violate any relativity principles since the energy travels with the group velocity and not the phase velocity. The group velocity should always be less than the speed of light in vacuum. (There are odd cases where this is apparently violated when there is attenuation, but this results mainly from absorbing/reflecting the back end of a pulse so that it appears that the peak is ahead of the back of the pulse—truly even this case does not have a violation, but I leave it to the interested reader to settle this for themselves [4]).

This is a complex quantity for mathematical simplicity; however, we take the real part when we are considering an actual field (with units volts/meter).

The corresponding magnetic field is found by substitution into Eq. (1.3):

$$\mathbf{H} = \hat{y}\sqrt{\frac{\epsilon_r}{\mu_r}}\frac{E_0}{c\mu_0} \exp\left(-i\omega t + ikz\right). \tag{1.8}$$

It is orthogonal to the electric field and both are orthogonal to the direction of propagation along z.

1.2 Fresnel Relations

Armed with the plane wave solution, it is possible to derive the reflection and transmission coefficients for transverse magnetic (TM, or p-polarized, from a German word which means parallel) and transverse electric (TE, or s-polarized, from the German word senkrecht) waves by the boundary conditions. (Transverse in this context is with respect to the plane of incidence). For TM, we have:

$$R_{\mathrm{TM}} = \frac{H_1^-}{H_1^+} = \frac{\epsilon_2 k_{1z} - \epsilon_1 k_{2z}}{\epsilon_2 k_{1z} + \epsilon_1 k_{2z}} \tag{1.9}$$

where $H_1^{\pm}$ are the magnetic fields of the incident and reflected waves on side 1 with:

$$k_{mz}^2 + k_x^2 = \mu_m \epsilon_m \left(\frac{\omega}{c}\right)^2 \tag{1.10}$$

where k_x is a constant of the problem (must be the same in all the materials) and m is the integer for the side. For the transmission:

$$T_{\mathrm{TM}} = \frac{H_2^+}{H_1^+} = \frac{2\epsilon_2 k_{1z}}{\epsilon_2 k_{1z} + \epsilon_1 k_{2z}}. \tag{1.11}$$

Similarly for TE, we have:

$$R_{\mathrm{TE}} = \frac{E_1^-}{E_1^+} = \frac{\mu_2 k_{1z} - \mu_1 k_{2z}}{\mu_2 k_{1z} + \mu_1 k_{2z}} \tag{1.12}$$

and

$$T_{\mathrm{TE}} = \frac{E_2^+}{E_1^+} = \frac{2\mu_2 k_{1z}}{\mu_2 k_{1z} + \mu_1 k_{2z}}. \tag{1.13}$$

1.3 Transfer Matrix

Using the Fresnel relations, we can solve for the general case where field is incident from both sides for TE:

$$\begin{bmatrix} E_2^+ \\ E_2^- \end{bmatrix} = \frac{1}{2}\begin{bmatrix} 1+\kappa_{21}\eta_{21} & 1-\kappa_{21}\eta_{21} \\ 1-\kappa_{21}\eta_{21} & 1+\kappa_{21}\eta_{21} \end{bmatrix} \begin{bmatrix} E_1^+ \\ E_1^- \end{bmatrix} = \overline{L}_{21} \begin{bmatrix} E_1^+ \\ E_1^- \end{bmatrix} \tag{1.14}$$

where $\kappa_{21} = k_{z2}/k_{z1}$ and $\eta_{21} = \mu_1/\mu_2$. For TM, we have:

$$\begin{bmatrix} H_2^+ \\ H_2^- \end{bmatrix} = \frac{1}{2}\begin{bmatrix} 1+\kappa_{21}\eta_{21} & 1-\kappa_{21}\eta_{21} \\ 1-\kappa_{21}\eta_{21} & 1+\kappa_{21}\eta_{21} \end{bmatrix} \begin{bmatrix} H_1^+ \\ H_1^- \end{bmatrix} = \overline{M}_{21} \begin{bmatrix} H_1^+ \\ H_1^- \end{bmatrix} \tag{1.15}$$

where $\kappa_{21} = k_{z2}/k_{z1}$ and $\eta_{21} = \epsilon_1/\epsilon_2$.

When propagating through a uniform region, the forward and backward waves are decoupled and we can define the transfer matrix as:

$$\overline{\Phi}(l) = \begin{bmatrix} \exp(ik_{mz}l) & 0 \\ 0 & \exp(-ik_{m}l) \end{bmatrix} \tag{1.16}$$

where l is the length of region m.

1.3.1 Example: Surface Plasmon Polariton

The simplest TM example is a single interface where k_{1z} and k_{2z} are imaginary, and we seek a solution that is exponentially bound to the surface so that the exponentially growing solutions are set to zero:

$$\begin{bmatrix} H_1^+ \\ H_1^- \end{bmatrix} = \begin{bmatrix} 0 \\ 1 \end{bmatrix} \tag{1.17}$$

$$\begin{bmatrix} H_2^+ \\ H_2^- \end{bmatrix} = \begin{bmatrix} 1 \\ 0 \end{bmatrix} \tag{1.18}$$

Substituting into Eq. (1.15) gives:

$$k_x = \frac{\omega}{c}\sqrt{\frac{\epsilon_1\epsilon_2}{\epsilon_1+\epsilon_2}} \tag{1.19}$$

This is the dispersion relation for a surface plasmon polariton, which exists at a surface of a metal (medium 2) and a dielectric (medium 1) with $\Re(\epsilon_2) < -\Re(\epsilon_1)$. Including the imaginary part of the permittivity accounts for absorption, which can also be found from Eq. (1.19). $\frac{\omega}{c}$ is sometimes referred to as k_0, the free-space propagation constant.

1.3.2 Example: Short-Range and Long-Range Surface Plasmons

While the surface plasmon polariton allows for an exponentially bound wave, more confinement can be obtained by using metal-insulator-metal (MIM) or insulator-metal-insulator (IMI) structures. These waveguide modes have tight confinement and high losses and are referred to as short-range surface plasmons (SRSPs). It is also possible to have less confinement with lower loss for long-range surface plasmons (LRSPs).

We consider a symmetric situation here:

$$\begin{bmatrix}1\\0\end{bmatrix} = \overline{M}_{12}\overline{\Phi}(l)\overline{M}_{21}\begin{bmatrix}0\\1\end{bmatrix} \tag{1.20}$$

Multiplying out these matrices gives the solution:

$$\tanh\left(\sqrt{k_x^2 - \epsilon_2 k_0^2}\frac{l}{2}\right) = -\frac{\epsilon_2\sqrt{k_x^2 - \epsilon_1 k_0^2}}{\epsilon_1\sqrt{k_x^2 - \epsilon_2 k_0^2}} \tag{1.21}$$

If the middle layer with ϵ_2 is an insulator (dielectric, not metal), then we can verify that as $l \to 0$, $k_x \to A/l$ where $A = \ln\frac{\epsilon_1-\epsilon_2}{\epsilon_1+\epsilon_2}$. Therefore, the magnitude of the propagation constant becomes larger and the wave is more tightly confined in the gap with greater losses. This is a SRSP.

If the middle layer with ϵ_2 is the metal, then we can verify that as $l \to 0$, a permitted solution is $k_x \to \sqrt{\epsilon_1}k_0$. This is the long-range surface plasmon where the wave extends into the dielectric and propagates with low loss.

It is also possible to have a SRSP in the case where the middle layer is a metal and this is found with the odd solution for the magnetic field:

$$\begin{bmatrix}1\\0\end{bmatrix} = \overline{M}_{12}\overline{\Phi}(l)\overline{M}_{21}\begin{bmatrix}0\\-1\end{bmatrix} \tag{1.22}$$

Then the solution becomes:

$$\tanh\left(\sqrt{k_x^2 - \epsilon_2 k_0^2}\frac{l}{2}\right) = -\frac{\epsilon_1\sqrt{k_x^2 - \epsilon_2 k_0^2}}{\epsilon_2\sqrt{k_x^2 - \epsilon_1 k_0^2}} \tag{1.23}$$

When the metal layer becomes thinner, then we can verify that as $l \to 0$, $k_x \to A/l$ where $A = \ln\frac{\epsilon_2-\epsilon_1}{\epsilon_1+\epsilon_2}$, which is the same limit as above since the metal and dielectric regions are switched.

Figure 1.1 shows the relation between ω and k for the SPP, SRSP, and LRSP, including the imaginary part of k which corresponds to attenuation. This is also known as the dispersion relation between energy and moment since photon energy is $\hbar\omega$ and photon momentum scales is $\hbar k$, where $\hbar$ is the reduced Planck constant.

1.3.3 Example: Surface Plasmon Resonance Sensing

Surface plasmon resonance (SPR) sensing considers the leaky wave solution where the magnetic field is only bound on the upper layer, but has propagating parts in

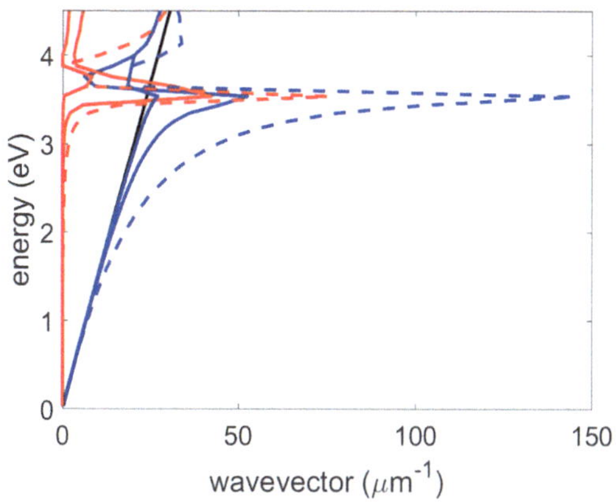

Fig. 1.1 Surface plasmon polariton dispersion relation, where blue is the real part and red is the imaginary part and black is the light line for propagation in water (refractive index 1.33). The solid line is the surface plasmon polariton on a semi-infinite silver slab in water, as given by Eq. (1.19). The dashed and dotted lines are for short-range (Eq. 1.23) and long-range (Eq. 1.21) surface plasmons for silver (permittivity values found in literature [5]) for a 20-nm-thick layer surrounded by material with refractive index of 1.33

the lower layer. The reflection changes as a function of angle incident from the lower layer and typically has a minimum. The reflection minimum can be thought of physically as the light coupling through the gold film and into a surface mode, which is lossy. It is essential that the metal is lossy to have a minimum in reflection. By tracking the change in the curve minimum with angle and/or wavelength, changes in the refractive index and thickness of a layer at the top surface can be sensed.

The reflection is found by solving:

$$B\begin{bmatrix} r \\ 1 \end{bmatrix} = \overline{M}_{43}\overline{\Phi_3}(l_3)\overline{M}_{32}\overline{\Phi_2}(l_2)\overline{M}_{21}\begin{bmatrix} 0 \\ 1 \end{bmatrix} \tag{1.24}$$

where $\overline{M}_{(j+1)j}$ is the generalization of the matrix in Eq. (1.15) between layers with index j and $j+1$ and $\overline{\Phi_j}(l_j)$ is the propogation matrix through layer j with length l_j. B is a constant that is divided out and so not relevant to the solution for the reflection r. Layer $j = 1$ is the overlayer (typically aqueous solution for an SPR sensor), layer $j = 2$ is the adlayer, layer $j = 3$ is the metal layer, and layer $j = 4$ is the substrate layer (typically glass with refractive index around 1.5). For each incidence angle in the substrate layer (θ), we can formulate a $k_x = k_0\sqrt{\epsilon_4}\sin\theta$ and solve for the magnitude of r. The solution is shown in Fig. 1.2, where an angle shift in the minimum of reflection occurs when a thin layer is added to the surface. The 50 -nm-thick gold layer is chosen to minimize the reflection (critical coupling).

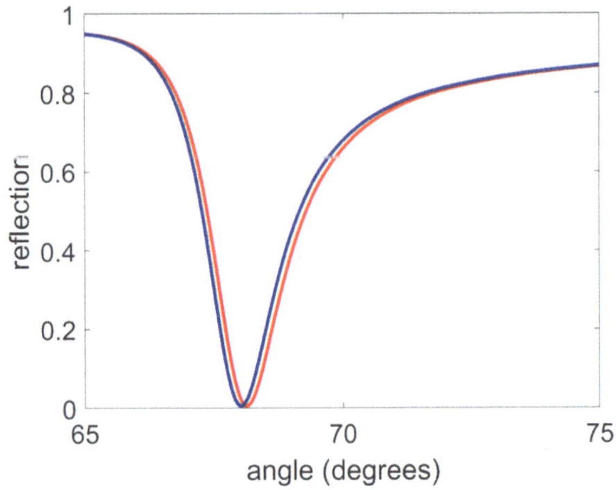

Fig. 1.2 Surface plasmon resonance sensing showing change in reflection intensity (from blue to red) with angle when an adlayer of 1 nm and refractive index of 1.5 is added to a 50 thick gold layer on a 1.5 refractive index substrate and aqueous (refractive index 1.33) overlayer. The permittivity of gold at 760 nm free-space wavelength is taken to be $-20.9125 + 1.2923i$

1.4 Effective Index Method

The transfer matrix approach is suitable for situations where there is translational invariance in two dimensions. This can also be applied to other symmetrical invariance (e.g., rotational invariance) with versions of the transmission matrix approach for cylindrical or spherical systems. If the index varies spatially in two dimensions and it is not possible to separate the variables (e.g., by using a different coordinate system like cylindrical or elliptical coordinates), then in general, computational methods are required to obtain highly accurate results.

The effective index method is used to obtain approximate solutions that provide insight rather than resorting to full numerical approaches. In this method, the refractive index distribution is divided into slices where the propagation constant of each slice is found by assuming translational invariance. As a result, the transfer matrix approach (or equivalent versions for cylindrical or other coordinate systems) can be used. This provides an "effective" index by dividing the propagation constant in each section by the free-space propagation constant. The procedure is then repeated, for the orthogonal polarization, using the effective index values and assuming translational invariance in each slice. The method is approximate because the boundary conditions are not satisfied along the length of each slice; however, so long as the highest index variation and smallest feature directions are solved first, the approximation can be quite accurate. To understand the use of this technique, an example is most instructive.

1.4.1 Example: Rectangular Plasmonic Waveguide

The metal clad dielectric rectangular waveguide propagation can be approximated using the effective index method. The approximation has two parts. In the first part, the TM short-range surface plasmon mode solution above is used to find an effective permittivity for the middle region. In the second part, that effective permittivity is used to approximate the entire region and the waveguide mode is solved for the TE mode.

The metal has a relative permittivity ϵ_1 and the dielectric has a relative permittivity ϵ_2. The narrowest gap is solved first for the TM mode, using Eq. (1.20). The effective relative permittivity of this middle region is given by $\epsilon_{\text{eff}} = k_x^2/k_0^2$, where k_x is found from Eq. (1.20). The effective permittivity of the outer regions are just the actual permittivity of those regions since they are uniform and the lowest order mode is just the normal plane wave. Next, we rotate the problem (so the orthogonal polarization is used) and solve for the TE mode replacing the middle region with a dielectric of relative permittivity $\epsilon_{\text{eff}} = \epsilon_3$. Using this with the matching of Eq. (1.14) gives:

$$\begin{bmatrix} 1 \\ 0 \end{bmatrix} = \overline{L}_{13}\overline{\Phi}(l)\overline{L}_{31}\begin{bmatrix} 0 \\ 1 \end{bmatrix} \tag{1.25}$$

Multiplying out these matrices gives the solution:

$$\tan\left(\frac{\sqrt{\epsilon_3 k_0^2 - \beta^2 l}}{2}\right) = -\frac{\mu_2\sqrt{\beta^2 - \epsilon_1 k_0^2}}{\mu_1\sqrt{\beta^2 - \epsilon_3 k_0^2}} \tag{1.26}$$

where β is the effective propagation constant of the lowest order mode in the x direction.

Figure 1.3 shows the effective index method applied to a silver rectangular aperture. The geometric parameters are chosen to be comparable with fabricated structures that showed a peak in transmission spectra close to the cutoff, as shown in Fig. 1.4 [6]. The resonance peaks are modified by the phase of reflection, which can be calculated by single mode matching [7].

1.5 Single Mode Matching

Mode matching is a generalization of the transfer matrix approach when there is not just a single mode in each region (forward and backward propagating), which is generally true if the permittivity or permeability varies laterally as well. In this case, the modes in each region are solved and matched at the boundary using the orthogonality relation to provide a system of equations for the coefficients of each mode. This is a general and rigorous computational method; however, in structures with subwavelength confinement coupled to an open region, it is often possible to approximate the situation with just a single mode in the region that provides subwavelength confinement, thereby giving an analytic or semi-analytic solution

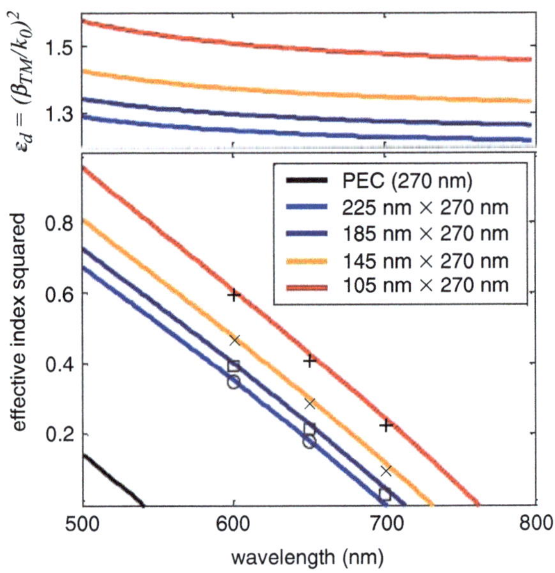

Fig. 1.3 Neglecting losses given by the imaginary part of the permittivity, the effective dielectric constant (top) associated with an MIM structure is then used to calculate the effective index squared of the entire structure (bottom) for different geometries in silver. Numerically calculated values are shown as symbols, as well as the perfect electric conductor (PEC) case. Reprinted with permission from Optica [8]

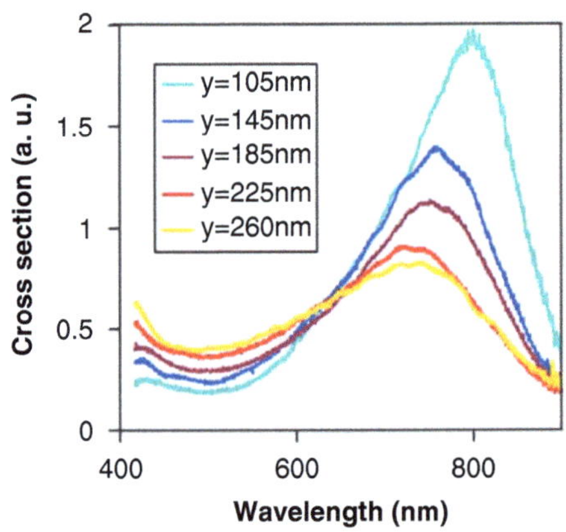

Fig. 1.4 Transmission through a 270-nm-wide aperture in a 300-nm-thick film silver for various widths. Reprinted with permission from Elsevier [6]

(involving solution to a single numerical integration). This gives accurate results if that mode couples predominantly to itself upon reflection and it is the main mode excited by any incident wave.

1.5.1 Orthogonality

The orthogonality relation allows mathematical separation of the modes. In cases where there are losses or gains (expressed by an imaginary part of the permittivity or permeability), the unconjugated orthogonality relation should be used:

$$\iint (\mathbf{E_{tm}} \times \mathbf{H_{tn}}) \cdot d\mathbf{a} = 0 \tag{1.27}$$

If the conjugated form of the orthogonality is used, the overlap integral can be nonzero and not lead to separation of the different modes. In general, the following may be true [9]:

$$\iint \left(\mathbf{E_{tm}} \times \mathbf{H^*_{tn}}\right) \cdot d\mathbf{a} \neq 0 \tag{1.28}$$

For situations of purely real permittivity and permeability, the conjugated orthogonality relation is often used and it will yield the correct results.

1.5.2 Example: Light in a Slit

Consider a slit of width a in a perfect conductor (this may be generalized to other geometries and materials). Figure 1.5 shows the geometry. We consider a plane wave normally incident from $z < 0$ in the plane $z = 0$ where the transverse electric field is matched on the left and right sides:

$$1 + \int_{-\infty}^{\infty} r(k_x) \exp(ik_x x)\, dk_x \simeq th(x) \tag{1.29}$$

where the left-hand side has unity incident plane field and the reflected field is an integral over all the different angles of plane waves written in terms of the x

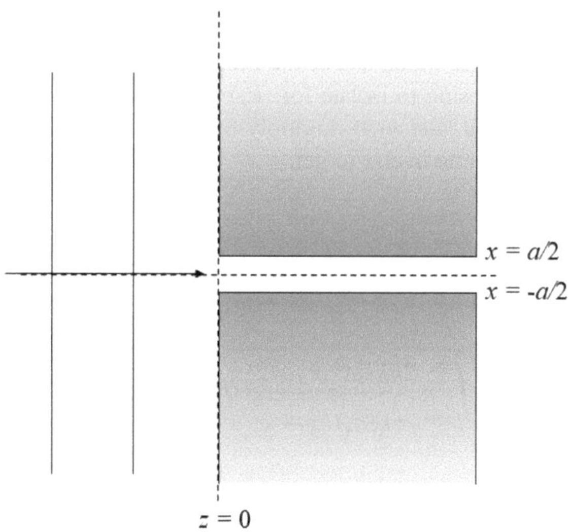

Fig. 1.5 Schematic representation of a slit in a metal, with plane wave normally incident from the left side

component of the field amplitude $r(k_x)$, where k_x is the transverse wavevector. The right-hand side assumes that there is only a single mode with transmission coefficient t and field distribution $h(x)$, which for the case of perfect electric conductor is given by:

$$h(x) = \begin{cases} 1, & \text{if } |x| \leq a/2 \\ 0 & \text{otherwise} \end{cases} \tag{1.30}$$

Similarly, there is an equation for the magnetic field:

$$\frac{1}{Z_0} - \int_{-\infty}^{\infty} \frac{k_0 r(k_x)}{Z_0\sqrt{k_0^2 - k_x^2}} \exp(ik_x x)\, dk_x \simeq \frac{t}{Z_0} h(x) \tag{1.31}$$

where $Z_0 = \sqrt{\frac{\mu_0}{\epsilon_0}}$ is the free-space impedance and the minus sign in front of the integral shows that the direction of propagation has flipped upon reflection (which is chosen to flip for the magnetic field by convention).

To use orthogonality to find an expression for $r(k_x)$, we multiply Eq. (1.29) by the magnetic field distribution $\exp\left(ik'_x x\right)/Z_0$ and integrate over x, switching the order of the integration and using the property:

$$\int_{-\infty}^{\infty} \exp\left(i(k_x + k'_x)x\right) dx = 2\pi\delta(k_x + k'_x) \tag{1.32}$$

where $\delta(k_x)$ is the Dirac delta function, which gives:

$$r(k'_x) = \frac{2t \sin\left(k'_x a/2\right)}{2\pi k'_x} - \delta(k'_x) \tag{1.33}$$

So we have made use of the orthogonality using the magnetic plane waves on the left-hand side to isolate for (k'_x). Now we use the electric field distribution on the right-hand side $h(x)$, multiply by Eq. (1.31) with the expression for $r(k'_x)$ above, and integrate over x to get:

$$t = \frac{2}{1 + I} \tag{1.34}$$

where

$$I = \int_{-\infty}^{\infty} \frac{2k_0 \sin^2(k_x a/2)}{\pi k_x^2 a\sqrt{k_0^2 - k_x^2}} dk_x \simeq \pi(a/\lambda) + 2(a/\lambda)i\left(\ln(2\pi(a/\lambda)) - 3/2\right) \tag{1.35}$$

where λ is the free-space wavelength and the approximation to the integral is for the subwavelength regime [10].

A similar approach is used to derive the reflection of the mode in the slit at the boundary semi-infinite free space:

$$r = \frac{1 - I}{1 + I} \tag{1.36}$$

The total transmission cross section of a film of length l is given by the coherent sum over multiple reflections to yield the Fabry-Perot result:

$$\sigma = a \frac{|t^2|(1 - |r^2|)}{|1 - r^2 \exp{(i2k_s l)}\,|^2} \tag{1.37}$$

where k_s is the propagation constant of the slit, but for the perfect electric conductor, this is just k_0.

1.5.2.1 Single Channel Limit

We consider the limit of a very narrow slit and observe that the cross section $\sigma \rightarrow \lambda/\pi$, which gives half the single channel limit for power transmission. The single channel limit is a general property of scatterers that says the maximum of the total scattering of each channel (i.e., waveguide mode in this case) is given by the wavelength rather than the geometric size. For this extreme subwavelength slit, the single channel is all that is permitted. The reason it is half the expected value is because the back-scattered light also gives the same contribution due to symmetry, and here we have only considered the forward scattering (transmission).

It is possible to exceed the single channel limit by introducing another channel (waveguide mode), with a straightforward extension of this theory [11].

1.5.2.2 Resonance Condition

The reflection coefficient r in Eq. (1.36) has a finite phase which shifts the Fabry-Perot resonances. Furthermore, in a real metal, we should consider changes to the mode shape and k_s. In the situation where the metal is sufficiently good (i.e., the magnitude of the permittivity is sufficiently large to have a field distribution given approximately by $h(x)$), we can approximate the changes to the Fabry-Perot resonances with the approximation to Eq. (1.21). As an example, Fig. 1.6 shows this calculation as compared with experimental data published in the literature.

1.5.2.3 Field Enhancement

It is possible to estimate the near-field intensity in a slit as σ/a, where again losses from k_s should be considered, as well as coupling in and reflection after

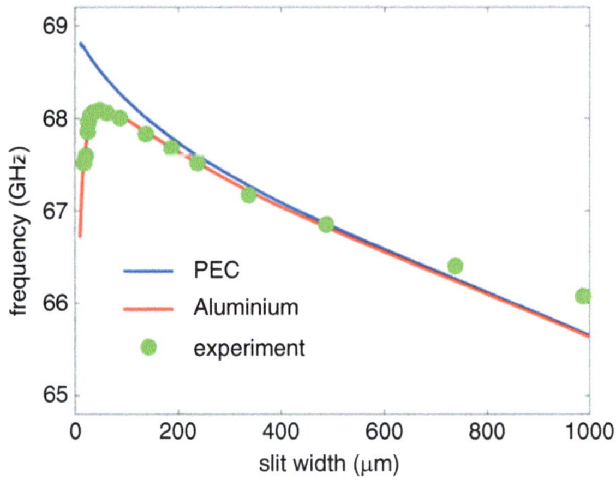

Fig. 1.6 Shift in transmission resonance through slit including finite conductivity of aluminum and comparing with experimental data [12]. Perfect electric conductor case is shown as well to show the impact of loss. Adapted from Ref. [11], ©Optica 2014 with permission

multiple round-trips. Considering the finite permittivity of gold and for a slit length of 150 nm, with alumina in the slit and for frequency of 0.3 THz gives a field enhancement of 1162, which is in good agreement with values found by comprehensive simulations [13].

1.6 Quasistatic Methods

In the case where the structure is much smaller than the wavelength, the spatial derivatives are mainly determined by the material boundaries since they vary faster in space than the wavelength itself. As a result, the time derivatives are negligible with respect to the spatial derivatives and the following is true:

$$\nabla \times \mathbf{E} \simeq 0 \tag{1.38}$$

or

$$\nabla \times \mathbf{H} \simeq 0. \tag{1.39}$$

1.6.1 Rayleigh Particle

Considering Eq. (1.38) for the extreme subwavelength limit, we can write $\mathbf{E} = -\nabla V$, where V is a potential. For a spheroid ellipsoid particle, we can calculate the field as a solution to Laplace's equation, which gives a constant field inside the particle [14, 15]:

$$E = \frac{\epsilon_1 E_0}{A\left(\epsilon_2 - \epsilon_1\right) + \epsilon_1} \tag{1.40}$$

where E_0 is the applied field (assumed to be along the long axis of the spheroid), $\epsilon_1$is the external permittivity, ϵ_2 is the permittivity in the spheroid, and for a prolate spheroid:

$$A = \frac{1-\xi^2}{2\xi^3}\left(\ln\frac{1+\xi}{1-\xi} - 2\xi\right) \simeq \frac{1}{3} - \frac{2\xi^2}{15} - \frac{2\xi^4}{35} + O(\xi^6) \tag{1.41}$$

where $\xi = \sqrt{1 - \frac{b^2}{a^2}}$ and a and b are the major and minor axes and the Taylor expansion is given for small deformations from a perfect sphere.

For an oblate spheroid:

$$A = \frac{1+\eta^2}{\eta^3}\left(\eta - \arctan\eta\right) \simeq \frac{1}{3} + \frac{2\eta^2}{15} - \frac{2\eta^4}{35} + O(\xi^6) \tag{1.42}$$

where $\eta = \sqrt{\frac{a^2}{b^2} - 1}$ and a and b are the major and minor axes and the Taylor expansion is given for small deformations from a perfect sphere.

1.6.1.1 Maximum Local Field at Plasmonic Resonance

The maximum local field of a metal spheroid occurs when the Fröhlich condition is satisfied such that the real part of the denominator of Eq. (1.40) is zero. Assuming the surrounding dielectric is lossless, the field enhancement at the tip of the spheroid is given by:

$$E = \frac{\epsilon_2\left[\epsilon_1 - \Re(\epsilon_2)\right]}{\Im(\epsilon_2)} E_0 \simeq -\frac{\epsilon_2 \Re(\epsilon_2)}{\Im(\epsilon_2)} E_0 \tag{1.43}$$

where the approximation assumes that the permittivity of the dielectric is negligible compared to that of the metal.

Figure 1.7 shows the field intensity enhancement at the plasmonic resonance for silver and gold. It is clear from this figure that the enhancement increases as the wavelength increases, but a very large aspect ratio is also required. A similar result is found in the next chapter for the fundamental material enhancement factor.

1.6.2 Dipole Polarization and Circuit Theory

In the quasistatic limit, the field outside of the spheroid is given by the sum of an incident field (constant field assumed over the region of the subwavelength particle)

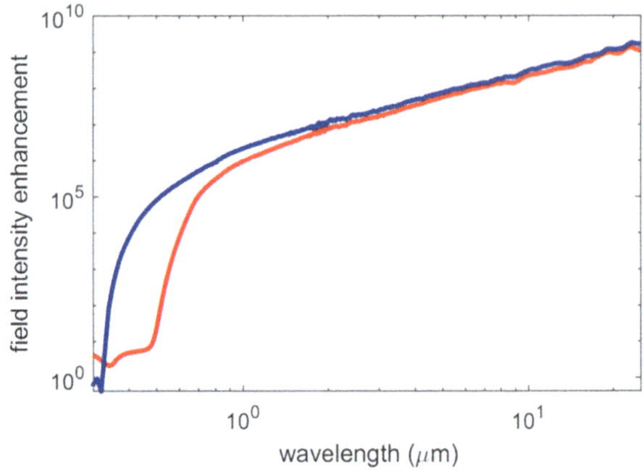

Fig. 1.7 Field intensity enhancement at the plasmonic resonance of a subwavelength spheroid for silver (blue) and gold (red)

and a dipole. Matching the field of the dipole at the boundary gives the polarizability (dipole moment divided by the incident field):

$$\alpha = 2\pi\epsilon_0\epsilon_1 a^3 \frac{(1-A)(\epsilon_2-\epsilon_1)}{A(\epsilon_2-\epsilon_1)+\epsilon_1} \tag{1.44}$$

Considering the field inside the sphere and the field outside the sphere, it is possible to calculate the total displacement current and relate this to the average potential difference. For a sphere, where $A = 1/3$, this gives the effective impedance of [16]:

$$Z_{\mathrm{sph}} = (-i\omega\epsilon_2\epsilon_0\pi a)^{-1} \tag{1.45}$$

and of the fringe fields:

$$Z_{\mathrm{fringe}} = (-i2\omega\epsilon_1\epsilon_0\pi a)^{-1} \tag{1.46}$$

This gives a simple interpretation of the plasmonic resonance as coming from an effective inductance when the permittivity is negative that cancels the capacitance of the fringe field when $\Re\epsilon_2 = -2\epsilon_1$. This may also be used to find the response of Janus particles [16], or the phase of reflection at the end of a nanorod with spherical ends [17].

1.6.3 Retardation and Single Channel Limit

The polarizability derived above was for the quasistatic limit; however, in the absence of losses (no imaginary part for $\epsilon_{1,2}$), the polarizability diverges at the Fröhlich condition and this would result in infinite scattering. As a result, some retardation needs to be added which amounts to a correction [18]:

$$\alpha' = \frac{\alpha}{1 - i\frac{2}{3}k^3\alpha}. \tag{1.47}$$

In the presence of absorption, this is usually negligible. Alternatively, if the loss is negligible, then α diverges, $\alpha' \to i\frac{3}{2k^3}$. The extinction cross section for an isotropic scatterer is $4\pi k \Im\left(\alpha'\right) \to \frac{3\lambda^2}{2\pi}$ where λ is the wavelength in the material. This is the single channel limit for free space, so the maximum scattering of a single channel object is given by the wavelength and not the geometry of the object itself. It is possible to obtain more scattering by adding additional channels, or additional resonances [11, 19, 20].

1.7 Formal Perturbation Theory

Perturbation theory in electromagnetics attempts to solve for the change in the electromagnetic response with a change in the geometry. This theory makes use of known solutions to calculate the change; however, it can be made more accurate by knowledge of, or a good guess for, the local field within the region of perturbation [21].

If we denote the original fields before perturbation as **E** and **H**, and those after perturbation as $\mathbf{E}'$ and $\mathbf{H}'$, we consider the quantity:

$$\mathbf{F} = \mathbf{E}' \times \mathbf{H} - \mathbf{E} \times \mathbf{H}' \tag{1.48}$$

and integrate the divergence of **F** over a volume, and using the divergence theorem [21–23]:

$$\frac{\delta\omega}{\omega'} = \frac{\frac{-i}{\omega'}\iint\left(\mathbf{E}' \times \mathbf{H} - \mathbf{E} \times \mathbf{H}'\right) d\Sigma - \iiint\left(\Delta\epsilon(\omega')\mathbf{E}' \cdot \mathbf{E} - \Delta\mu(\omega')\mathbf{H}' \cdot \mathbf{H}\right) dV}{\iiint\left(\frac{\partial\omega\epsilon(\omega)}{\partial\omega}\mathbf{E}' \cdot \mathbf{E} - \frac{\partial\omega\mu(\omega)}{\partial\omega}\mathbf{H}' \cdot \mathbf{H}\right) dV} \tag{1.49}$$

where Σ is the surface integral around the volume of interest, the second integral in the numerator is only nonzero over the regions of perturbation (where $\Delta\epsilon$ and $\Delta\mu$ are nonzero), and the derivatives in the denominator are evaluated at $(\omega+\omega')/2$. The expression may be generalized to include anisotropic materials in a straightforward way. An imaginary part to ω and $\delta\omega$ allows for losses.

This expression is exact to the first order in $\delta\omega$; however, it may accurately be approximated for small perturbations with $\omega \simeq \omega'$, and $\mathbf{E}'$, $\mathbf{H}'$ can be replaced with **E**, **H** in the denominator only [21, 23]. In many practical cases, the surface integral can be set to zero by use of perfectly matched boundaries terminating in a perfect conductor (so there is no field at the surface):

$$\frac{\delta\omega}{\omega} \simeq \frac{-\iiint \left(\Delta\epsilon(\omega')\mathbf{E}' \cdot \mathbf{E} - \Delta\mu(\omega')\mathbf{H}' \cdot \mathbf{H}\right) dV}{\iiint \left(\frac{\partial\omega\epsilon(\omega)}{\partial\omega}\mathbf{E} \cdot \mathbf{E} - \frac{\partial\omega\mu(\omega)}{\partial\omega}\mathbf{H} \cdot \mathbf{H}\right) dV} \tag{1.50}$$

Key to making Eq. (1.50) accurate is suitable choice for $\mathbf{E}'$ and $\mathbf{H}'$. For small perturbations, these may be replaced with the field calculated by the quasistatic methods of the last section, as has been demonstrated previously [21]. Figure 1.8 shows a calculation done for a plasmonic particle in a photonic crystal waveguide where the near field around the particle was calculated with a fine mesh that would make the entire simulation 12 times faster. Another way to approximate perturbations at surfaces is by making use of continuity so that the tangential field is the same as the unperturbed field, $\mathbf{E}'_{\parallel} \simeq \mathbf{E}_{\parallel}$; however, the normal field is scaled by the ratio of the permittivity values $\mathbf{E}'_{\perp} \simeq \frac{\epsilon}{\epsilon+\Delta\epsilon}\mathbf{E}_{\perp}$. Similar arguments can be made for the magnetic field by the equivalence principle.

We may apply perturbation theory of cavities to waveguides by making a cavity out of a waveguide with two perfect reflectors on either end and a standing wave in the middle. Therefore, there is a relation between the change in the complex propagation constant of the waveguide and the complex frequency of the cavity:

$$\frac{\delta\beta}{\beta} = -\frac{v_p}{v_g}\frac{\delta\omega}{\omega} \tag{1.51}$$

where the group velocity $v_g = \frac{\partial\omega}{\partial k}$ and the phase velocity $v_p = \frac{\omega}{k}$ naturally removes the influence of dispersion (since the waveguide is considered to be at a single frequency). The integrals in Eq. (1.50) extend over the cross section of the waveguide.

1.8 Coupled Mode Theory

1.8.1 Complex Coupled Mode Equations

Coupled mode theory is used to formulate the coupling between known orthogonal modes due to a perturbation $\Delta\epsilon$ (which can be generalized to include anisotropy and magnetic response perturbations as well). Using the unconjugated form of the orthogonality relation allows for the inclusion of complex permittivity, which includes lossy and gain materials that may be used to include perfectly matched layers and thereby truncate the calculation domain [9]. The transverse electric and magnetic field at each position along the propagation direction z may be expressed as:

$$\mathbf{E}_{\perp} = \sum_{m} \left(a_m(z) + b_m(z)\right) \mathbf{e}_{\perp,m}(x, y) \tag{1.52}$$

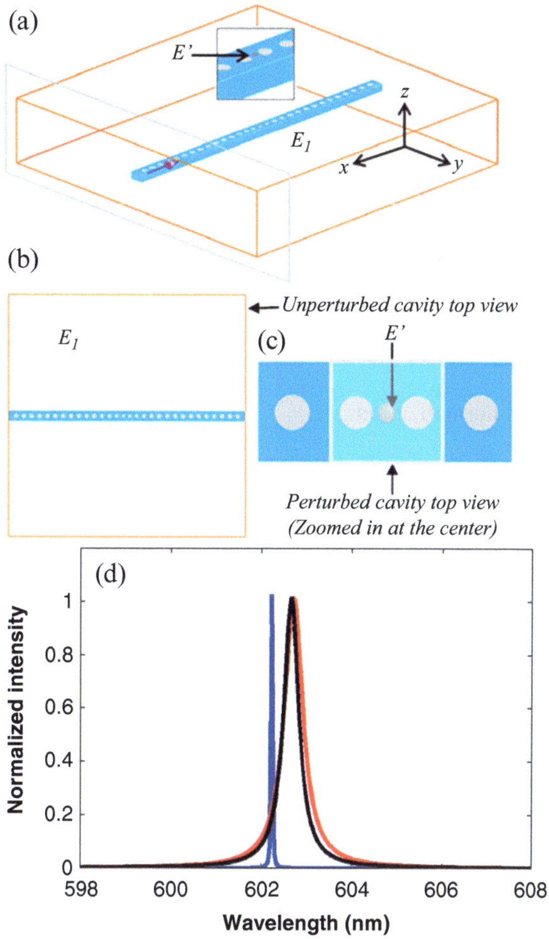

Fig. 1.8 (**a**) A 1D photonic crystal cavity in silicon nitride is perturbed by adding a spheroid silver nanoparticle at the center. (**b**) The calculation is first done for the unperturbed cavity using numerical methods. (**c**) The perturbed region is then calculated to obtain an accurate expression for the near field using very fine mesh for high accuracy. (**d**) A comparison of the unperturbed cavity (blue), the perturbed cavity with full numerical calculation (black), and the perturbed cavity using perturbation theory (red). While the fields in the perturbation region were calculated numerically, the simulation time was reduced by a factor of 12 by using perturbation theory. Mukherjee and Gordon [22] ©Optica 2012

and

$$\mathbf{H}_{\perp} = \sum_m (a_m(z) - b_m(z))\, \mathbf{h}_{\perp,m}(x, y). \tag{1.53}$$

Using orthogonality (Eq. (1.27)) and reciprocity (Eq. (1.48)), the coupled mode equations can be derived as:

$$\frac{\partial a_n}{\partial z} - i\beta_n a_n = i \sum_m (\kappa_{mn} b_n + \chi_{mn} a_n)\,/N_n \tag{1.54}$$

and

$$\frac{\partial b_n}{\partial z} + i\beta_n b_n = -i\sum_m (\kappa_{mn} a_n + \chi_{mn} b_n)/N_n \tag{1.55}$$

where

$$\kappa_{mn} = \frac{\omega}{2}\iint \Delta\epsilon \left(\mathbf{e}_{\perp,\mathbf{m}} \cdot \mathbf{e}_{\perp,\mathbf{n}} - \frac{\epsilon}{\epsilon + \Delta\epsilon} e_{\parallel,m} e_{\parallel,n}\right) dA \tag{1.56}$$

and

$$\chi_{mn} = \frac{\omega}{2}\iint \Delta\epsilon \left(\mathbf{e}_{\perp,\mathbf{m}} \cdot \mathbf{e}_{\perp,\mathbf{n}} + \frac{\epsilon}{\epsilon + \Delta\epsilon} e_{\parallel,m} e_{\parallel,n}\right) dA \tag{1.57}$$

with normalization

$$N_n = \int \hat{z} \cdot \left(\mathbf{e}_{\perp,\mathbf{n}} \times \mathbf{h}_{\perp,\mathbf{n}}\right) dA. \tag{1.58}$$

1.8.2 Example: Counter and Codirectional Coupling

The coupled mode equations are usually employed to solve counter and codirectional coupling between only two modes, neglecting the contributions from other modes by the rotating wave approximation [24]. The rotating wave approximation ignores terms evolving rapidly along z since this coupling will quickly go in and out of phase and so there will be no field buildup; any exchanged energy is quickly transferred back.

As an example of counter-directional coupling, we consider a periodic grating that couples a transverse electric forward-going mode to the same backward-going mode, and we normalize to set $N = 1$, dropping the subscript and writing:

$$\kappa = \chi = 2\kappa_0 \cos(Kz) \tag{1.59}$$

where $K = 2\pi/\Lambda$ and Λ is the period of the grating.

Then the solutions for grating length L become:

$$a(L) = \frac{\sqrt{\kappa_0^2 - \delta^2} \exp\left(i(\delta + \beta)L\right)}{\left(\sqrt{\kappa_0^2 - \delta^2} \cosh\sqrt{\kappa_0^2 - \delta^2} L - i\delta \sinh\sqrt{\kappa_0^2 - \delta^2} L\right)} \tag{1.60}$$

and

$$b(0) = \frac{i\kappa}{\sqrt{\kappa_0^2 - \delta^2}\cosh\sqrt{\kappa_0^2 - \delta^2}L - i\delta} \tag{1.61}$$

where $\delta = \beta - K/2$, and the Bragg resonance corresponds to $\delta = 0$. A similar approach can be applied for codirectional coupling between modes (here we choose forward-going modes with subscripts 1 and 2), where the solutions are:

$$a_1(z) = \exp\left(i(\beta_1 + \delta)z\right)\left(\cos\left(\sqrt{\kappa_0^2 + \delta^2 z}\right) - \frac{i\delta}{\sqrt{\kappa_0^2 + \delta^2}}\sin\left(\sqrt{\kappa_0^2 + \delta^2 z}\right)\right) \tag{1.62}$$

and

$$a_2(z) = \exp\left(i(\beta_2 - \delta)z\right)\left(\frac{i\kappa_0}{\sqrt{\kappa_0^2 + \delta^2}}\sin\left(\sqrt{\kappa_0^2 + \delta^2 z}\right)\right) \tag{1.63}$$

when the field is initially contained in mode 1 and $\delta = \beta_2 - \beta_1 + K$.

1.8.3 Example: Uniform Dielectric Discontinuity

Coupled mode theory provides an exact solution when all the participating modes are considered. For example, we consider a step in the index that would give the Fresnel relations discussed above. If we consider the coupled mode Eqs. (1.54) and (1.55) for the case of forward and backward plane waves traveling normal to a uniform discontinuity $\delta\epsilon$, we can solve for the eigenvectors and find the amplitude of the reflected wave to be $(n_1 - n_2)/(n_1 + n_2)$ where $n_1 = \sqrt{\epsilon_1}$ and $n_2 = \sqrt{\epsilon_1 + \Delta\epsilon}$ for unity forward-going wave. This is the expected result for the Fresnel equations, which shows that a single forward-going wave in the perturbed dielectric is made up of forward- and backward-propagating modes that correspond to the incident and reflected waves at the boundary.

1.8.4 Example: Numerical Method

Using any convenient set of basis modes (e.g., the modes of a cylindrical metal waveguide), it is possible to use the coupled mode equations to represent the modes of a perturbed structure in terms of the basis modes. This is done by diagonalizing the coupled mode equations to generate new eigenvectors (the new modes written in terms of the basis modes) and eigenvalues (the corresponding propagation constants). Figure 1.9 shows an example calculation for a six hollow

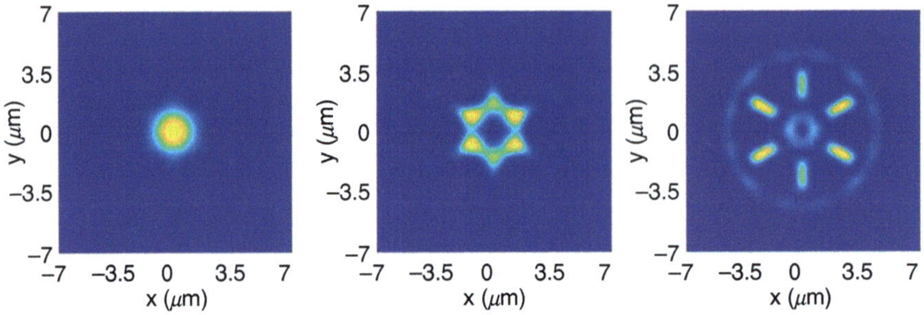

(a) Mode 1; $n_{\text{eff}} = 1.4938$. (b) Mode 19; $n_{\text{eff}} = 1.4745$. (c) Mode 39; $n_{\text{eff}} = 1.4617$.

Fig. 1.9 Electric field intensity of three modes obtained by coupled mode theory for a microstructured dielectric optical fiber and their corresponding effective index values [25]

cylinder microstructured fiber, where the results agree to the last digit with finite difference mode solver solutions.

Mode matching at the boundaries between two regions of different scattering profiles is simplified by using a common basis for each side, since orthogonality may be used to speed up the overlap integrals (where an integral is replaced with a summation). This is a generalization of techniques that use a Fourier basis, like Fourier modal method, or rigorous coupled wave analysis, since the approach can be extended to different bases [26].

1.9 Summary and Outlook

This chapter presented several analytical methods that can be used to obtain physical insight into the optical response of structures, particularly focusing on subwavelength structures. While computational power is rapidly advancing and there has been tremendous success in machine learning approaches to optical design that make this the first stop for many researchers, good physical intuition will always be of value to interpret the results and be wary of pathological results.

References

1. R. Merlin, Metamaterials and the Landau–Lifshitz permeability argument: large permittivity begets high-frequency magnetism. Proc. Natl. Acad. Sci. **106**(6), 1693–1698 (2009)
2. V.G. Veselago, Electrodynamics of substances with simultaneously negative and. Usp. fiz. nauk **92**(7), 517 (1967)
3. J.B. Pendry, Negative refraction makes a perfect lens. Phys. Rev. Lett. **85**(18), 3966 (2000)
4. A.M. Steinberg, P.G. Kwiat, R.Y. Chiao, Measurement of the single-photon tunneling time. Phys. Rev. Lett. **71**(5), 708 (1993)

5. R.L. Olmon, B. Slovick, T.W. Johnson, D. Shelton, S.-H. Oh, G.D. Boreman, M.B. Raschke, Optical dielectric function of gold. Phys. Rev. B **86**(23), 235147 (2012)
6. A. Degiron, H.J. Lezec, N. Yamamoto, T.W. Ebbesen, Optical transmission properties of a single subwavelength aperture in a real metal. Opt. Commun. **239**(1–3), 61–66 (2004)
7. F.J. García-Vidal, L. Martín-Moreno, E. Moreno, L.K.S. Kumar, R. Gordon, Transmission of light through a single rectangular hole in a real metal. Phys. Rev. B **74**(15), 153411 (2006)
8. R. Gordon, A.G. Brolo, Increased cut-off wavelength for a subwavelength hole in a real metal. Opt. Express **13**(6), 1933–1938 (2005)
9. W.-P. Huang, J. Mu, Complex coupled-mode theory for optical waveguides. Opt. Express **17**(21), 19134–19152 (2009)
10. R. Gordon, Angle-dependent optical transmission through a narrow slit in a thick metal film. Phys. Rev. B **75**(19), 193401 (2007)
11. S. Chen, S. Jin, R. Gordon, Super-transmission from a finite subwavelength arrangement of slits in a metal film. Opt. Express **22**(11), 13418–13426 (2014)
12. J.R. Suckling, A.P. Hibbins, M.J. Lockyear, T.W. Preist, J.R. Sambles, C.R. Lawrence, Finite conductance governs the resonance transmission of thin metal slits at microwave frequencies. Phys. Rev. Lett. **92**(14), 147401 (2004)
13. X. Chen, H.-R. Park, N.C. Lindquist, J. Shaver, M. Pelton, S.-H. Oh, Squeezing millimeter waves through a single, nanometer-wide, centimeter-long slit. Sci. Rep. **4**(1), 1–5 (2014)
14. J.A. Stratton, *Electromagnetic Theory*, vol. 33. (John Wiley & Sons, London, 2007)
15. L.D. Landau, E.M. Lifshitz, *Electrodynamics of Continuous Media*. (Pergamon, New York, 1984)
16. N. Engheta, A. Salandrino, A. Alù, Circuit elements at optical frequencies: nanoinductors, nanocapacitors, and nanoresistors. Phys. Rev. Lett. 95, 095504 (2005)
17. W. Su, X. Li, J. Bornemann, R. Gordon, Theory of nanorod antenna resonances including end-reflection phase. Phys. Rev. B **91**, 165401 (2015)
18. B.T. Draine The discrete-dipole approximation and its application to interstellar graphite grains. Astron. J. **333**, 848–872 (1988)
19. Z. Ruan, S. Fan, Superscattering of light from subwavelength nanostructures. Phys. Rev. Lett. **105**(1), 013901 (2010)
20. C. Qian, X. Lin, Y. Yang, X. Xiong, H. Wang, E. Li, I. Kaminer, B. Zhang, H. Chen, Experimental observation of superscattering. Phys. Rev. Lett. **122**(6), 063901 (2019)
21. R.A. Waldron, Perturbation theory of resonant cavities. Proc. IEE Part C Monogr. **107**(12), 272–274 (1960)
22. I. Mukherjee, R. Gordon, Analysis of hybrid plasmonic-photonic crystal structures using perturbation theory. Opt. Express **20**(15), 16992–17000 (2012)
23. J. Yang, H. Giessen, P. Lalanne, Simple analytical expression for the peak-frequency shifts of plasmonic resonances for sensing. Nano Lett. **15**(5), 3439–3444 (2015)
24. H. Kogelnik, 2. theory of dielectric waveguides, in *Integrated Optics* (Springer, Berlin,1975), pp. 13–81
25. T. DeWolf, R. Gordon, Complex coupled mode theory electromagnetic mode solver. Opt. Express **25**(23), 28337–28351 (2017)
26. A. Ahmed, M. Liscidini, R. Gordon, Design and analysis of high-index-contrast gratings using coupled mode theory. IEEE Photonics J. **2**(6), 884–893 (2010)

Chapter 2
Fundamental Limits to Near-Field Optical Response

Owen D. Miller

Abstract Near-field optics is an exciting frontier of photonics and plasmonics. The tandem of strongly localized fields and enhanced emission rates offers significant opportunities for wide-ranging applications while also creating the following basic questions: How large can such enhancements be? To what extent do material losses inhibit optimal response? Over what bandwidths can these effects be sustained? This chapter surveys theoretical techniques for answering these questions. We start with physical intuition and mathematical definitions of the response functions of interest (LDOS, CDOS, SERS, NFRHT, etc.), after which we describe the general theoretical techniques for bounding such functions. Finally, we apply those techniques specifically to near-field optics, for which we describe known bounds, optimal designs, and open questions.

2.1 Introduction

Near-field optics is an exciting frontier of photonics and plasmonics. The near field is the region of space within much less than one electromagnetic wavelength of a source, and "near-field optics" refers to the phenomena that arise when optical-frequency sources interact with material structures in their near field. Free-space waves exhibit negligible variations over such small length scales, which might lead one to think this regime simply reduces to classical electrostatics and circuit theory. A new twist in the optical near field is the emergence of **polaritons**, modes that arise near the interfaces between negative- and positive-permittivity materials [1]. Polaritons emerge from an interplay of geometry and material susceptibility, instead

O. D. Miller (✉)
Department of Applied Physics, Yale University, New Haven, CT, USA
e-mail: owen.miller@yale.edu

R. Gordon (ed.), *Advances in Near-Field Optics*, Springer Series in Optical Sciences 244, https://doi.org/10.1007/978-3-031-34742-9_2

of geometry and wave interference, to confine optical waves. Freedom from wave-interference requirements leads to a striking possibility: resonant fields whose size (spatial confinement) is *decoupled* from its wavelength. Highly confined polaritons enable two reciprocal effects: incoming free-space waves can be concentrated to spatial regions much smaller than the electromagnetic wavelength (well below the diffraction limit) and, conversely, that patterned materials close to a dipolar emitter can significantly amplify outgoing radiation.

The tandem of strongly localized fields and enhanced emission rates offers significant opportunities for applications including spectroscopy [2, 3], nanolasers [4], coherent plasmon generation [5], and broadband single-photon sources [6]. It also generates the following fundamental questions: How large can such enhancements be? Are there limits to field localization? All known polaritonic materials have significant or at least nontrivial amounts of material loss; to what extent does the loss affect these quantities? Over what bandwidths can these effects be sustained?

This chapter surveys theoretical techniques for answering these questions. The same features that make the near field appealing also make it theoretically challenging: there are not fixed photon flows, modal descriptions require exquisite care, and analytical descriptions are not possible except in the simplest high-symmetry scenarios. Over the past decade, thankfully, there has been a surge of interest in identifying what is possible in these systems. One key to the success of these approaches is to not attempt to develop models that apply to every possible instance of a given scattering scenario, but instead to develop techniques that identify *bounds* to the *extreme possibilities* of each scattering scenario. In this chapter, we describe these techniques in detail. We start with physical intuition and mathematical definitions of the response functions of interest (Sect. 2.2), after which we describe the general theoretical techniques for bounding such functions (Sect. 2.3). Finally, we apply those techniques specifically to near-field optics, for which we describe known bounds, optimal designs, and open questions (Sect. 2.4).

2.2 Near-Field Optical Response Functions

In this section, we summarize the background intuition and mathematical equations describing six key near-field optical response functions: local density of states (Sect. 2.2.1), which is proportional to the radiation of a single dipolar current; free-electron radiation (Sect. 2.2.2), which is the collective radiation of a line of current created by an electron beam; the cross density of states (Sect. 2.2.3), which measures modal or emission correlations across different spatial locations; surface-enhanced Raman scattering (Sect. 2.2.4), which is the simultaneous enhancement of incident radiation and outgoing luminescence, typically for imaging or sensing applications; near-field radiative heat transfer (Sect. 2.2.5), which is the transfer of radiative energy from a hot body to a cold one, at near-field separations; and mode volume (Sect. 2.2.6), which refers to the spatial confinement of a resonant mode. Many of these response functions are depicted in Fig. 2.1.

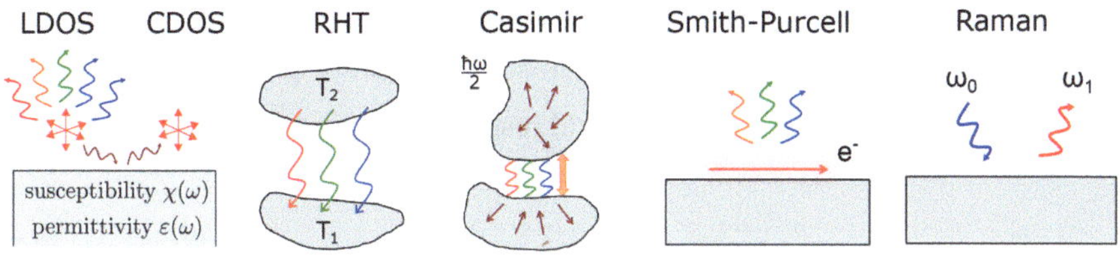

Fig. 2.1 An array of near-field optical response functions of broad interest (Adapted from Ref. [7])

2.2.1 LDOS

The first and arguably most important near-field response quantity is the *local density of states* (LDOS). The central role of LDOS is a result of the extent to which it underpins many connected ideas in near-field optics [8].

The first connection is to the power radiated by a dipole. In general, the work per time done by a field $\mathbf{E}$ on a current $\mathbf{J}$ in a volume V is given by $(1/2)\,\mathrm{Re}\int_V \mathbf{J}^* \cdot \mathbf{E}$. This is a generalized version of Watt's law in circuit theory, and it encodes the work done by the electric field mediating the electric force on the charges in the current, across a distance traveled by the charges given by the product of their speed and the time interval of interest. By Newton's second law, the work per time done by a current $\mathbf{J}$ on a field $\mathbf{E}$ is the negative of the expression above, $-(1/2)\,\mathrm{Re}\int_V \mathbf{J}^* \cdot \mathbf{E}$. We can convert the current density $\mathbf{J}$ to a dipole density $\mathbf{P}$ by the relation $\mathbf{J} = \partial \mathbf{P}/\partial t = -i\omega\mathbf{P}$ for harmonic frequency ω ($e^{-i\omega t}$ convention). Then the power radiated by a dipole at $\mathbf{x}_0$ with dipole moment $\mathbf{p}$ (and therefore dipole density $\mathbf{P} = \mathbf{p}\delta(\mathbf{x} - \mathbf{x}_0)$) is

$$
\begin{aligned}
P_{\text{rad}} &= -\frac{1}{2}\,\mathrm{Re}\int_V \mathbf{J}^* \cdot \mathbf{E}\,\mathrm{d}\mathbf{x} \\
&= \frac{\omega}{2}\,\mathrm{Im}\int_V \mathbf{P}^* \cdot \mathbf{E}\,\mathrm{d}\mathbf{x} \\
&= \frac{\omega}{2}\,\mathrm{Im}\left[\mathbf{p}^* \cdot \mathbf{E}(\mathbf{x}_0)\right].
\end{aligned}
$$

The electric field at $\mathbf{x}_0$, $\mathbf{E}(\mathbf{x}_0)$, is the field produced by a delta-function dipole source, which exactly coincides with the dyadic Green's function (GF) $\mathbb{G}$, evaluated at $\mathbf{x}_0$ from a source at $\mathbf{x}_0$, multiplied by the dipole moment $\mathbf{p}$, giving:

$$
P_{\text{rad}} = \frac{\omega}{2}\,\mathrm{Im}\left[\mathbf{p}^* \cdot \mathbb{G}\left(\mathbf{x}_0, \mathbf{x}_0\right)\mathbf{p}\right].
$$

The imaginary part of a complex number of the form $z^\dagger A z$ is $\mathrm{Im}(z^\dagger A z) = z^\dagger(\mathrm{Im}\,A)z$ by symmetry, where $\mathrm{Im}\,A$ refers to the anti-Hermitian part of A ($\mathrm{Im}\,A = (A - A^\dagger)/2i$). So we have

$$P_{\text{rad}} = \frac{\omega}{2}\mathbf{p}^{\dagger}\left[\operatorname{Im}\mathbb{G}\left(\mathbf{x}_0,\mathbf{x}_0\right)\right]\mathbf{p}. \tag{2.1}$$

This result gives us the first key near-field response function, the imaginary part of the Green's function evaluated at the source position,

$$\operatorname{Im}\mathbb{G}(\mathbf{x}_0,\mathbf{x}_0), \tag{2.2}$$

which is proportional to the radiation rate of an electric dipole into any environment.

Spontaneous emission typically occurs via electric-dipole transitions in atomic or molecular systems, so the rate of spontaneous emission is governed by the imaginary part of the GF. It has been recognized for many decades that this rate is not an immutable constant, but a function of the environment. Just as specifying the amplitude of a current or voltage source in a circuit does not dictate the power delivered by the source, which depends on the impedance of the load, specifying the amplitude of a dipole moment does not dictate the power it delivers to its electromagnetic environment. This fact inspired the concept of a photonic bandgap [9] and photonic crystals [10, 11], with the goal of *inhibiting* spontaneous emission, originally to avoid laser power loss. It has conversely inspired significant effort toward *amplifying* spontaneous emission, for applications such as single-molecule imaging [2, 3]. An early recognition of this fact came from Purcell, who noted that an emitter radiating into a single-photonic-mode environment would have an altered spontaneous-emission rate [12]. Purcell recognized that for a single-mode resonator with quality factor Q and mode volume V, the density of states (per unit volume and per unit frequency) becomes $(Q/\omega)/V$. The relative change of the spontaneous-emission rate is the **Purcell factor**, which is proportional to $\lambda^3 Q/V$.

Purcell derived this expression in the context of enhancing magnetic-dipole transitions in spin systems, but exactly the same argument applies to electric-dipole transitions, where it is most used today. This expression drives many modern investigations of high-quality-factor and/or small-mode-volume cavity design [13–19], to reach the largest Purcell enhancement possible. It can be generalized to multimode, high-Q systems: if each mode has mode field $\mathbf{E}_i$, center frequency ω_i, and half-width-at-half-maximum γ_i (corresponding to a *mode lifetime* of $1/(2\gamma_i)$), the power radiated by a dipole with moment $\mathbf{p}$ located at position $\mathbf{x}_0$ is [20]

$$P_{\text{rad}} \approx \frac{\omega^2}{4}\sum_i \frac{\gamma_i\left|\mathbf{E}_i^{\dagger}(\mathbf{x}_0)\mathbf{p}\right|^2}{(\omega-\omega_i)^2+\gamma_i^2} \tag{2.3}$$

In the limit of infinite Q, the Lorentzian lineshapes become delta functions, and the summation simplifies to delta functions multiplied by the overlap of modal fields with the dipole moment. The overlap of each mode with the dipole is a measure of the relative modal energy concentration at that particular point in space. Hence, the overall summation can be understood as a **local density of states**, or LDOS (with appropriate prefactors). The power radiated by a dipole into an electromagnetic environment, then, is directly proportional to the local density of electromagnetic modes; inserting the correct prefactors leads to an LDOS expression in terms of $\operatorname{Im}\mathbb{G}$ [8, 21–23]:

$$\mathrm{LDOS}(\omega, \mathbf{x}) = \frac{1}{\pi\omega} \operatorname{Tr} \operatorname{Im} \mathbb{G}(\mathbf{x}_0, \mathbf{x}_0), \tag{2.4}$$

where the trace encodes a summation over all independent polarizations. (Note that, e.g., Ref. [8] defines the Green's function with an extra $1/\omega^2$ factor, which leads to ω in the numerator of their analog to Eq. (2.4).) In free space, the LDOS coincides with the density of states (as there are no spatial variations) and is given by $\mathrm{LDOS}(\omega) = \omega^2/2\pi^2c^3$. Technically, the expression of Eq. (2.4) is the *electric* LDOS; one can similarly define a magnetic LDOS through a summation over the relative magnetic-field strengths, or more generally by the power radiated by a magnetic dipole. For a magnetic Green's function $\mathbb{G}^{(HM)}$, denoting the magnetic field from a magnetic-dipole source, the magnetic LDOS is [8]

$$\mathrm{LDOS}^{(m)}(\omega, \mathbf{x}) = \frac{1}{\pi\omega} \operatorname{Tr} \operatorname{Im} \mathbb{G}^{(HM)}(\mathbf{x}_0, \mathbf{x}_0). \tag{2.5}$$

The sum of Eqs. (2.4) and (2.5) is referred to as the *total* LDOS, representing the totality of electric- and magnetic-field energy localized to a point $\mathbf{x}_0$, at frequency ω, over all modes. (Significant alterations to the modal-decomposition expressions are needed, e.g., in plasmonic (and polaritonic) systems [24, 25].) Such descriptions are mathematically accurate only in the high-quality-factor, nonoverlapping-mode limit [26, 27], but the dipole-radiation interpretation generalizes to any linear scattering scenario.

To summarize, the imaginary part of the Green's function, $\operatorname{Im}\mathbb{G}(\mathbf{x}_0, \mathbf{x}_0)$, is a measure of the power radiated by electric and/or magnetic dipoles in an arbitrary environment, which is proportional to the spontaneous-emission rate of a dipolar emitter, and it encapsulates the Purcell factor, particularly the ratio Q/V, of high-quality-factor modes that concentrate energy at that point. We have extensively described LDOS due to its versatility and cross-cutting nature. The following quantities have more focused and niche applications and can be described more concisely.

2.2.2 *Free-Electron Radiation*

Radiation by a free-electron beam is closely related to LDOS, with the key distinction being that the current distribution is now a line source. An electron (charge $-e$) propagating through free space at constant velocity $v\hat{\mathbf{x}}$ comprises a free current density $\mathbf{J}(\mathbf{r}, t) = -\hat{\mathbf{x}}ev\delta(y)\delta(z)\delta(x - vt)$, which generates a frequency-dependent incident field [28]:

$$\mathbf{E}_{\mathrm{inc}} = \frac{e\kappa_p e^{ik_v x}}{2\pi\omega\varepsilon_0}\left[\hat{\mathbf{x}} i\kappa_\rho K_0(\kappa_\rho\rho) - \hat{\rho} k_v K_1(\kappa_\rho\rho)\right], \tag{2.6}$$

written in cylindrical coordinates (x, ρ, θ), where K_n is the modified Bessel function of the second kind, $k_v = \omega/v$, and $\kappa_\rho = \sqrt{k_v^2 - k^2} = k/\beta\gamma$ ($k = \omega/c$, free-space wavevector; $\gamma = 1/\sqrt{1-\beta^2}$, Lorentz factor). When there is no back-action on the source from its interactions with a scatterer, then the free-electron photon emission and energy loss is a typical scattering problem, with Eq. (2.6) as the incident field.

An important feature of Eq. (2.6) is that the incident field is entirely evanescent (the asymptotic decay of the special function K_n is given by e^{-kr}/kr in the far field). This is expected on physical grounds, as an electron moving at constant velocity cannot radiate. Once a scattering body is brought close to the electron beam, however, the situation changes: the evanescent incident field can excite modes in the scatterer that couple to far-field radiation. (Physically, the electromagnetic-field-mediated interaction of the electron beam with the scatterer can lead to deceleration and therefore radiation.) The radiated power can be computed by an LDOS-like expression, $\frac{1}{2}\operatorname{Re}\int \mathbf{J}^* \cdot \mathbf{E}$, where $\mathbf{J}$ is the free-electron current density, but the bound techniques developed below for scattering bodies are most easily applied to the polarization fields $\mathbf{P}$ within the scatterer, so we prefer an equivalent expression in terms of $\mathbf{P}$. One option would be a linear combination of a direct-radiation term with a scatterer-interaction-radiation term, but the evanescent-only nature of the incident field implies that the direct-radiation term is zero. Instead, the only power lost by the electron beam is that which is extinguished by the scatterer, into absorption losses or far-field radiation. As we discuss more thoroughly in Sect. 2.3.1, the extinction of a scattering body V is given by

$$P_{\mathrm{ext}} = \frac{\omega}{2}\operatorname{Im}\int_V \mathbf{E}_{\mathrm{inc}}^*(\mathbf{x}) \cdot \mathbf{P}(\mathbf{x})\,\mathrm{d}\mathbf{x}, \tag{2.7}$$

which we will use to analyze the free-electron loss, as $P_{\mathrm{loss}} = P_{\mathrm{ext}}$.

When the beam passes by the scatterer without intersecting it, the resulting radiation is referred to as **Smith–Purcell radiation**. When the beam passes through the scatterer, causing radiation, it is referred to as **transition radiation**. And when the beam radiates while propagating *inside* a refractive medium (within which the modified speed of light can be smaller than the electron speed), it is referred to as **Cherenkov radiation**. The Smith–Purcell process resides squarely in the realm of near-field electromagnetism.

2.2.3 CDOS

In Sect. 2.2.1, we showed that the power radiated by a single dipole at position $\mathbf{x}$ is proportional to the LDOS at that point, which itself is proportional to $\operatorname{Im}\mathbb{G}(\mathbf{x}, \mathbf{x})$.

Consider now the power radiated by *two* dipoles, $\mathbf{p}_1$ and $\mathbf{p}_2$, at positions $\mathbf{x}_1$ and $\mathbf{x}_2$, for a total dipole density of $\mathbf{P}(\mathbf{x}) = \mathbf{p}_1\delta(\mathbf{x}-\mathbf{x}_1) + \mathbf{p}_2\delta(\mathbf{x}-\mathbf{x}_2)$. The power they jointly radiate is given by

$$\begin{aligned} P_{\text{rad}} &= \frac{\omega}{2}\int_V\int_{V'} \mathbf{P}(\mathbf{x})\,\text{Im}\,\mathbb{G}(\mathbf{x},\mathbf{x}')\mathbf{P}(\mathbf{x}')\,\text{d}\mathbf{x}\,\text{d}\mathbf{x}' \\ &= \frac{\omega}{2}\Big\{\mathbf{p}_1^\dagger\left[\text{Im}\,\mathbb{G}(\mathbf{x}_1,\mathbf{x}_1)\right]\mathbf{p}_1 + \mathbf{p}_2^\dagger\left[\text{Im}\,\mathbb{G}(\mathbf{x}_2,\mathbf{x}_2)\right]\mathbf{p}_2 \\ &\quad + \mathbf{p}_1^\dagger\left[\text{Im}\,\mathbb{G}(\mathbf{x}_1,\mathbf{x}_2)\right]\mathbf{p}_2 + \mathbf{p}_2^\dagger\left[\text{Im}\,\mathbb{G}(\mathbf{x}_2,\mathbf{x}_1)\right]\mathbf{p}_1\Big\}. \end{aligned} \tag{2.8}$$

The first two terms are the powers radiated by the two dipoles in isolation (or when incoherently excited); the second pair of terms is the positive or negative contribution that arises for constructive or destructive (coherent) interference between the two dipoles. For reciprocal media (of arbitrary patterning), the third and fourth terms are complex conjugates of each other, such that we can just consider one of them (say, the third term) in determining the two-dipole interference. By analogy with Eq. (2.4), we can define a **cross density of states (CDOS)** by the expression:

$$\text{CDOS}_{ij}(\omega,\mathbf{x}_1,\mathbf{x}_2) = \frac{1}{\pi\omega}\,\text{Im}\,\mathbb{G}_{ij}(\mathbf{x}_1,\mathbf{x}_2), \tag{2.9}$$

which differs from Ref. [29] only by the absence of a 2 in the prefactor. The sign of the CDOS indicates the sign of the interference term, while its magnitude is a field-correlation strength between the two points of interest in a given electromagnetic environment. The amplification of emission that can occur when the sign is positive is an example of superradiance, while the reduction of emission when the sign is negative is an example of subradiance, in each case mediated by the local CDOS [30]. Because the CDOS is the off-diagonal part of a positive-definite matrix, it is straightforward to show that its magnitude is bounded above by the square root of the product of the diagonal terms in the matrix, i.e., the local densities of states of the two dipoles in isolation [31].

In systems that are closed, or approximately closed, there is another interesting interpretation of the CDOS [29, 31]. Just as the LDOS can be interpreted as a local modal density, the CDOS can be interpreted as a local modal *connectivity*—it is a measure of spatial coherence between two points. In Ref. [29], it was shown the one can compute local coherence lengths from spatial integrals of the CDOS. From these local coherence lengths, it was unambiguously demonstrated that "spatial squeezing" of eigenmodes occurs in systems of disordered plasmonic nanoparticles. This plausibly explains surprising experimental results when probing the local response of such disordered films [32], showing the value of CDOS as an independent concept from LDOS.

There are two other areas in which CDOS emerges as a key metric: Forster energy transfer [33–35] and quantum entanglement and super-radiative coupling between qubits [36–40]. The general idea in each case is a dipole $\mathbf{p}_1$ transferring energy to a second dipole $\mathbf{p}_2$. In this scenario, $\mathbf{p}_1$ and $\mathbf{p}_2$ are considered *fixed*. By Poynting's theorem, the energy flux into a small bounding surface of $\mathbf{p}_2$, for a field $\mathbf{E}_1$ generated by $\mathbf{p}_1$, is

$$\frac{\omega}{2}\operatorname{Im}\left[\mathbf{p}_2^\dagger \mathbf{E}_1(\mathbf{x}_2)\right] = \frac{\omega}{2}\operatorname{Im}\left[\mathbf{p}_2^\dagger \mathbb{G}(\mathbf{x}_2, \mathbf{x}_1)\mathbf{p}_1\right], \tag{2.10}$$

which is a form of the CDOS. The fixed nature of the second dipole, $\mathbf{p}_2$, is crucial for the CDOS metric to be the correct one. If the second dipole is *induced by the field emanating from the first dipole*, then $\mathbf{p}_2 = \alpha_2 \mathbf{E}_1(\mathbf{x}_2)$, and the correct energy-transfer expression would be the imaginary part of the polarizability multiplied by the squared absolute value of the Green's function.

2.2.4 Surface-Enhanced Raman Scattering (SERS)

Surface-enhanced Raman scattering is a technique whereby molecules are excited by a pump field, subsequently emitting Stokes- (or anti-Stokes-) shifted radiation (fluorescence) that can be used for imaging or identification [41–44]. The small cross sections of most chemical molecules result in very low pump and emission efficiencies in conventional Raman spectroscopy [45], but one can engineer the near-field environment to enhance both the concentration of the pump field and the emission rate. Efficiency improvements of up to 12 orders of magnitude have been demonstrated, enabling single-molecule detection and a variety of applications.

SERS is a nonlinear process, in which a single dipolar molecular sees both a pump enhancement and a spontaneous-emission enhancement. A key insight for understanding SERS is that the weakness of the nonlinearities of the individual molecules means that the nonlinear process can be treated as the *composition* of linear processes, in which the pump first enhances the excited-population densities (or, classically, the dipole amplitudes), and then the spontaneous-emission enhancements can be treated as a second step, essentially independent of the first.

We can write the key metric of SERS by considering these two steps in sequence, following a procedure outlined in Ref. [46]. First, an illumination field at frequency ω_0 impinges upon the molecule and its environment; in tandem, a total field of $\mathbf{E}_{\omega_0}(\mathbf{x}_0)$ is generated at the molecule. The Raman process generates a dipole moment at frequency ω_1 given by

$$\mathbf{p}_{\omega_1} = \boldsymbol{\alpha}_{\text{Raman}} \mathbf{E}_{\omega_0}(\mathbf{x}_0) \tag{2.11}$$

where $\boldsymbol{\alpha}_{\text{Raman}}$ is the molecular polarizability. Next, the power radiated at ω_1 by this dipole is given, per Eq. (2.1), by

$$P_{\text{rad},\omega_1} = \mathbf{p}_{\omega_1}^\dagger \left[\operatorname{Im} \mathbb{G}_{\omega_1}(\mathbf{x}_0, \mathbf{x}_0)\right] \mathbf{p}_{\omega_1}. \tag{2.12}$$

Hence, we see that there are two opportunities for amplification of SERS: concentrating the incoming field $\mathbf{E}_{\omega 0}$ that determines the dipole amplitude and enhancing the outgoing radiation by maximizing the LDOS, proportional to $\operatorname{Im} \mathbb{G}_{\omega 1}(\mathbf{x}_0, \mathbf{x}_0)$, at the location of the dipole. To separate the two contributions, we can write the dipole moment as $\mathbf{p} = \|\boldsymbol{\alpha}\mathbf{E}\| (\boldsymbol{\alpha}\mathbf{E}/\|\boldsymbol{\alpha}\mathbf{E}\|)$, i.e., an amplitude multiplied by a unit vector. If we denote the unit vector as $\hat{\mathbf{p}}_{\omega 1}$, then we can write

$$P_{\text{rad},\omega 1} = \|\boldsymbol{\alpha}_{\text{Raman}} \mathbf{E}_{\omega 0}\|^2 \hat{\mathbf{p}}_{\omega 1}^{\dagger} \left[\operatorname{Im} \mathbb{G}_{\omega 1}(\mathbf{x}_0, \mathbf{x}_0) \right] \hat{\mathbf{p}}_{\omega 1}, \tag{2.13}$$

where now the first term encapsulates ω_0-frequency concentration, and the second term encapsulates ω_1-frequency LDOS-enhancement. Straightforward arguments lead to a net SERS enhancement, relative to a base rate P_0 without any nearby surface, given by

$$\frac{P_{\text{rad},\omega 1}}{P_0} = \left(\frac{\|\boldsymbol{\alpha}_{\text{Raman}} \mathbf{E}_{\omega 0}\|^2}{\|\boldsymbol{\alpha}_{\text{Raman}}\|^2 \|\mathbf{E}_{\text{inc},\omega 0}\|^2} \right) \left(\frac{\rho_{\hat{\mathbf{p}},\omega 1}}{\rho_{0,\omega 1}} \right), \tag{2.14}$$

where $\|\boldsymbol{\alpha}\|$ refers to the induced matrix norm of $\boldsymbol{\alpha}$, $\rho_{\hat{\mathbf{p}},\omega 1}$ is the ω_1-frequency LDOS for a $\hat{\mathbf{p}}$-polarized dipole, and $\rho_{0,\omega 1}$ in this expression is the background ω_1-frequency LDOS of a $\hat{\mathbf{p}}$-polarized dipole (not the typical summation over all polarizations). The two parenthetical terms in Eq. (2.14) must both be bounded to identify fundamental limits to SERS enhancements.

2.2.5 *Near-Field Radiative Heat Transfer*

The warming of the cold earth by the hot sun is mediated by radiative transfer, i.e., photons radiated from the sun to the earth. The maximum rate at which such a process could occur is of course given by the **blackbody** rate, which is determined only by the solid angle subtended by the earth from the sun (or vice versa). Determination of this blackbody rate requires no knowledge of multiple-scattering processes between the two bodies. In the far field, the only "channels" (carriers of power into and out of a scattering region) are propagating-wave channels; by Kirchhoff's law [47], one need only know the absorption or emission rates of the two bodies in isolation to know their maximum radiative-exchange rate. A more general viewpoint of far-field radiation, via the idea of communication channels, is discussed in Sect. 2.3.2.

It has been known for 75 years [48, 49] that two bodies separated by less than a thermal wavelength can exchange radiative heat at significantly larger rates than their far-field counterparts. Once in the near field, the bodies can exchange photons not only through radiative channels but also through evanescent channels; moreover, as the separation distance d is reduced, the number of evanescent channels that can be accessed increases dramatically, scaling as $1/d^2$. These channels can be accessed via any mechanism that produces strong near fields. Polaritonic surface waves, via either plasmons or phonon–polariton materials, are a natural choice, and hyperbolic metamaterials (whose strongest effect is not surface waves but instead high-wavenumber bulk modes with nonzero evanescent tails) can provide similar performance [50, 51]. Photonic crystals can also support surface waves, but the confinement of those waves is typically related to the size of their bandgap [11], thereby scaling with frequency, yielding surface waves with significantly less confinement than their metallic counterparts.

The complexity of near-field radiative heat transfer (NFRHT) is daunting, both experimentally and theoretically. The first experimental demonstrations of enhancements in NFRHT via near-field coupling were not achieved until the 2000s [52–54], many decades after the original predictions [48, 49], and measurements in the extreme near field were not achieved until 2015 [55]. There are a number of technical hurdles to experimental measurements, especially maintaining consistent, nanometer-scale gap separations over large-scale device diameters while simultaneously measuring miniscule heat currents [55].

The theoretical challenge has been no less severe. NFRHT involves rapidly decaying near fields (requiring high resolution), typically over large-area surfaces (requiring a large simulation region), for spatially incoherent and broadband thermal sources (such that the equivalent of very many simulations is needed). The computational complexity of this endeavor has limited the analysis of NFRHT almost exclusively to high-symmetry structures (planar/spherical bodies, metamaterials, etc.) [56–61], small resonators [58, 62], two-dimensional systems [63], and the like. We review the planar-body interaction, which is informative, while emphasizing the need (and opportunity) for new theoretical tools to understand what is possible when exchanging radiative heat in the near field. (Casimir forces [64–66] are theoretical brethren of NFRHT, the only difference being that they arise from zero-point instead of thermal fluctuations. Though we do not review them here, there is at least one recent work developing theoretical bounds/upper limits using techniques similar to those described throughout this chapter [67].)

Consider two near-field bodies with temperatures T_1 and T_2, respectively. By the fluctuation–dissipation theorem, the incoherent currents in body 1, $\mathbf{J}_1$, have ensemble averages (denoted $\langle\rangle$) given by [58]

$$\langle \mathbf{J}_1(\mathbf{x}, \omega)\mathbf{J}_1^\dagger(\mathbf{x}', \omega)\rangle = \frac{4\varepsilon_0\omega}{\pi} \operatorname{Im}\left[\chi_1(\mathbf{x}, \omega)\right] \Theta(\omega, T_1)\delta(\mathbf{x} - \mathbf{x}')\mathcal{I}, \tag{2.15}$$

where $\chi_1(\mathbf{x}, \omega)$ is the material susceptibility of body 1, $\mathcal{I}$ is the 3×3 identity matrix, and $\Theta(\omega, T)$ is the Planck distribution:

$$\Theta(\omega, T) = \frac{\hbar\omega}{e^{\hbar\omega/kT} - 1}. \tag{2.16}$$

These currents radiate to body 2, at each frequency ω, at a rate that we denote $\Phi_{21}(\omega)$. The rate $\Phi_{21}(\omega)$ is given by the ensemble average of the flux into body 2, i.e., $\langle -\frac{1}{2} \operatorname{Re} \int_{S_2} \mathbf{E} \times \mathbf{H}^* \cdot \hat{\mathbf{n}}\rangle$, where S_2 is a bounding surface of V_2, $\hat{\mathbf{n}}$ is the outward normal, and the field sources are given by Eq. (2.15), except without the Planck function. The Planck function is separated so that $\Phi_{21}(\omega)$ is independent of temperature and depends only on the electromagnetic environment. Then the radiative heat transfer rate into 2 from currents in 1, denoted H_{21}, is given by

$$H_{21} = \int \Phi_{21}(\omega)\Theta(\omega, T_1)\,\mathrm{d}\omega. \tag{2.17}$$

Similarly, the rate of transfer from body 2 to body 1, H_{12}, is given by

$$H_{12} = \int \Phi_{12}(\omega)\Theta(\omega, T_2)\,\mathrm{d}\omega, \tag{2.18}$$

and the net transfer rate is the difference between the two. For reciprocal bodies, the rates $\Phi_{12}(\omega)$ and $\Phi_{21}(\omega)$ are always equal (by exchanging the source and "measurement" locations), but this is also true more generally: for *two* bodies exchanging radiative heat in the near field, $\Phi_{12}(\omega)$ and $\Phi_{21}(\omega)$ must be equal, or else one could have net energy exchange with both bodies at equal temperatures, in violation of the second law of thermodynamics. Note that if three bodies are present, or either body radiates significant amounts of energy into the far field, this relation need not hold in nonreciprocal systems, and indeed "persistent currents" have been predicted in three-body systems in the near field [68]. Throughout this chapter, we will focus on the prototypical two-body case, so we can take

$$\Phi_{12}(\omega) = \Phi_{21}(\omega) = \Phi(\omega), \tag{2.19}$$

without assuming reciprocity. Hence, the *net* NFRHT rate between the two bodies is given by

$$H_{2\leftarrow 1} = \int \Phi(\omega)\left[\Theta(\omega, T_1) - \Theta(\omega, T_2)\right]\mathrm{d}\omega. \tag{2.20}$$

Often, it is illuminating to reduce the problem to a single temperature T and study the differential heat transfer for a temperature differential ΔT. The net heat exchange divided by this temperature differential is the **heat transfer coefficient**, or HTC, which is given by Eq. (2.20), except that the temperature difference is replaced by a single derivative of $\Theta(\omega, T)$ with respect to temperature:

$$\mathrm{HTC} = \int \Phi(\omega)\frac{\partial\Theta(\omega, T)}{\partial T}\,\mathrm{d}\omega. \tag{2.21}$$

Hence, the quantity $\Phi(\omega)$ is the designable quantity in NFRHT and is the focus of the NFRHT bounds appearing across Sect. 2.4.

2.2.6 Mode Volume

Finally, we turn to a unique near-field quantity: mode volume. Intuitively, **mode volume** encapsulates an "amount of space" occupied by an electromagnetic mode. Obviously, defining the volume of a continuous density is necessarily subjective. But we can develop an intuitive approach to the common volume definition. The energy density of a mode m at any point $\mathbf{x}$ is proportional to $\varepsilon(\mathbf{x})|\mathbf{E}_m(\mathbf{x})|^2$. If the maximum energy density occurs at a point $\mathbf{x}_0$, we can define the volume of the mode as follows: let us redistribute the energy into a binary pattern in which at every point in space, it can only take the values 0 or $\varepsilon(\mathbf{x}_0)|\mathbf{E}_m(\mathbf{x}_0)|^2$. Let us also require that the total energy of the mode will not change in this binarization, i.e., $\int \varepsilon(\mathbf{x})|\mathbf{E}(\mathbf{x})|^2$ remains fixed. Then the corresponding redistributed field will occupy the volume:

$$V_m = \frac{\int \varepsilon(\mathbf{x})|\mathbf{E}_m(\mathbf{x})|^2}{\varepsilon(\mathbf{x}_0)|\mathbf{E}_m(\mathbf{x}_0)|^2}. \tag{2.22}$$

Typical modes of interest, which have strong field concentration and Gaussian- or Lorentzian-like energy decay, are well-suited to such an interpretation.

More rigorously, per Eq. (2.3), the modal field intensity is the quantity that determines the interaction of a dipole with a specific mode, and the contribution of that mode to the spontaneous emission of the dipole. Then an alternative interpretation of the quantity in Eq. (2.22) is that the numerator can be taken to be 1, for a normalized mode, and the denominator is the relevant coupling term in the Hamiltonian that is to be maximized. This alternative approach explains why a common mathematical objective is to minimize the expression in Eq. (2.22), without reference to any physical concept of volume.

A critical question around mode volume is whether such a concept is even valid. For closed (or periodic) systems with nondispersive, real-valued permittivities, the Maxwell operator is Hermitian, and there is an orthogonal basis of modal fields that can be orthonormalized. Dispersion in the material systems makes the eigenproblem nonlinear, but for Drude–Lorentz-like dispersions, one can introduce auxiliary variables, and in this higher-dimensional space, there is again a linear, Hermitian eigenproblem [69]. But once losses are introduced, either through open boundary conditions or material dissipation, the operator is no longer Hermitian, and the modes cannot be orthonormalized with an energy-related inner product [25]. Instead, one must work with *quasinormal modes* (QNMs), for which two issues arise.

If material losses are the dominant loss mechanism, as is typical in plasmonics, then the key new subtlety often is the modification of orthogonality: the modes are orthogonal in an *unconjugated* "inner product" (e.g., $\int \varepsilon\mathbf{E}_1\cdot\mathbf{E}_2$ instead of $\int \varepsilon\mathbf{E}_1^*\cdot\mathbf{E}_2$), which then replaces the standard conjugated inner product in modal expansions such as Eq. (2.3). While this is mathematically convenient, it can stymie our typical intuition. A beautiful example is demonstrated in Ref. [24]. There, it is shown that the spontaneous emission near a two-resonator antenna can be dominated by two

QNMs, as expected. However, if one tries to attribute individual contributions from each QNM, one of the QNMs appears to contribute *negative* spontaneous emission. This is attributable to the modified inner product: modes that are orthogonal in the unconjugated inner product are not orthogonal in an energy inner product, and their contributions to a positive energy flow (such as spontaneous emission) are invariably linked; one can no longer separate a power quantity such as LDOS into individual contributions from constituent modes. Ultimately, one can define mode volume as a complex-valued quantity [70], in which case it no longer becomes an independent quantity of interest to minimize or maximize, but rather an ingredient for other scattering quantities of interest.

If radiation losses are the dominant loss mechanism, one faces a hurdle even before orthogonality: just *normalizing* the modal fields becomes tricky. If the modal fields eventually radiate in free space, they will asymptotically scale as $e^{ik_m r}/r$, where $k_m = \omega_m/c$ is the wavenumber of the mode and r is a distance from the scatterer. But the losses to radiation transform the resonant eigenvalues to poles in the lower-half of the complex-frequency plane, i.e., $\omega_m \rightarrow \omega_m^{(r)} - i\omega_m^{(i)}$, where $\omega_m^{(i)} > 0$. Hence, the modal fields grow exponentially, $\sim e^{\omega_m^{(i)} r/c}$, such that any integrals of the form $\int \mathbf{E}^2$ or $\int |\mathbf{E}|^2$ diverge. There are a few resolutions to this issue (cf. Sec. 4 of Ref. [71]). Perhaps the simplest is to use computational perfectly matched layers (PMLs) to confine the fields to a finite region. Then, for any accurate discretization of the Maxwell operator, one is simply left with a finite-sized, non-Hermitian matrix, whose eigenvectors will generically be orthonormalizable under the unconjugated inner product. (Exceptions to this occur at aptly named *exceptional points*, where modes coalesce, and one needs Jordan vectors to complete the basis [72, 73].) Orthogonalization of these modes requires the same modification of the inner product discussed above (even without material loss [27], as radiation is a loss mechanism). Moreover, there is one further difficulty: sometimes important contributions to energy expression can come from fields that *primarily reside in the PML region*. One can think of this as a condition for a complete basis. It is difficult to attribute physical intuition or meaning to such contributions.

In Sect. 2.4.4, where we develop bounds for mode volume, we will only deal with cases of lossless dielectric materials, we assume the quality factors are sufficiently high that the system is approximately closed, and we assume isolated resonances. This is the limit in which the mode volume as defined by Eq. (2.22) is exactly the quantity that enters the LDOS expression of Eq. (2.3), which is typically the underlying goal of minimizing mode volume in the first place. In scenarios where one must use quasinormal modes, it is probably better to eschew them altogether (if one wants a bound) and to instead work directly with the scattering quantity (e.g., LDOS) of interest.

2.3 Analytical and Computational Bound Approaches

Across many areas of science and technology, "fundamental limits" or "bounds" play an important role in technological selection, theoretical understanding, and optimal design. Examples abound:

- The **Shockley–Queisser limits** for solar-cell energy conversion efficiency. Originally developed for single-cell, all-angle solar absorption and energy conversion [74], the basic framework they developed identifies two required loss mechanisms in any solar cell: radiation back to the sun (at the open circuit condition [75]), and thermalization losses in the establishment of quasi-Fermi levels in each band. Almost any proposed solar-energy-conversion technique must be put through a Shockley–Queisser analysis to earn serious consideration as a technology.
- The **Yablonovitch** $4n^2$ **limit**, for the maximum broadband, all-angle absorption enhancement in any optically thick material [76]. The factor $4n^2$, for a refractive index n, arises from the density-of-states enhancement in a high-index material, a 2X enhancement from mirrors on the rear surface, and a 2X enhancement from the reorientation of mostly vertical rays into random angles.
- The **Wheeler–Chu limit** to antenna quality factor, Q [77, 78]. It is difficult for a subwavelength antenna (such as a cell phone antenna) to operate over a wide bandwidth, and the Wheeler–Chu (sometimes Harrington is also given credit [79]) limit imposes a bound on the maximum operational bandwidth. Most state-of-the-art antenna designs operate very close to the Wheeler–Chu limit [80].
- The **Bergman–Milton bounds** on the effective properties of a composite material [81–86].
- The **Abbe diffraction limit** on the maximum focusing of an optical beam. This limit can be circumvented in the near field [87, 88], or even in the far field if one is willing to tolerate side lobes [89–95].
- The **Shannon bounds** [96], a foundational idea in information theory [97].

Many of these examples involve electromagnetism, but typically only for noninteracting waves and simplified physical regimes. The Yablonovitch $4n^2$ limit applies in geometric (ray) optics, the Wheeler–Chu limit only arises in highly subwavelength structures, and the diffraction limit applies only to free-space (or homogeneous-medium) propagation. Is it possible to create an analogous theoretical framework for the full Maxwell equations, identifying fundamental spectral response bounds while accounting for the exceptional points [98, 99], speckle patterns [100], bound states in the continuum [101], and other exotic phenomena permitted by the wave equation? A flurry of work over the past decade suggests that in many scenarios, the answer should be "yes." In the following subsections, we outline the key new ideas that have been developed.

2.3.1 Global Conservation Laws

One approach particularly well-suited to formulating bounds is to replace the complexity of the full Maxwell-equation design constraints with a single constraint that encodes some type of conservation law. The Yablonovitch limit, discussed in the previous section, offers a powerful example: to identify maximum absorption enhancement in a geometric-optics setting, one can replace the complexity of ray-tracing dynamics with a single density-of-states constraint. Unfortunately, one cannot extend such density-of-states arguments to full-Maxwell and near-field settings, but other types of "conservation laws" can be identified. A **global conservation law** that has been particularly fruitful for nanophotonics is the **optical theorem**. The optical theorem [102–104] is a statement of global power conservation: the total power extinguished from an incident beam by a scattering body (or bodies) equals the sum of the powers scattered and absorbed by that body. Writing the extinguished, scattered, and absorbed powers as P_{ext}, P_{scat}, and P_{abs}, respectively, the optical theorem can be expressed as

$$P_{\text{ext}} = P_{\text{scat}} + P_{\text{abs}}. \tag{2.23}$$

Conventionally, the optical theorem is specified in terms of the far-field scattering amplitudes of a scattering body [102], in which case the extinction is shown to be directly proportional to the imaginary part of the forward-scattering amplitude. This expression can be interpreted as a mathematical statement of the physical intuition that the total power taken from an incident beam can be detected in the phase and amplitude of its shadow. The analysis does not have to be done in the far field; another common version is to relate the extinguished-, scattered-, and absorbed-power fluxes via surface integrals of the relevant Poynting fluxes [103]. Still one more version of the optical theorem, and the one that turns out to be most useful for wide-ranging bound applications, is to use the divergence theorem to relate the surface fluxes to the fields within the volume of the scatterer and write all powers in terms of the polarization currents and fields induced in those scatterers [104]. As we briefly alluded to in the discussion of free-electron radiation in Sect. 2.2.2, the work done by a field $\mathbf{E}$ on a polarization field $\mathbf{P}$ in a volume V is given by $\left(\frac{\omega}{2}\right) \operatorname{Im} \int_V \mathbf{E}^* \cdot \mathbf{P} = \left(\frac{\omega}{2}\right) \int_V \mathbf{P}^* \left[\operatorname{Im} \chi / |\chi|^2\right] \mathbf{P}$, where χ is the material susceptibility. (We assume throughout scalar, electric material susceptibilities χ. The generalizations to magnetic, anisotropic, and bianisotropic materials are straightforward in every case.) Extinction is the work done by the incident field on the induced polarization field, scattered power is the work done by that polarization field on the scattered fields $\mathbf{E}_{\text{scat}}$, and absorbed power is the work done by the total field on the polarization field. Hence, the optical theorem reads:

$$\operatorname{Im} \int_V \mathbf{E}^*_{\text{inc}}(\mathbf{x}) \cdot \mathbf{P}(\mathbf{x})\, d\mathbf{x} = \operatorname{Im} \int_V \int_V \mathbf{P}^*(\mathbf{x}) \cdot \mathbb{G}_0(\mathbf{x}, \mathbf{x}')\mathbf{P}(\mathbf{x}')\, d\mathbf{x}\, d\mathbf{x}' + \int_V \mathbf{P}^*(\mathbf{x}) \cdot \frac{\operatorname{Im} \chi(\mathbf{x})}{|\chi(\mathbf{x})|^2} \mathbf{P}(\mathbf{x})\, d\mathbf{x}, \tag{2.24}$$

where we have substituted $\mathbf{E}_{\text{scat}}(\mathbf{x}) = \int_V \mathbb{G}_0(\mathbf{x}, \mathbf{x}')\mathbf{P}(\mathbf{x}')\, d\mathbf{x}'$ for the scattered field and dropped the constant factor $(\omega/2)$ preceding every integral. Equation (2.24) relates extinction on the left-hand side to the sum of scattered and absorbed powers on the right-hand side. For intuition and compactness, it is helpful to rewrite equations like Eq. (2.24) in a matrix/vector form. We can assume any arbitrarily high-resolution discretization in which $\mathbf{P}(\mathbf{x})$ becomes a vector $\mathbf{p}$, the integral operator $\int_V \mathbb{G}(\mathbf{x}, \mathbf{x}')\, d\mathbf{x}'$ becomes a matrix $\mathbb{G}_0$, and integrals of the conjugate of a field $\mathbf{a}(\mathbf{x})$ with another $\mathbf{b}(\mathbf{x})$ are replaced with vector inner products $\mathbf{a}^\dagger \mathbf{b}$. It is also helpful to define a material parameter $\xi(\mathbf{x}) = -1/\chi(\mathbf{x})$, and a corresponding (diagonal) matrix $\xi = -\chi^{-1}$. With these notational changes, Eq. (2.24) can be rewritten as

$$\operatorname{Im}\left(\mathbf{e}^\dagger_{\text{inc}} \mathbf{p}\right) = \mathbf{p}^\dagger \left[\operatorname{Im} \mathbb{G}_0 + \operatorname{Im} \xi\right] \mathbf{p}. \tag{2.25}$$

This is the vectorized version of the optical theorem, and it illuminates some of the mathematical structure embedded in this particular version of power conservation. The left-hand side is a linear function of the polarization field $\mathbf{p}$, while the right-hand side is a quadratic function. Moreover, in passive systems, the absorbed and scattered powers are nonnegative quantities. This nonnegativity is embedded in the matrices (operators) $\operatorname{Im} \mathbb{G}_0$ and $\operatorname{Im} \xi$, both of which are positive semidefinite (denoted by "≥ 0") in passive systems:

$$\operatorname{Im} \mathbb{G}_0 \geq 0, \tag{2.26}$$

$$\operatorname{Im} \xi \geq 0. \tag{2.27}$$

The positive semidefinite nature of these matrices implies that the right-hand side of Eq. (2.25) is a convex quadratic functional of $\mathbf{p}$. Hence, Eq. (2.25) can be interpreted as an ellipsoid (as opposed to a hyperboloid) in the high-dimensional space occupied by $\mathbf{p}$.

A key feature of Eq. (2.25), and the conservation laws to follow, is that it is **domain oblivious** [105]. Suppose we enforce that constraint on a high-symmetry domain, such as a sphere or half-space, where the operator $\mathbb{G}_0$ might be easy to construct. Of course, enforcing Eq. (2.25) will enforce power conservation on the sphere itself. But it *also enforces power conservation on all sub-domains of the sphere*. This is not obvious—the operator $\mathbb{G}_0$ is different for every choice of domain and range, and once we have chosen a sphere for both, it seems that we are stuck with only the sphere domain. The key, however, is the appearance of $\mathbf{p}$ in each term of Eq. (2.25), and twice on the right-hand side. To enforce Eq. (2.25) on a smaller sub-domain, instead of changing the domain and range of the operator, we can instead enforce the polarization $\mathbf{p}$ to be zero at each point outside the sub-domain but inside the enclosing domain. On the right-hand side, this effectively changes both the domain and range of $\mathbb{G}_0$, while on the left-hand side, it nulls any extinction contribution from outside the sub-domain. Hence, the conservation law of Eq. (2.25), and all of the volume-integral-based conservation laws to follow, is domain oblivious.

Power conservation via the optical theorem has led to a surprisingly wide array of bounds and fundamental limits in electromagnetic systems. The key idea is to drop the full Maxwell-equation constraint that is implicit in any design problem and replace it with only the power-conservation expression of Eq. (2.25). Even with just this single constraint, surprisingly good bounds can be attained. As an example, consider systems where absorptive losses are more important than radiation/scattering losses. In such systems, we can drop the $\operatorname{Im}\mathbb{G}_0$ term in the optical theorem of Eq. (2.25) and use its positivity to write a constraint that absorbed power, be less than or equal to extinction:

$$\mathbf{p}^\dagger (\operatorname{Im}\xi)\,\mathbf{p} \le \operatorname{Im}\left(\mathbf{e}_{\text{inc}}^\dagger \mathbf{p}\right). \tag{2.28}$$

This constraint implies a bound on the strength of the polarization field, because the left-hand-side term is quadratic (and positive-definite) in $\mathbf{p}$, while the right-hand side is linear in $\mathbf{p}$. A few steps of variational calculus [106] can identify the largest polarization-field strength that can be induced in a scatterer:

$$\|\mathbf{p}\|^2 = \mathbf{p}^\dagger \mathbf{p} = \int_V |\mathbf{P}(\mathbf{x})|^2 \, \mathrm{d}\mathbf{x} \le \frac{\|\mathbf{e}_{\text{inc}}\|^2}{\operatorname{Im}\xi} = \frac{|\chi|^2}{\operatorname{Im}\chi} \int_V |\mathbf{E}_{\text{inc}}(\mathbf{x})|^2 \, \mathrm{d}\mathbf{x}. \tag{2.29}$$

We have a first bound: in a lossy material, wherein $\operatorname{Im}\chi > 0$, there is a bound on the largest polarization currents that can be induced in a scatterer, based only on the material properties and the energy of the *incident* wave in the scattering region. Polarization currents beyond this strength would have absorbed powers larger than their extinction, implying an unphysical negative scattered power.

Beyond the strength of the polarization field itself, one can use similar variational-calculus arguments to identify bounds on wide-ranging quantities: extinction, absorption, and scattering; in bulk materials [106], 2D materials [107], and lossy environments [108, 109]; high-radiative-efficiency scatterers [110]; and even near-field quantities such as local density of states [46, 106], near-field radiative heat transfer [107, 111], and Smith–Purcell radiation [112]. As a canonical example, let us consider the extinction, absorption, and scattering cross sections of a scattering body with volume V, susceptibility χ, and a plane-wave incident field. Cross sections $\sigma_{\text{ext,abs,scat}}$ are the relevant powers divided by the intensity of the incident wave; the corresponding bounds are

$$\frac{\sigma_{\text{abs,scat,ext}}}{V} \leq \frac{\beta\omega}{c} \frac{|\chi|^2}{\operatorname{Im}\chi} \qquad \beta_{\text{abs,ext}} = 1, \beta_{\text{scat}} = \frac{1}{4}. \tag{2.30}$$

Per-volume cross sections are bounded above by the frequency of the incoming waves and the material susceptibilities. Plasmonic nanoparticles can approach these bounds [106, 107, 113].

One subtlety that arises in the near field (whose bounds are discussed in depth in Sect. 2.4) is which conservation laws to use. The absorption- and extinction-based constraint of Eq. (2.28) may not be ideal for local density of states, for example, as the power radiated by a dipole is not exactly the same as the power extinguished by a nearby scatterer. (There is a separate pathway for the dipole to radiate directly to the far field, and this radiation can destructively/constructively interfere with waves scattered by the scatterer.) The optical theorem of Eq. (2.25) arises from equating fluxes through a surface surrounding the scatterer. Instead, in the near field, one can draw a surface around the dipolar source itself. Then one can identify new conservation laws, which now relate the total power radiated by the dipole (the LDOS) to the sum of power absorbed in the scatterer and power radiated to the far field.

In some systems, radiation losses are the limiting factor rather than absorption losses. Prominent examples include metals at low frequencies and low-loss dielectrics. In these systems, the key component of the optical theorem of Eq. (2.25) is the radiation-loss term with $\operatorname{Im}\mathbb{G}_0$, not the absorption-loss term. Of course, absorption must be positive, so we can drop it and replace the optical theorem with a second inequality version:

$$\mathbf{p}^\dagger \left(\operatorname{Im}\mathbb{G}_0\right) \mathbf{p} \leq \operatorname{Im}\left(\mathbf{e}_{\text{inc}}^\dagger \mathbf{p}\right). \tag{2.31}$$

Although the $\operatorname{Im}\mathbb{G}_0$ matrix may appear daunting, we typically use high-symmetry volumes for our designable domains, and we can use analytical or semi-analytical forms of $\operatorname{Im}\mathbb{G}_0$ in those domains. (Such usage does not restrict the validity of the bound to only the high-symmetry domain; as discussed above, this expression is domain oblivious.) One common high-symmetry domain is a sphere, in which case $\operatorname{Im}\mathbb{G}_0$ can be written in a basis of vector spherical waves [114–116]. Application of this approach to the question of maximum cross sections yields different bounds from the ones of Eq. (2.30). One must limit the number of spherical waves that can contribute to the scattering process; allowing only the first N electric multipole leads to maximum cross sections proportional to the square of the wavelength, λ:

$$\sigma_{\text{abs,scat,ext}} \leq \frac{\beta\lambda^2}{\pi}\left(N^2 + 2N\right) \qquad \beta_{\text{scat,ext}} = 1, \beta_{\text{abs}} = \frac{1}{4}, \tag{2.32}$$

with double the value if the magnetic vector spherical waves can be equally excited. Note the different values of β for absorption and scattering in the absorption-limited case of Eq. (2.30) versus the radiation-limited case of Eq. (2.32). The different coefficients arise because of the different conditions under which maximum extinction occurs. In an absorption-dominated system, arbitrarily small scattering is possible (in principle), such that the maximum for extinction and absorption coincide, while the scattered-power maximum requires a reduction in absorption relative to extinction and a $1/4$ coefficient to account for the matching that must occur. The opposite occurs in scattering-limited systems, where absorption can be arbitrarily small (in principle), the maximum for extinction and scattering coincide, and an extra factor of $1/4$ is introduced when absorption is to be maximized. The bound of Eq. (2.32) was originally derived for antenna applications or spherically symmetry scatterers via long and/or restrictive arguments [117–121]; the single conservation law of Eq. (2.31) is sufficient to derive Eq. (2.32) in quite general settings [122, 123]. (An interesting precursor to the global-conservation-law approach is Ref. [124], which identifies metrics that intrinsically have bounded optima over polarization currents, even *without* any constraints.)

Of course, in some settings, both absorption and radiation losses will be important to capture what is possible, and the bounds of Eqs. (2.30, 2.32) may not be sufficient. It is possible to capture both loss mechanisms in a single bound by using the entirety of the optical theorem, Eq. (2.25), without dropping either term. This was first recognized in Refs. [116, 125, 126]. Ref. [116] used this approach to derive bounds on the thinnest possible perfect absorber. (Or, conversely, the maximum absorption of an arbitrarily patterned thin film with a given maximum thickness.) Cross-sectional bounds given in Ref. [116, 125, 126] are generalizations of the two bounds listed above, Eqs. (2.30,2.32), containing each as separate asymptotic limits. At normal incidence, one can derive a simple transcendental equation for the minimum thickness, h_{min}, of a perfect absorber with material parameter $\xi = -1/\chi$:

$$h_{\text{min}} = \left(\frac{2\lambda}{\pi}\right) \frac{\operatorname{Im}\xi(\omega)}{1 - \operatorname{sinc}^2(\omega h_{\text{min}}/c)}. \tag{2.33}$$

This approach has been successfully applied to the identification of the minimum thickness of a metasurface reflector [127].

Finally, at the global-conservation level, one can go one step further, as first recognized in Refs. [125, 126]. The optical theorem of Eq. (2.25) represents the conservation of *real* power across the volume of a scatterer, which can be understood as the conservation of the real part of the Poynting vector through any bounding surface. Additionally, the imaginary part of the Poynting vector corresponds to what is known as *reactive* power [103]. The complex-valued version of the optical theorem is essentially the same as Eq. (2.25) but without the imaginary part in any of the terms; a careful analysis leads to the generalized optical theorem:

$$-\mathbf{p}^\dagger \mathbf{e}_{\text{inc}} = \mathbf{p}^\dagger \left[\mathbb{G}_0 + \xi\right] \mathbf{p}. \tag{2.34}$$

The real and imaginary parts of Eq. (2.34) now offer *two* global conservation laws that must be satisfied in any scatterer. The real-power conservation law accounts for absorption- and radiation-loss pathways, while the reactive-power conservation law accounts for resonance conditions in real materials. The latter has been shown to be beneficial for tightening bounds in plasmonic materials that are relatively large (wavelength-scale sizes are quite large for plasmonic resonators) or which have very large negative real susceptibilities and/or very small imaginary susceptibilities [126]. This approach has been applied to bounds in cloaks [128] and focusing efficiency [129]. Equation (2.34) can be derived in one step from the volume-integral equation [130] (or Lippmann–Schwinger equation), which in this notation reads $[\mathbb{G}_0 + \xi]\,\mathbf{p} = -\mathbf{e}_{\text{inc}}$, simply by taking the inner product of that equation with $\mathbf{p}$.

In this section, we have seen that the optical theorem, written over the volume polarization fields induced in a scatterer, offers a single (or two) global conservation laws that can be used to identify bounds in wide-ranging applications. In Sect. 2.3.3, we show that it is also a starting point for generating an infinite number of "local" conservation laws. First, however, we will explore an approach that is closely related to global conservation laws: so-called "channel" bounds.

2.3.2 *Channel Bounds*

In this section, we explore another technique for identifying bounds to what is possible: decomposing power transfer into a set of independent or orthogonal power-carrying "channels." Then the upper limits distill to the maximum power (or alternative objective) per channel multiplied by the number of possible channels.

A particularly elegant formulation of channels was proposed by D. A. B. Miller and colleagues in the early 2000s [131–134]. Consider a transmitter region that wants to communicate (i.e., send information/energy) to a receiver region, and a vacuum (or background) Green's function operator $\mathbb{G}_0$ comprising the fields in the receiver from sources in the transmitter. How many communication channels are possible? There is a simple, rigorous mathematical answer to this question: if one decomposes the $\mathbb{G}_0$ operator via a singular value decomposition (SVD) [135],

$$\mathbb{G}_0 = \mathbb{U}\mathbb{S}\mathbb{V}^\dagger, \tag{2.35}$$

then each pair of singular vectors forms an independent channel. The singular value decomposition encodes orthogonality and normalization. For example, the first right singular value, which we can call $\mathbf{v}_1$, radiates *only* to the first left singular vector $\mathbf{u}_1$ in the receiver region, and the strength of this connection is given exactly by the first singular value, which we can call s_1. This triplet $(\mathbf{v}_1, \mathbf{u}_1, s_1)$ mathematically define a **communication channel**, as are all the pairs in the SVD. There cannot be an infinite number of such channels with arbitrarily large strengths, as the channel strengths obey a simple sum rule related to the integral of the Green's function over the transmitter and receiver volumes:

$$\sum_i |s_i|^2 = \text{Tr}\left(\mathbb{S}^\dagger \mathbb{S}\right) = \text{Tr}\left(\mathbb{G}_0^\dagger \mathbb{G}_0\right) = \int_{V_T} \int_{V_R} \|\mathbb{G}_0(\mathbf{x}_T, \mathbf{x}_S)\|^2 \, \text{d}\mathbf{x}_T \, \text{d}\mathbf{x}_R. \tag{2.36}$$

One can define more granular bounds as well: for any transmitter/receiver regions enclosed within high-symmetry bounding domains, one can identify upper limits for each individual singular value [136]. The singular values must decay exponentially in two-dimensional systems, whereas in three dimensions, their decay can be sub-exponential. This SVD-based decomposition of Eq. (2.35) implicitly uses a field-energy normalization; one can alternatively use power-transfer normalizations and arrive at related bounds for the communication strength between two volumes [137–139]. Each of these is a powerful approach for free-space communication systems such as MIMO [140, 141]. More generally, they capture a general truth about free-space propagation: it can always be decomposed into orthogonal, power-carrying channels.

In the near field, however, evanescent waves do *not* offer an equivalent set of power-carrying channels. Evanescent waves obey different mathematical orthonormalization rules, which are consistent with the following fact: evanescent waves decaying (or growing) in one direction *cannot* carry power and power can be transmitted only in the presence of oppositely directed evanescent waves [142]. A prototypical example: a single interface can only exhibit total internal reflection alongside evanescent-wave excitation, whereas the introduction of second interface, and counter-propagating evanescent waves, can lead to the tunneling of power through a "barrier."

In lieu of the general SVD approach, in high-symmetry scenarios, it is often possible to decompose power transfer in a high-symmetry basis. For example, a spherically symmetric scatterer preserves the quantum numbers of incoming vector spherical waves and cannot scatter into waves of different quantum numbers, which implies that each vector spherical wave comprises a "channel" for incoming and outgoing radiation. Similarly, in planar systems, the in-plane (parallel) wavevector $\mathbf{k}$ is a conserved quantity, in which case one can isolate the scattering process into each $\mathbf{k}$-dependent propagating and evanescent plane wave. One cannot define free-space evanescent-wave channels, per the orthonormalization discussion above, but a more complete analysis can lead to $\mathbf{k}$-dependent transfer coefficients that are readily interpretable as a channel-based power decomposition. We discuss the successful application of these ideas to near-field radiative heat transfer in Sect. 2.4.1.4. A word of caution is important, however: the assumption of a high-symmetry structure dramatically limits the set of structures to which such bounds apply, and in many scenarios, it has been found that the symmetry-independent approaches of global conservation laws (previous section) and local conservation laws (next section) yield both tighter *and* more general bounds.

2.3.3 Local Conservation Laws

In the global-conservation-law section of Sect. 2.3.1, we discussed that one or two conservation-of-power constraints are already sufficient for bounds in many scenarios of interest. Of course, one or two constraints cannot capture every objective of interest: if, for example, one wanted to know the largest average response over multiple incident fields, certainly more constraints are needed. Thankfully, it turns out that there is a systematic way to generate a large number of conservation-law constraints for any nanophotonic design problem of interest.

The key is to identify ***local*** **conservation laws** that apply at every point within the scatterer [105, 115]. These conservation laws can be "built" from a volume-integral formulation of the underlying governing dynamics, but we will use a more intuitive approach to develop them. The "generalized optical theorem" is written in Eq. (2.34) in vector/matrix notation; the equivalent integral expression is

$$\int_V \int_V \mathbf{P}^*(\mathbf{x})\mathbb{G}_0(\mathbf{x},\mathbf{x}')\mathbf{P}(\mathbf{x}')\,\mathrm{d}\mathbf{x}\,\mathrm{d}\mathbf{x}' + \int_V \mathbf{P}^*(\mathbf{x})\xi(\mathbf{x})\mathbf{P}(\mathbf{x})\,\mathrm{d}\mathbf{x} = -\int_V \mathbf{P}^*(\mathbf{x})\mathbf{E}_{\mathrm{inc}}(\mathbf{x})\,\mathrm{d}\mathbf{x}. \tag{2.37}$$

To formulate local conservation laws, we simply recognize the following: for the first integral over the entire scatterer V that appears in every term, we can replace V with $V_{\mathbf{x}}$, where $V_{\mathbf{x}}$ is an infinitesimal volume centered around any point $\mathbf{x}$ within the scatterer. With this replacement, the dependence on $\mathbf{x}$ of each integrand becomes approximately constant (exactly constant in the zero-volume limit), and the integral simplifies to just multiplication by the volume $V_{\mathbf{x}}$, which appears in every term and can be cancelled, leaving

$$\int_V \mathbf{P}^*(\mathbf{x})\mathbb{G}_0(\mathbf{x},\mathbf{x}')\mathbf{P}(\mathbf{x}')\,\mathrm{d}\mathbf{x}' + \mathbf{P}^*(\mathbf{x})\xi(\mathbf{x})\mathbf{P}(\mathbf{x}) = -\mathbf{P}^*(\mathbf{x})\mathbf{E}_{\mathrm{inc}}(\mathbf{x}). \tag{2.38}$$

More rigorous justifications are given in Refs. [105, 115] and can proceed either from the volume-integral formulation or, with equal validity, by converting the volume integrals around $V_{\mathbf{x}}$ into surface integrals (via the divergence theorem), in which case Eq. (2.38) is interpreted simply as flux conservation through the surface of $V_{\mathbf{x}}$. To convert Eq. (2.38) to the more compact vector notation, we denote new matrices $\mathbb{D}_i$ as diagonal matrices of all zeros except a single 1 at diagonal entry i, in which case Eq. (2.38) can be written as

$$\mathbf{p}^\dagger\mathbb{D}_i\left(\mathbb{G}_0 + \xi\right)\mathbf{p} = -\mathbf{e}_{\mathrm{inc}}^\dagger\mathbb{D}_i\mathbf{x}, \tag{2.39}$$

which must hold for all spatial locations' index by i. Equation (2.39) offers an infinite set of local conservation laws that must be satisfied for any (linear) scattering body. Moreover, just as for the global conservation laws, Eq. (2.39) is domain oblivious. Hence, if the constraints of Eq. (2.39) lead to a bound, then that bound will apply to all sub-domains (or "patterns") contained therein.

There is a systematic procedure that one can follow for identifying fundamental limits using the constraints of Eq. (2.39). If one discards the Maxwell differential (or integral) equations, and only imposes the constraints of Eq. (2.39), the resulting optimization problem has the form of a **quadratically constrained quadratic program**, or **QCQP**. QCQPs arise across many areas of science and engineering [143–148], and there are many mathematical approaches for solving them. One in particular is useful for identifying bounds: one can relax a QCQP to a **semidefinite program (SDP)** in a higher-dimensional space [145, 149], which can be solved for its global optimum by standard algorithms in polynomial time [150, 151]. The solution of the SDP is guaranteed to be a bound, or fundamental limit, on the solution of the problem of interest. (The semidefinite program can also be regarded as the "dual" [151] of the dual of the QCQP [152], which is another way to see that it leads to bounds.)

Thus, local conservation laws lead to a systematic procedure for identifying bounds, or fundamental limits, to electromagnetic quantities of interest. One replaces the governing Maxwell equations with the domain-oblivious conservation-law constraints of Eq. (2.39), forms a semidefinite program from the objective and constraints, and solves the SDP to find a bound. To avoid the computational complexity of using all of the constraints, one can iteratively select only the "maximally violated" constraints, for rapid convergence to the bound of interest [105]. A mathematically oriented review of bounds related to Eq. (2.39) is given in Ref. [153]. Extensions of various types are given in Ref. [154] (multi-functionality), Ref. [155] (quantum optimal control), Ref. [156] (efficiency metrics), and Ref. [157] (other physical equations).

2.3.4 Sum Rules

Whereas the three previous sections primarily emphasized fundamental limits across spatial degrees of freedom, at a single frequency, **sum rules** center around *spectral* degrees of freedom and constraints related to bandwidth. Sum rules are a prime example of applied complex analysis. Most often, they are taught and discussed in the context of material susceptibilities, so we will start there, before focusing on our key interest, scattering problems. In the Appendix Sect. 2.6, we provide a short review of key results from complex analysis, and the intuition behind their derivations, culminating in the Cauchy residue theorem that is used for all sum rules. Cauchy's residue theorem, for our purposes, can be distilled to the following statement. Consider a function $f(z)$ that is **analytic** (has no poles) in some domain D in the complex z plane. (Below, the analytic variable z will be the frequency ω.) Then the function $f(z)/(z - z_0)$ has a simple pole at z_0, for z_0 in D, and any integral of this function along a closed contour in D containing z_0 simplifies to the value of the function at the pole:

$$\oint_{\gamma} \frac{f(z)}{z - z_0} = 2\pi i f(z_0), \tag{2.40}$$

where $f(z_0)$ is the "residue" of the function $f(z)/(z - z_0)$. Now let us put Cauchy's residue theorem to use.

Consider a material susceptibility χ that relates an electric field $\mathbf{E}$ to an induced polarization field $\mathbf{P}$. Typically, we might directly consider the frequency-domain relationship of these variables:

$$\mathbf{P}(\omega) = \chi(\omega)\mathbf{E}(\omega), \tag{2.41}$$

where we are suppressing spatial dependencies in these expressions for simplicity. (All of the position dependencies are straightforward.) This multiplicative frequency-domain relation arises from a *convolutional* time-domain relationship: the polarization field at a given field is related to the electric field at all other times convolved with the susceptibility function (as a function of time):

$$\mathbf{P}(t) = \int \chi(t - t')\mathbf{E}(t')\,\mathrm{d}t'. \tag{2.42}$$

(We do not use different variables for the time- and frequency-domain definitions; the domain should be clear in each context.) **Causality** is the formal specification that *cause precedes effect*. Material susceptibilities are causal: the polarization field cannot arise before the electric field has arrived, which means that for some origin of time, the susceptibility function is identically zero at all preceding times:

$$\chi(t - t') = 0 \qquad \text{for } t < t'. \tag{2.43}$$

In the usual Fourier-transform relation between the time- and frequency-domain susceptibility functions, then, one can set the lower limit of the time-domain integral to be 0:

$$\chi(\omega) = \frac{1}{2\pi}\int_{-\infty}^{\infty} \chi(t)e^{i\omega t}\,\mathrm{d}t = \frac{1}{2\pi}\int_{0}^{\infty} \chi(t)e^{i\omega t}\,\mathrm{d}t. \tag{2.44}$$

Setting the lower limit of the integral to 0 has an important ramification. Let us assume the susceptibility takes a finite value for all real frequencies. (Metals are an exception, with divergent susceptibilities at zero frequencies, but known modifications to the rules below can be developed to account for this singularity [158, 159].) This implies that the integral of Eq. (2.44) converges to the correct finite value at each frequency. Now let us consider a **complex-valued** frequency $\omega = \omega_0 + i\Delta\omega$. If we insert this frequency into Eq. (2.44), we find

$$\chi(\omega_0 + i\Delta\omega) = \frac{1}{2\pi}\int_{0}^{\infty} \chi(t)e^{i\omega_0 t}e^{-\Delta\omega t}\,\mathrm{d}t, \tag{2.45}$$

which is equivalent to the integral of Eq. (2.44), except now that there is the additional exponential decay term $e^{-\Delta\omega t}$ in the integrand. This exponential decay term can only aid in convergence, and under appropriate technical assumptions (e.g., Titchmarsh's theorem [160]), one can prove the intuitive idea that Eq. (2.45) cannot diverge for any $\Delta\omega$. This implies that the material susceptibility $\chi(\omega)$ is analytic in the upper-half of the complex-frequency plane. (Conversely, frequencies in the lower half would have the exponentially diverging term $e^{\Delta\omega t}$ in their integrands, leading to divergences at certain frequencies, which is where the system *resonances* are located.) Hence, we can use Cauchy's integral theorem of Eq. (2.40) with $\chi(\omega)$ as the analytic function in the numerator of the integrand. The typical usage of the integral theorem is to select a pole on the real axis (or, technically, in the limit of approaching the real axis from above), and to use a contour C that follows the real line, includes a semicircular deformation around ω', and then closes along a semicircle approaching infinity in the upper-half plane. This contour actually does not enclose *any* poles, instead "side-stepping" the real-axis pole, at a frequency we denote by ω. Hence, we have

$$\oint_C \frac{\chi(\omega')}{\omega' - \omega}\, \mathrm{d}\omega' = 0. \tag{2.46}$$

The integral over C can be broken into three components: the principal-valued integral along the real axis from negative infinity to infinity (skipping ω'), the semicircular arc going into the upper-half plane, and the semicircular arc rotating clockwise around ω. The second of these terms is zero (for sufficiently rapid decay of $\chi(\omega)$), while the third term is simply $-i\pi\chi(\omega)$ (half of the typical Cauchy residue term since it is half of a circle, with a negative sign for the clockwise rotation). Equating the negative of the third term to the first, we have

$$i\pi\chi(\omega) = \int_{-\infty}^{\infty} \frac{\chi(\omega')}{\omega' - \omega}\, \mathrm{d}\omega'. \tag{2.47}$$

We can take the imaginary part of both sides, and use the symmetry of χ around the origin, $\chi(-\omega) = \chi^*(\omega)$, to arrive at one of the **Kramers–Kronig** (KK) relations for a material susceptibility:

$$\operatorname{Re}\chi(\omega) = \frac{2}{\pi}\int_0^{\infty} \frac{\omega' \operatorname{Im}\chi(\omega')}{(\omega')^2 - \omega^2}\, \mathrm{d}\omega'. \tag{2.48}$$

The counterpart KK relation relates the imaginary part of $\chi(\omega)$ to an integral involving the real part. These KK relations are the foundations of sum rules. There are two special pole frequencies ω at which we may have additional information about the material response: infinity frequency and zero frequency (statics). In the limit of infinitely large frequencies, all materials become transparent, with a susceptibility that must scale as

$$\chi(\omega) \to -\frac{\omega_p^2}{\omega^2} \qquad \text{as } \omega \to \infty, \tag{2.49}$$

where ω_p is a constant proportional to the total electron density of the material [158, 159]. Inserting this asymptotic limit into the KK relation of Eq. (2.48), we find our first example of a sum rule:

$$\int_0^\infty \omega \operatorname{Im} \chi(\omega) \, \mathrm{d}\omega = \frac{\pi \omega_p^2}{2}. \tag{2.50}$$

Equation (2.50) is known as either the **TRK sum rule** or the f **sum rule** [158, 159]. It relates the weighted integral of the imaginary part of the susceptibility to simple constants multiplied by the electron density of the material of interest. The quantity $\omega \operatorname{Im} \chi(\omega)$ is proportional to the *oscillator strengths* in single-electron susceptibility models [161]. Alternatively, in the low-frequency limit, one may know the static refractive index n_0 of a given material; inserting $\omega = 0$ in the KK relation of Eq. (2.48) gives the low-frequency sum rule:

$$\int_0^\infty \frac{\operatorname{Im} \chi(\omega)}{\omega} \, \mathrm{d}\omega = \frac{\pi}{2}\left(n_0^2 - 1\right). \tag{2.51}$$

The two sum rules of Eqs. (2.50,2.51) are well-known material sum rules that are useful for spectroscopy [158, 159] as well as for bounds on material properties [162–164]. We have repeated their well-known derivations to familiarize the reader with the machinery of KK relations and sum rules, which we apply next to scattering properties.

Just as the origin for material sum rules was recognition of material susceptibility as a causal (linear) response function, for scattering sum rules, we want to start by recognizing that the electromagnetic field $\mathbf{E}$ generated by a source (presumably current) is also a causal linear response function: $\mathbf{E}$ cannot be nonzero before the current $\mathbf{J}$ is nonzero. Hence, the electric field at all times before an origin must be zero, which again leads to analyticity in the upper-half of the complex-frequency plane. Yet we do not want KK relations for the electric field at specific points in space; we want KK relations (and sum rules) for relevant power quantities. Typical expressions of interest might be the field intensity, $|\mathbf{E}(\mathbf{x}, \omega)|^2$, or the Poynting flux $(1/2) \operatorname{Re}\left[\mathbf{E}^*(\mathbf{x}, \omega) \times \mathbf{H}(\mathbf{x}, \omega)\right]$, at a point $\mathbf{x}$, but *neither* of these quantities is analytic in the upper-half plane. The problematic term in each case is $\mathbf{E}^*(\omega)$. Analyticity is not preserved under complex conjugation, and indeed by symmetry, we know that $\mathbf{E}^*(\omega) = \mathbf{E}(-\omega)$ on the real line; if we try to continue ω into the upper-half plane, the $-\omega$ argument moves into the lower-half plane, where the resonances reside. Hence, $\mathbf{E}^*(\omega)$ can have poles, and the corresponding power terms do not have simple KK relations or sum rules.

We are rescued, again, by the optical theorem. Whereas absorbed and scattered powers always involve conjugated total fields, *extinction*, by virtue of the optical theorem, takes a different form (Eq. (2.7)), which is proportional to the overlap integral of the conjugate of the incident field with the induced polarization field, $\int_V \mathbf{E}_{\text{inc}}^* \cdot \mathbf{P}$. Many common incident fields, such as plane waves of the form $e^{i\omega x/c}$, are analytic *everywhere* in the complex plane, and their conjugates *can* be analytically continued. The polarization field is the product of the analytic material susceptibility with the analytic electric field, and thus is itself analytic. Hence, extinction expressions contain a term that will obey KK relations and sum rules, which we denote $s(\omega)$:

$$P_{\text{ext}}(\omega) = \frac{\omega}{2} \operatorname{Im} \underbrace{\int_V \mathbf{E}_{\text{inc}}^*(\mathbf{x}, \omega) \cdot \mathbf{P}(\mathbf{x}, \omega)\, \mathrm{d}\mathbf{x}}_{s(\omega)}. \tag{2.52}$$

By the arguments laid out above, the quantity $s(\omega)$ is analytic in the upper-half plane. It satisfies the other required assumptions as well (e.g., sufficient decay at infinity) for incident fields such as plane waves; we can immediately write a KK relation for it:

$$\operatorname{Re} s(\omega) = \frac{2}{\pi} \int_0^\infty \frac{\omega' \operatorname{Im} s(\omega')}{(\omega')^2 - \omega^2} \mathrm{d}\omega'. \tag{2.53}$$

Notice that the term in the numerator of the integrand is exactly proportional to extinction; hence, sum rules for the imaginary part of $s(\omega)$ (by analogy with the sum rules for $\operatorname{Im} \chi$) will necessarily be sum rules for extinction. Again paralleling the susceptibility analysis, we can take the limit as $\omega \to \infty$, in which case

$$\begin{aligned} s(\omega) &= \int \mathbf{E}_{\text{inc}}^*(\mathbf{x}, \omega) \cdot \mathbf{P}(\mathbf{x}, \omega)\, \mathrm{d}\mathbf{x} \\ &\to -\frac{\omega_p^2}{\omega^2} \int_V |\mathbf{E}_{\text{inc}}(\mathbf{x}, \omega)|^2\, \mathrm{d}\mathbf{x} \\ &= -\frac{\omega_p^2}{\omega^2} |\mathbf{E}_0|^2\, V, \end{aligned} \tag{2.54}$$

where $\mathbf{E}_0$ is the (constant) vector amplitude of the plane wave and V is the volume of the scatterer. Evaluating the KK relation for $s(\omega)$, Eq. (2.53), in the high-frequency limit gives a sum rule for the imaginary part of $s(\omega)$:

$$\int_0^\infty \omega \operatorname{Im} s(\omega)\, \mathrm{d}\omega = \frac{\pi \omega_p^2}{2} |\mathbf{E}_0|^2\, V, \tag{2.55}$$

which in turn implies a sum rule for extinction (via Eq. (2.52)):

$$\int_0^\infty P_{\text{ext}}(\omega)\, \mathrm{d}\omega = \frac{\pi \omega_p^2}{4} |\mathbf{E}_0|^2\, V. \tag{2.56}$$

Equation (2.56) dictates that the total integrated extinction of any scattering body is fixed by the amplitude of the incident plane wave and the total number of electrons in the scatterer (from the product of ω_p^2 with V) and is otherwise independent of the shape, resonance profile, and any other characteristics of the scattering body.

Just as for a material susceptibility, one can also derive a sum rule for P_{ext} by setting $\omega = 0$ in the KK relation for $s(\omega)$, Eq. (2.53). The key low-frequency information we can utilize is that the induced dipole moment of the scatterer is related to the incident field via a polarizability tensor $\boldsymbol{\alpha}$. Following a few algebraic steps [165] paralleling the low-frequency material sum rule, one similarly finds a sum rule for the integral of $P_{\text{ext}}(\omega)/\omega^2$. The term $(1/\omega^2)\,\mathrm{d}\omega$ is exactly proportional to $\mathrm{d}\lambda$, where $\lambda = 2\pi c/\omega$ is the wavelength, so this sum rule is often written as a sum rule over wavelength:

$$\int_0^\infty P_{\text{ext}}(\omega)\,\mathrm{d}\lambda = \pi^2 \mathbf{E}_0 \cdot \boldsymbol{\alpha}\mathbf{E}_0. \tag{2.57}$$

There is an additional magnetic polarizability term in materials with a nonzero magnetostatic response [165]. Interestingly, Eq. (2.57) has different dependencies than Eq. (2.56): the polarizability has a weak dependence on material, but a strong dependence on shape. The low-frequency sum rule implies that scattering bodies with the same size and shape, but made of different materials, can have nearly identical wavelength-integrated extinctions. Moreover, electrostatic polarizabilities obey "domain monotonicity" bounds that dictate that the quantity $\mathbf{E}_0 \cdot \boldsymbol{\alpha}\mathbf{E}_0$ must increase as the scatterer domain increases in size, such that one can bound integrated extinction via high-symmetry enclosures for which the right-hand side of Eq. (2.57) often takes a simplified analytical form. Taken together, the high- and low-frequency sum rules of Eqs. (2.56, 2.57) comprise strong constraints on the possible scattering lineshapes of arbitrary scatterers.

Equations (2.56, 2.57) are classical sum rules with a long history. The high-frequency sum rule, Eq. (2.56), was known at least as early as 1963 [166], when the connection to material-susceptibility sum rules was first made. A specialized version of the low-frequency sum rule, Eq. (2.57), was first proposed by Purcell in 1969 [167], in order to bound the minimum volume occupied by interstellar dust. It was generalized to arbitrary scattering bodies in Ref. [165], where the monotonicity bounds (originally developed by Jones [168]) were connected to the low-frequency sum rules. For many years, it seemed that plane-wave extinction might be the *only* scattering quantity for which sum rules can be derived. In recent years, however, it has been recognized that near-field local density of states has a similar form—it is the real or imaginary part of an amplitude, instead of the squared magnitude of an amplitude—for which sum rules can also be derived. We describe this sum rule and its implications in Sect. 2.4.2.

2.4 Fundamental Limits in the Near Field

We have set the stage: we have introduced near-field optics, defined many of the response functions of interest, and described tools formulated for electromagnetic-response bounds. In this section, we describe how these ingredients come together for bounds and fundamental limits to near-field response. We identify different bounds–and the different techniques required to derive them–based on the frequency range of interest: a single frequency (Sect. 2.4.1), all frequencies (Sect. 2.4.2), and finite, nonzero bandwidths (Sect. 2.4.3). We leave bounds for mode volume, which seemingly requires very different techniques, to the final section of the chapter (Sect. 2.4.4).

2.4.1 Single-Frequency Bounds

In Sect. 2.3, we described two techniques that can be used to identify single-frequency bounds to any linear-electromagnetic response function of interest: conservation laws and channel decompositions. In this subsection, we summarize how one can adapt, specialize, and/or combine those approaches in the near field, for spontaneous-emission and CDOS engineering, Smith–Purcell radiation enhancements, and spectral NFRHT response.

2.4.1.1 Spontaneous Emission

The canonical near-field quantity is LDOS, which as discussed in Sect. 2.2.1 is proportional to the spontaneous-emission rate of an electric dipole at a given location. In a closed system, the LDOS is a sum of delta functions over the modes of the system, in which case the LDOS diverges at the modal frequencies. In an open system, however, the modal intuition no longer applies, leading to the more general Green's function expression of Eq. (2.4). This scattering quantity lends itself well to the conservation-law-based scattering-response bounds described in Sect. 2.3.3.

We can repeat here the Green's function expression for LDOS, which we will denote in this section by $\rho(\mathbf{x}, \omega)$:

$$\rho(\mathbf{x}, \omega) = \frac{1}{\pi \omega} \operatorname{Tr} \operatorname{Im} \mathbb{G}(\mathbf{x}, \mathbf{x}, \omega). \tag{2.58}$$

The trace of the Green's function can be computed with a summation over three orthogonal unit vectors $\mathbf{s}_j$, for $j = 1, 2, 3$, in which case the trace can be interpreted as the incoherent summation of the fields from three dipoles with amplitudes $\varepsilon_0 \mathbf{s}_j$. There is an initial impediment to applying the conservation-law framework to this expression: it is not written explicitly as a function of the polarization fields, whose constraints are critical to meaningful bounds. This impediment is easily hurdled: one can decompose the Green's function into its incident and scattered components. The scattered fields are the convolutions of the free-space Green's function matrix $\mathbb{G}_0$ from the scattering domain to the dipole point; by reciprocity, the overlap of $\mathbf{s}_j$ with $\mathbb{G}_0$ is the field incident upon the scattering body V. By this line of reasoning, for a scalar isotropic medium (the general bianisotropic case is derived in Ref. [106]), one can rewrite LDOS as

$$\rho(\mathbf{x}, \omega) = \rho_0(\omega) + \frac{1}{\pi\omega} \operatorname{Im} \sum_j \int_V \mathbf{E}_{\text{inc},\mathbf{s}_j} \cdot \mathbf{P}_{s_j} \, dV, \tag{2.59}$$

where $\rho_0(\omega)$ is the free-space LDOS (which is position-independent and given below Eq. (2.4)), and the $\mathbf{s}_j$ subscript encodes the three dipole orientations. Using the same discretized vector/matrix notation as we initiated with Eq. (2.25), this expression can equivalently be written as

$$\rho(\mathbf{x}, \omega) = \rho_0(\omega) + \frac{1}{\pi\omega} \operatorname{Im} \sum_j \mathbf{e}_{\text{inc},\mathbf{s}_j}^T \mathbf{p}_{s_j}. \tag{2.60}$$

Now we see that LDOS is a linear function of the polarization fields induced in the scattering body. We want to know the largest possible value of LDOS, of Eq. (2.60), subject to the Maxwell equations, but of course the latter constraint contains all of the complexity of the design problem. Instead, we drop the Maxwell-equation constraint and impose only one of the conservation laws of Sect. 2.3. To start, we can impose the conservation law that absorbed power be smaller than extinguished power, of Eq. (2.28), which leads to the optimization problem:

$$\begin{aligned} \max_{\mathbf{p}_{\mathbf{s}_j}} \quad & \frac{1}{\pi\omega} \operatorname{Im} \sum_j \mathbf{e}_{\text{inc},\mathbf{s}_j}^T \mathbf{p}_{\mathbf{s}_j} \\ \text{s.t.} \quad & (\operatorname{Im}\xi)\, \mathbf{p}_{\mathbf{s}_j}^\dagger \mathbf{p}_{\mathbf{s}_j} \le \operatorname{Im}\left(\mathbf{e}_{\text{inc},\mathbf{s}_j}^\dagger \mathbf{p}_{\mathbf{s}_j}\right). \end{aligned} \tag{2.61}$$

Treating each dipole orientation $\mathbf{s}_j$ independently, one can find from a Lagrangian analysis that the optimal $\mathbf{p}_{\mathbf{s}_j}$ comprises a linear combination of $\mathbf{e}_{\text{inc},\mathbf{s}_j}$ and $\mathbf{e}_{\text{inc},\mathbf{s}_j}^*$; in the near field, where the incident field and its conjugate are nearly identical, and the LDOS is dominated by its scattered-field contribution, we ultimately find the following bound [106]:

$$\rho(\mathbf{x}, \omega) \le \frac{1}{\pi\omega} \frac{|\chi(\omega)|^2}{\operatorname{Im}\chi(\omega)} \sum_{\mathbf{s}_j} \left\|\mathbf{e}_{\text{inc},\mathbf{s}_j}\right\|^2 = \frac{1}{\pi\omega} \frac{|\chi(\omega)|^2}{\operatorname{Im}\chi(\omega)} \sum_{\mathbf{s}_j} \int_V \left|\mathbf{E}_{\text{inc},\mathbf{s}_j}\right|^2 \mathrm{d}\mathbf{x}. \tag{2.62}$$

Normalizing by the free-space electric LDOS $\rho_0(\omega)$, and performing the integral over an enclosing half-space (and keeping only the term that decreases most rapidly with separation distance d), one finds [106]

$$\frac{\rho(\mathbf{x}, \omega)}{\rho_0(\omega)} \le \frac{1}{8(kd)^3} \frac{|\chi(\omega)|^2}{\operatorname{Im} \chi(\omega)}, \tag{2.63}$$

where $k = \omega/c$ is the free-space wavenumber. Equation (2.63) represents our first near-field bound. This bound only depends on two parameters of the system: the separation distance d, relative to the wavenumber, and the **material enhancement factor**:

$$\frac{|\chi(\omega)|^2}{\operatorname{Im} \chi}. \tag{2.64}$$

The material enhancement factor encodes a key trade-off: a large susceptibility magnitude implies large possible polarization currents, while a large imaginary part of the susceptibility implies losses that necessarily restrict resonant enhancement. In Drude metals with $\chi = -\omega_p^2/(\omega^2 + i\gamma\omega)$, the material enhancement factor is given by $\omega_p^2/\gamma\omega$, showing that the largest possible single-frequency response is achievable in materials with large electron densities and small losses. The material enhancement factor is described in further detail in Refs. [106, 169].

The second key parameter is the distance d; the factor $1/d^3$ encodes the dramatic enhancements that are possible in the near field. These enhancements are typically achieved with plasmonic modes, and the factor $1/d^3$ arises from the most rapidly decaying component of the free-space Green's function, $\sim 1/r^3$; squaring this term and integrating over a three-dimensional volume leads to the inverse-cubic dependence. The last point also suggests an important caveat: systems with a different dimensionality must have different scaling laws as a function separation distance. Designing for 2D materials, for example, leads to integrals over 2D (or very thin) domains, leading to a $1/d^4$ near-field enhancement factor. There are also more slowly increasing terms that arise from the mid-field and far-field contributions to the free-space Green's function.

Finally, it should be noted that certain constraints of interest can be seamlessly integrated into the optimization problem of Eq. (2.61). Of particular importance in plasmonics applications is **radiative efficiency**. When one finds a bound on extinction or LDOS, the bound may suggest very large enhancements, but all of that enhancement could be going into material absorption rather than far-field radiation or scattering. Suppose a given application requires a certain radiative efficiency, such as some fraction η of the total emission going into the far field. This can be written mathematically as the constraint that absorption be smaller than $(1 - \eta)$ multiplied by the extinction, or $P_{\text{abs}} \le (1 - \eta)P_{\text{ext}}$. Absorption is quadratic in the polarization field, while extinction is linear in the polarization field, such that this expression represents an additional constraint that can be seamlessly incorporated into Eq. (2.61). Often the bound of interest, with this constraint, is analytically solvable. Ref. [110] identifies precisely such bounds on high-radiative-efficiency plasmonics, prescribing a trade-off between large response and radiative efficiency.

In Ref. [110], it is not only shown that high-radiative-efficiency bounds can be derived; it is also shown that hybrid dielectric-metal designs can approach the bounds *and* that they surpass the same fundamental limits evaluated for metal-only structures. This example showcases the power of using bounds to understand the broader landscape of a photonics application area of interest.

2.4.1.2 CDOS

Bounds to CDOS can be found along very similar lines to the LDOS bounds of above. We can define the trace of the CDOS via Eq. (2.9), taking

$$\rho(\mathbf{x}_1, \mathbf{x}_2, \omega) = \frac{1}{\pi\omega} \operatorname{Tr} \operatorname{Im} \mathbb{G}(\mathbf{x}_1, \mathbf{x}_2, \omega). \tag{2.65}$$

Then, we can separate out a scattered contribution coming from the polarization fields induced in the scatterer, just as for LDOS, and when this term dominates (i.e., the geometry primarily mediates the CDOS), we have

$$\rho(\mathbf{x}_1, \mathbf{x}_2, \omega) = \frac{1}{\pi\omega} \sum \mathbf{e}^T_{\text{inc},\mathbf{s}_j,\mathbf{x}_1} \mathbf{p}_{\mathbf{s}_j,\mathbf{x}_2}, \tag{2.66}$$

where the position subscripts on $\mathbf{e}_{\text{inc}}$ and $\mathbf{p}$ denote the source positions of the $\mathbf{s}_j$-polarized dipoles. Hence, in CDOS, the field incident from one position is overlapped with the polarization field induced by a source from a second position. The bound for CDOS will be identical to that of Eq. (2.62), but with $\|\mathbf{e}_{\text{inc},\mathbf{s}_j}\|^2$ replaced by $\|\mathbf{e}_{\text{inc},\mathbf{s}_j,\mathbf{x}_1}\| \|\mathbf{e}_{\text{inc},\mathbf{s}_j,\mathbf{x}_2}\|$. Finally, normalizing by free-space LDOS and dropping all except the most rapidly varying terms as a function of separation distances d_1, d_2, one arrives at the bound [7]:

$$\frac{\rho(\mathbf{x}_1, \mathbf{x}_2, \omega)}{\rho_0(\omega)} \le \frac{1}{4k^3\sqrt{d_1^3 d_2^3}} \frac{|\chi(\omega)|^2}{\operatorname{Im}\chi(\omega)}. \tag{2.67}$$

The discussion of the terms that appeared in the LDOS bound of Eq. (2.63) can be translated almost seamlessly here: the same material dependence shows up, corresponding to the same possibilities for plasmonic enhancement, and the same distance dependencies due to the same enhancements of the near fields of the two dipoles. There are likely two further enhancements that can be made to Eq. (2.67). First, Eq. (2.67) is a factor of 2 larger than Eq. (2.63), when the former is evaluated in the limit as $\mathbf{x}_1 \to \mathbf{x}_2$. This is almost certainly because the bound of Eq. (2.67) in Ref. [7] came from evaluating bounds for each diagonal element, simplifying, and *then* taking the trace. Taking the trace and then simplifying the bound would likely remove this factor of 2. Second, the bound of Eq. (2.67) does *not* depend on the distance between the two dipoles, d_{12}. This may be physical in certain limits, e.g., when a plasmon can maintain its amplitude in propagating from one dipole to the other, but may not be physical when such propagation is not possible, and one would expect improved bounds to capture this. It is likely true that applying the many-conservation-law approach of Sect. 2.3.3 would incorporate such effects. Nevertheless, Eq. (2.67) is a good starting point to understand the upper limits to engineering CDOS in photonic environments.

2.4.1.3 Smith–Purcell Radiation

Another exciting application area for the single-frequency bound approach is to Smith–Purcell radiation, which is the radiation that occurs when a free electron passes near a structured material. A constant-velocity free electron produces only a near field, with no far-field component, but when the evanescent wave interacts with grating-like structures, the gratings can couple the near fields to propagating far fields, leading to a release of energy from the electron in the form of electromagnetic radiation. The natural question, then, is how large this energy release can be?

Mathematically, this question is identical to the question of the work done by a dipole (i.e., LDOS), except that the incident field is different in this case, and is given by Eq. (2.6). Maximizing the overlap of *this* incident field with the induced polarization field, subject to the same constraint of Eq. (2.61), leads to a bound on the Smith–Purcell emission spectral probability given by [112]:

$$\Gamma(\omega) \leq \frac{\alpha}{2\pi c} \frac{|\chi|^2}{\operatorname{Im}\chi} \frac{L\theta}{\beta} \left[(\kappa_\rho d) K_0(\kappa_\rho d) K_1(\kappa_\rho d)\right], \tag{2.68}$$

where $\Gamma = P/\hbar\omega$ for emission power P, α is the fine-structure constant, $\beta = v/c$ is the normalized electron velocity, L and θ are the height and opening azimuthal angle of the cylindrical sector containing the patterned material, $\kappa_\rho = k/\beta\gamma$ is the wavenumber divided by β and the Lorentz factor γ, d is the distance of the beam from the surface, and K_n is the modified Bessel function of the second kind. Although the exact expression is somewhat complex, we see that Smith–Purcell radiation also directly benefits from the material enhancement factor $|\chi|^2/\operatorname{Im}\chi$. A seemingly surprising conclusion also emerged from Eq. (2.68): *slow* electrons, at small enough separations, can lead to *greater* radiation enhancements than *fast* (i.e., high-energy) electrons. All constant-velocity electrons do not radiate when their speed is smaller than the speed of light in the background medium, and emit only near fields. But high-speed electrons are closer to surpassing the Cherenkov threshold, and hence the fields they generate decay more slowly, out to larger distances. By contrast, low-speed electrons have very strong but very tightly confined near fields. But if one brings a patterned surface close enough, the strong very near fields of slow electrons have greater potential for radiation enhancements than the more moderate near fields of fast electrons.

Some of the general trends, and absolute numerical values, of the bound of Eq. (2.68) were validated theoretically and experimentally in Ref. [112]. In particular, Fig. 2.2 shows an experimental setup for measuring the Smith–Purcell

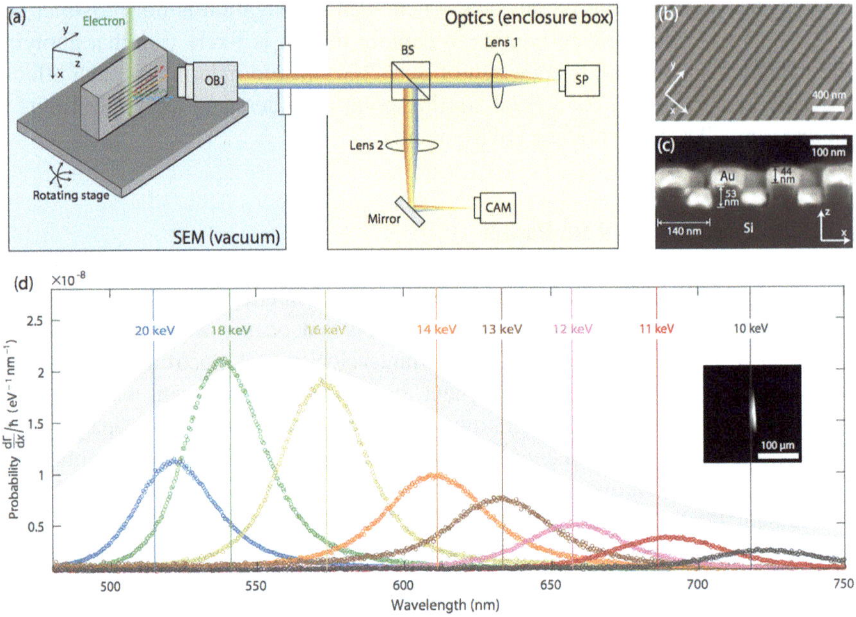

Fig. 2.2 The bounds of Eq. (2.68) dictate upper limits to Smith–Purcell emission rates. (**a**–**d**) The experiments of Ref. [112] quantitatively confirm that designed metallic gratings can approach the fundamental performance limits (Adapted from Ref. [112])

radiation for electron beams with varying energies, as well as designed gold-on-silicon gratings whose parameters were optimized for maximum response. The key result is shown in panel (d), where the gray region indicates the fundamental bounds, as a function of photon wavelength, with some width to account for experimental uncertainties. The colored data points are quantitatively measured probabilities (with no fitting parameters), showing that both the quantitative values of the bounds are nearly approachable and that the complex wavelength dependence (emerging from an interplay between the material enhancement factor and the optical near fields) correctly captures the response of high-performance designs.

2.4.1.4 Spectral NFRHT

Near-field radiative heat transfer, NFRHT, introduced in Sect. 2.2.5, offers an extraordinary challenge for fundamental limits. It comprises rapidly decaying, large-area, broadband thermal sources for which little has been understood about upper bounds for quite some time. While we tackle the question of broadband enhancements in Sect. 2.4.3, in this section, we describe the recent progress in understanding maximum NFRHT at a single frequency. There are three key results that we can highlight: channel bounds for planar bodies [57, 60], material-loss bounds [111], and an amalgamation of the two [170, 171].

Channel bounds to NFRHT are described as "Landauer bounds," due to their similarities with Landauer transport. For planar bodies with in-plane translational (and therefor rotational) symmetries, the in-plane wavenumber is a conserved quantity, and the energy flux from one body to another can be decomposed into propagating and evanescent plane-wave channels with no cross-channel scattering. One can decompose the fields emanating from the emitting body into normalized plane-wave modes, insert them into the fluctuation-averaged flux, i.e., the average of the integral $\frac{1}{2}\int_A \mathbf{E} \times \mathbf{H}^* \cdot \hat{\mathbf{n}}$, for separating plane A and normal vector $\hat{\mathbf{n}}$. This results in an expression for the flux rate $\Phi(\omega)$, of Eq. (2.20) and Eq. (2.21), given by

$$\Phi(\omega) = \sum_{j=s,p} \frac{1}{2\pi} \int \frac{\mathrm{d}^2\boldsymbol{\kappa}}{4\pi^2} T_j^{12}(\omega, \kappa, d), \tag{2.69}$$

where $\boldsymbol{\kappa}$ is the in-plane wave propagation constant (and κ its magnitude), j is a polarization index, k_0 is the free-space wavenumber, and the T_i are "transmission coefficients," which depend on the specific Fresnel reflection coefficients of the two interfaces [60]. This expression has an elegant interpretation: NFRHT is the composition of plane-wave fluxes, each contributing with a weight T_i. Moreover, the coefficients T_i are bounded above by 1, for both the propagating and evanescent waves [57, 60, 172]. Then, if there is a limit to the largest wavenumber across which a nonzero transmission can be achieved, one will have a bound on the maximum spectral RHT.

Hence, it is possible to identify a maximal rate of NFRHT which is given by power transferred with "Landauer" transmission unity over all possible plane waves [57, 172]. While intuitive, however, this bound has two serious drawbacks. The first is that if one literally computes the integral of Eq. (2.69) over all possible waves, the result is infinite, as there are an infinite number of plane-wave channels. Of course one cannot reasonably expect to achieve unity transmission over channels with infinitely large in-plane wavenumbers (as they decay exponentially fast), implying that there must be a maximal channel at which the sum should be terminated. But how to choose this value? One proposal, from Ref. [57], was that the maximal accessible channel should be proportional to $1/a$, where a is the lattice spacing of the material, the reasoning being that beyond this limit, the use of a continuum model of the materials would not be valid. Another proposal, from Ref. [172], is that the maximal accessible channel wavenumber is given by $k_{\max} = 1/d$, where d is the separation between the two bodies, the reasoning being that the exponential decay of the evanescent waves makes it difficult to achieve large transmission beyond $1/d$. Each of the resulting bounds (one from $k_{\max} = 1/a$ and the other from $k_{\max} = 1/d$) has shortcomings: the lattice-spacing-defined bound is extraordinarily high for any reasonable lattice constant, well beyond all other bounds discussed below. And the separation-defined bound is in fact not a true bound: it can be superseded with reasonable material parameters [111], which in fact do show nontrivial transmission beyond $1/d$. Hence, the two known versions of the bound are either far too large or surpassable.

The second serious drawback of using Eq. (2.69) is that it only applies to planar bodies with translational symmetry in all directions. The use of conservation laws for bounds, discussed next, leads to bounds that can be applied to planar bodies with any patterning while also being tighter than the channel bounds resulting from Eq. (2.69).

The first use of conservation laws for spectral NFRHT bounds appeared in Ref. [111]. The mathematical procedure is sufficiently complex that we will not go through it in detail here, but the intuition can be explained. The idea is to use the global conservation law requiring $P_{\text{abs}} \leq P_{\text{ext}}$ in the spectral NFRHT problem. The difficulty is that the sources are embedded *within* one of the scattering bodies, which leads to divergences if one blindly applies the constraint $P_{\text{abs}} \leq P_{\text{ext}}$. However, the radiative exchange of heat can be decomposed into two subsequent scattering problems, both of which have sources separated from scatterers. In the first step, the incident field is given by the field emanating from body 1 *in the presence of body 1*, with only the second body serving as the scatterer. The absorption in this second body is bounded by the extinction by this second body, which leaves a bound in terms of the second material and the "incident field" emanating from body 1. Of course, we do not know exactly what this field is for any pattern. At this point, however, we can use reciprocity to rewrite the field emanating from body 1 in terms of fields emanating from the free space of body 2's domain, being absorbed by body 1. The constraint $P_{\text{abs}} \leq P_{\text{ext}}$ can be applied to this scattering process again, ultimately yielding a single-frequency, flux-per-area A bound given by [111]

$$\frac{\Phi(\omega)}{A} \leq \frac{1}{16\pi^2 d^2} \frac{|\chi_1|^2}{\operatorname{Im} \chi_1} \frac{|\chi_2|^2}{\operatorname{Im} \chi_2}, \tag{2.70}$$

where d is the separation distance between the two bodies and χ_1 and χ_2 are their optical susceptibilities, respectively. This bound includes two key dependencies: the material enhancement factor $|\chi|^2/\operatorname{Im}\chi$ and a $1/d^2$ dependence arising from the rapidly decaying near fields in the electromagnetic Green's function. The bound of Eq. (2.70) is promising, as it suggests significant possible enhancements of spectral NFRHT, and it is plausible: the actual NFRHT of two planar bodies with equal susceptibilities, on resonance, is given by $\Phi(\omega)/A = 1/(4\pi^2 d^2) \ln\left[|\chi|^4/(4(\operatorname{Im}\chi)^2)\right]$, with nearly identical dependencies as Eq. (2.70), except for the logarithmic dependence on the material enhancement. Can this be overcome, with instead linear enhancements in $|\chi|^2/\operatorname{Im}\chi$? For some materials, the answer is "yes," as shown with computational inverse design in Ref. [173]. More generally, however, such linear enhancements are not generic, and one can further tighten the bound of Eq. (2.70).

Refs. [170, 171] showed that one can tighten the bound of Eq. (2.70) by combining the use of a global conservation law with that of a channel decomposition. If one decomposes the *general* (not specific to translation-symmetric) scattering response into plane waves, and further imposes conservation laws for absorption and extinction (of the bodies in tandem as well as in isolation), then a long mathematical process leads to a tighter bound. If we define $\mathbb{G}_{0,AB}$ to be the free-space Green's function matrix for sources in body A to measurement points in body B, and g_i the singular values of $\mathbb{G}_{0,AB}$, then the resulting bound is given by Molesky et al. [170]

$$\Phi(\omega) \le \sum_i \left[\frac{1}{2\pi} \Theta(\zeta_A \zeta_B g_i^2 - 1) + \frac{2}{\pi} \frac{\zeta_A \zeta_B g_i^2}{(1 + \zeta_A \zeta_B g_i^2)^2} \Theta(1 - \zeta_A \zeta_B g_i^2) \right], \tag{2.71}$$

where $\zeta_{A,B} = |\chi_{A,B}|^2 / \operatorname{Im} \chi_{A,B}$. One can see that the expression of Eq. (2.71) has components of both material response (in $\zeta_{A,B}$) and channels (in the g_i factors) in it. Strikingly, in the near-field limit, expression Eq. (2.71) is given by Venkataram et al. [171]

$$\Phi(\omega) \frac{d^2}{A} \le \frac{1}{4\pi^2} \ln\left(1 + \frac{\zeta_A \zeta_B}{4}\right) + \frac{\Theta(\zeta_A \zeta_B - 4)}{8\pi^2} \left\{ \ln(\zeta_A \zeta_B) + \frac{1}{4}\left[\ln\left(\frac{\zeta_A \zeta_B}{4}\right)\right]^2 - 2\ln\left(1 + \frac{\zeta_A \zeta_B}{4}\right) \right\}, \tag{2.72}$$

which correctly captures the *logarithmic* material dependence that is seen in planar bodies. This significantly tightens the bound of Eq. (2.70) for plasmonic materials such as silver or gold which have large material enhancement factors $|\chi|^2 / \operatorname{Im} \chi$. The genesis and utility of the bounds of Eqs. (2.70)–(2.72) are illustrated in Fig. 2.3, which contains the derivation of the conservation-law bounds of Eq. (2.70) in Fig. 2.3a, the design of structures showing the material dependence of Eq. (2.70) in Fig. 2.3b, and the more general combination of conservation law and channel-decomposition approach of Eq. (2.72) in Fig. 2.3c.

Generically, it is not possible to find "tighter" single-frequency dependencies than those that arise in Eq. (2.72), as both the distance and material enhancement dependencies are achievable in realistic-material planar designs. The only possible improvements are the coefficient prefactors, as well as the correct material dependence away from the surface-plasmon frequency, suggesting that Eq. (2.72) indeed captures the key trade-offs in single-frequency NFRHT. A key remaining question, then, is what is possible over a broad bandwidth? This question is resolved in Sect. 2.4.3.

2.4.2 All-Frequency Sum Rules

In Sect. 2.3.4, we developed the key elements needed for sum rules: a causal linear response function, an objective that does not involve the conjugate of that function,

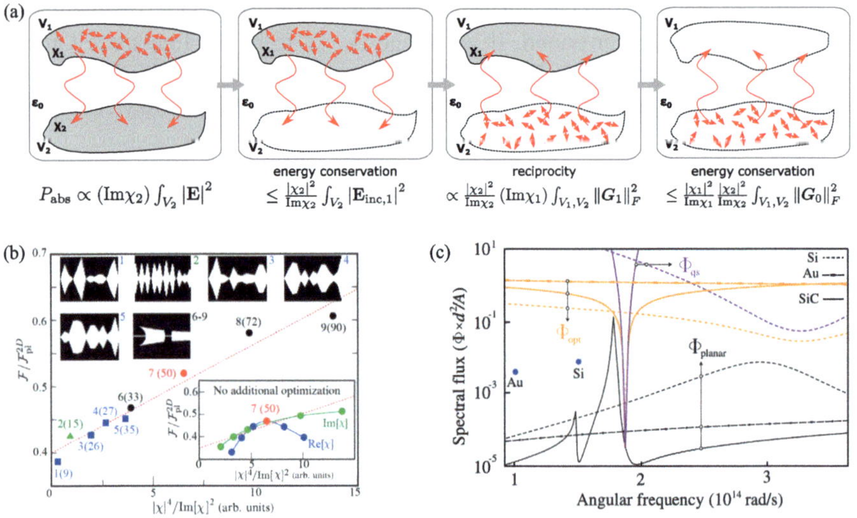

Fig. 2.3 A collection of bounds on single-frequency near-field radiative heat transfer. (**a**) The approach of Ref. [111] using material loss as the only constraint, exploiting reciprocity to bound the response given that the sources are embedded *within* one of the arbitrarily patterned scattering bodies. (Adapted from Ref. [111].) (**b**) Bounds and designs from Ref. [173] showing the feasibility, in specific regimes, of achieving enhancements proportional to the square of the material enhancement factor $|\chi|^2/\operatorname{Im}\chi$. (Adapted from Ref. [173].) (**c**) Tightened bounds from Refs. [170, 171], precluding the possibility of extraordinary response at frequencies away from the surface-polariton frequency of a material of interest (Adapted from Ref. [171])

and certain technical conditions (e.g., sufficient decay). Optical extinction is the prototype example, as the optical theorem prescribes that extinction should be proportional to the imaginary part of the overlap of the incident field with the induced polarization field, a quantity that is analytic (for plane-wave incident fields) in the upper-half plane. Within the past few years [7, 174], it has been realized that there is a near-field analog of extinction: the local density of states, or LDOS. As derived in Sect. 2.2.1, (electric) LDOS is given by the trace of the imaginary part of the (electric) Green's function, evaluated at the source location:

$$\mathrm{LDOS}(\mathbf{x}, \omega) = \operatorname{Im}\operatorname{Tr}\left[\frac{1}{\pi\omega}\mathbb{G}(\mathbf{x}, \mathbf{x}, \omega)\right]. \tag{2.73}$$

The key similarity with extinction is that LDOS is the imaginary part of an amplitude, rather than a squared norm (which depends on the complex conjugate of that amplitude). At first blush, then, it would appear that one can port exactly the derivation used for extinction to derive sum rules for LDOS. However, there are three obstacles that must be overcome.

First, LDOS diverges at high frequencies. Ignoring the effects of a scatterer (which are effectively infinitely far away at infinitely large frequencies), and as seen in Eq. (2.4), the free-space photon density of states scales as ω^2 as frequency goes to infinity. A diverging LDOS violates the asymptotic-decay requirement of KK relations, prohibiting a sum rule. The resolution, however, is straightforward: one should subtract the free-space LDOS $\rho_0(\omega)$ from the total LDOS, leaving only the scatterer-based contribution $\rho_s(\omega)$:

$$\begin{aligned}\rho_s(\mathbf{x},\omega) = \rho(\mathbf{x},\omega) - \rho_0(\omega) &= \operatorname{Im}\operatorname{Tr}\left[\frac{1}{\pi\omega}\left(\mathbb{G}(\mathbf{x},\mathbf{x},\omega) - \mathbb{G}_0(\mathbf{x},\mathbf{x},\omega)\right)\right] \\ &= \operatorname{Im}\operatorname{Tr}\left[\frac{1}{\pi\omega}\mathbb{G}_s(\mathbf{x},\mathbf{x},\omega)\right], \end{aligned} \tag{2.74}$$

where we define $\mathbb{G}_s$ as the scattered-field part of the Green's function. After isolating the scatterer's contribution to the LDOS, one can verify that the "scattered LDOS" indeed decays sufficiently quickly at high frequencies [7]. Hence, this approach of subtracting the free-space LDOS, an approach generalized in "dispersion relations with one subtraction" [160], resolves the first issue of diverging LDOS.

The second issue is that one is not free to arbitrarily choose the pole frequency for a KK relation involving the scattered LDOS. The Green's function itself is finite and generically nonzero at every real frequency, but by definition, the LDOS includes a factor of $1/\omega$, as in Eq. (2.73). (This does not correspond to a divergent LDOS at zero frequency, as the *imaginary* part of the Green's function goes to zero at frequency, but the real part does not generically go to 0.) This function, then, already has a pole at the origin. One could try to move the pole to infinite frequency, for example, by multiplying by $\omega/(\omega - \omega_0)$ and taking the limit as $\omega_0 \to \infty$, but the high-frequency asymptotic behavior of LDOS is quite complicated. Hence, there is likely only a single meaningful sum rule for near-field LDOS, which arises from the intrinsic pole at zero frequency.

The third issue is that the real part of the Green's function diverges, since the source and measurement locations coincide; sum rules relate the integral of the imaginary part to the real part (or vice versa), which leads to the impermissible evaluation of an infinite quantity. (Such an integral *should* diverge; the free-space LDOS increases with frequency, meaning that any integral over all frequencies will of course diverge.) One resolution to this issue was proposed in Ref. [175]: to remove the longitudinal contribution to the Green's function, which removes the singularity and suggests that over all frequencies there can be no net change in spontaneous-emission enhancements. But this removal thereby precludes the possibility for near-to-far-field coupling that is crucial for spontaneous-emission engineering, which is why a conventional refractive-index sum rule is recovered. Instead, it was recognized in Refs. [7, 174] that there is an alternative mechanism for overcoming this obstacle: to subtract out the free-space LDOS term from the total term. The free-space term is the one responsible for the diverging real part, yet the free-space LDOS is exactly known and hence there is no need for a KK relation for that part anyhow. Hence, this obstacle is resolved by the same procedure as the first one, and we can proceed to deriving a scattered-LDOS sum rule.

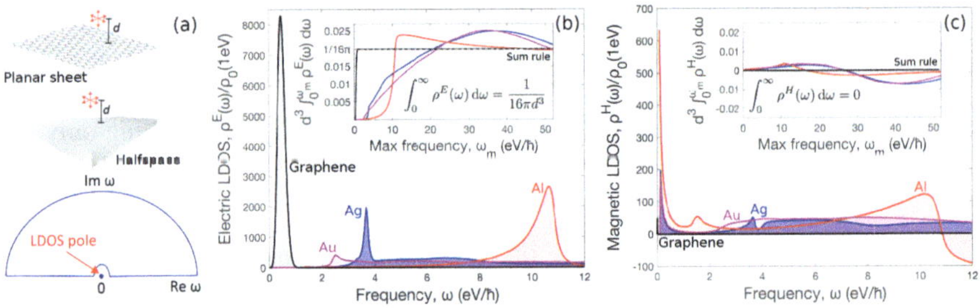

Fig. 2.4 (**a**) Sum rules, derived using the techniques of Sect. 2.3.4 and the contour on the lower left, impose strong constraints on LDOS lineshapes. (**b**) Electric LDOS of various material half-spaces and 2D sheets, with different resonance peaks and bandwidths. The inset, however, shows that the integral converges to identical values for each scenario. (**c**) Similarly with magnetic LDOS, whose sum rule is now zero. The sum rules are for the scattered-field contributions to the LDOS, which can be negative at frequencies where spontaneous emission is suppressed by the presence of a scatterer (Adapted from Ref. [7])

The hemispherical contour (with hemispherical bump at the origin), in tandem with the same Cauchy-residue arguments for far-field sum rules in Sect. 2.3.4, leads to a sum rule for $\rho - \rho_0$ analogous to the far-field case [7]:

$$\int_0^\infty \rho_s(\omega, \mathbf{x})\, \mathrm{d}\omega = \frac{1}{2} \operatorname{Re} \operatorname{Tr} \mathbb{G}_s(\mathbf{x}, \mathbf{x})\big|_{\omega=0} = \alpha_{\text{LDOS}}. \tag{2.75}$$

Now we have connected the all-frequency scattered-field component of electric LDOS to its electrostatic Green's function. Is that informative? It turns out to be quite informative, because there are near-field "domain monotonicity" theorems [7] that ensure that this shape-dependent Green's function term is bounded above by its form in any enclosure, and we can choose high-symmetry enclosures where it has a simple analytical form. For example, for a planar half-space, the near-field electrostatic constant is simply

$$\alpha_{\text{LDOS,plane}} = \frac{1}{16\pi d^3} \left[\frac{\varepsilon(0) - 1}{\varepsilon(0) + 1} \right], \tag{2.76}$$

where $\varepsilon(0)$ is the zero-frequency (electrostatic) permittivity. For conductive materials whose permittivity diverges at zero frequency, the corresponding fraction in Eq. (2.76) is simply 1, which can also be used as a general bound for any material. Notably, for the *magnetic* LDOS above an electric material, the right-hand side of the counterpart to Eq. (2.76) is *zero*: the scattering contribution to the magnetic LDOS must average out to zero (i.e., it provides suppression and enhancement of the free-space LDOS in equal amounts).

An example of the utility of the LDOS sum rule is given in Fig. 2.4. The electric LDOS is shown for three typical metals: gold (Au), silver (Ag), and aluminum (Al), as well as for a single graphene sheet (with Fermi level 0.6 eV). These four systems show LDOS peaks at quite different frequencies, from below 1 eV to beyond 10 eV, with very different quality factors leading to quite different "spreads" in their spectral response. Yet as is made clear by the inset of Fig. 2.4, the integrated response is exactly equal for each of these systems, as must be true from Eq. (2.76) (the material constant α for each system is exactly 1). Sum rules illuminate unifying principles that must apply across seemingly disparate systems.

2.4.3 *Finite, Nonzero Bandwidth*

The techniques of the previous two sections apply to single-frequency and all-frequency scenarios. In this section, we probe an intermediate regime: finite, nonzero bandwidth. Techniques that work for any arbitrary bandwidth would be tantalizingly powerful, as they would incorporate the single- and all-frequency results as asymptotic limits of a more general theory. Yet the techniques of the previous section would seem incapable of extension to nonzero, finite bandwidths: there is no single scattering problem for which power-conservation laws can be imposed, nor can the contour integrals of the sum-rule approaches be easily modified to a finite bandwidth. In this section, we describe two recently developed approaches to tackle finite-bandwidth bounds: first, transforming bandwidth-averaged response to a complex frequency (largely following Ref. [7]), and second, identifying an oscillator-based representation of any scattering matrix (largely following Ref. [176]).

2.4.3.1 Complex-Frequency Bounds

Ref. [7] recognized an intermediate route that utilized both techniques in one fell swoop. The idea can be summarized succinctly: finite-bandwidth average response can be transformed to a scattering problem at a single, *complex*-valued frequency, where quadratic constraints analogous to power conservation can be imposed. The complex frequency accounts for bandwidth, while the power-conservation analog imposes a finite bound. We now develop this intuition mathematically.

To compute the bandwidth average of a response function such as LDOS, one must define a "window function" that encodes the center frequency, the bandwidth, and the nature of the averaging. A common choice is a linear combination of step functions, but this choice turns out to be mathematically treacherous. A simple (and mathematically serendipitous) choice is a Lorentzian function. Uses of tailored window functions for bandwidth averaging were first proposed in Refs. [17, 177]; in the first, bandwidth-averaged extinction was analyzed for scaling laws for optical cloaking, while in the second, they were used to regularize the computational inverse design of maximum LDOS. Our quantity of interest, the frequency-averaged LDOS, $\langle \rho \rangle$, can be written as [7]

$$\langle \rho \rangle = \int_{-\infty}^{\infty} \rho(\omega) H_{\omega_0,\Delta\omega}(\omega)\, d\omega, \tag{2.77}$$

where $H_{\omega_0,\Delta\omega}(\omega)$ is the Lorentzian window function:

$$H_{\omega_0,\Delta\omega}(\omega) = \frac{\Delta\omega/\pi}{(\omega-\omega_0)^2 + (\Delta\omega)^2}, \tag{2.78}$$

where ω_0 is the center frequency and $\Delta\omega$ is the bandwidth of interest. In Eq. (2.77), we define the frequency integral from $-\infty$ instead of 0 for smoothness; typically, the window function will be narrow enough to render this difference negligible; conversely, in the all-frequency limit, the symmetry of the LDOS around zero frequency ensures we are working with the correct quantity. We are interested only in the near-field enhancements of ρ, so we will drop the free-space LDOS, as was useful in the sum-rule section to avoid spatial and spectral divergences. Then, consider the integral of Eq. (2.77): it already covers the entire real line, and we can imagine adding to it a hemispherical contour in the UHP that will contribute infinitesimally. Then the integral is a closed contour, and we can use complex-analytic techniques based on the analyticity of the integrand and the locations of the poles of the integrand. The integrand is not analytic, but the LDOS can be written as $\rho(\omega) = \operatorname{Im} s(\omega)$, where $s(\omega)$, proportional to the trace of the imaginary part of the scattered component of the Green's function, *is* analytic. Taking the imaginary part outside the integral, the remainder of the integrand of Eq. (2.77) has two poles away from the lower-half plane: one at zero, thanks to the $1/\omega$ term in the LDOS, and a second at $\omega_0 + i\Delta\omega$. Then, a few lines of algebra give the frequency average of $\rho(\omega)$ as [7]

$$\langle \rho \rangle = \operatorname{Im} s(\omega_0 + i\Delta\omega) + 2H_{\omega_0,\Delta\omega}(0)\alpha_{\text{LDOS}}. \tag{2.79}$$

The second term comes from the contribution of the sum rule at a given frequency and ensures that the ultimate expression will give the sum rule in the asymptotic limit $\Delta\omega \to \infty$. Here, for simplicity and pedagogy, we will assume a sufficiently narrow bandwidth that the second term can be ignored. (It can always be reintroduced in the final expression.) The first term is the imaginary part of the LDOS scattering amplitude, evaluated at the *complex* frequency $\tilde{\omega} = \omega_0 + i\Delta\omega$. What is the largest this term can be?

To bound the complex-frequency term, we can develop a generalization of the real-frequency conservation-law approach. In Ref. [7], we developed such a generalization via a somewhat complicated line of differential-equation reasoning; here, we develop a simpler (but no less general) integral-equation form. The starting point is the complex-valued integral equation:

$$\left[\mathbb{G}_0(\tilde{\omega}) + \xi(\tilde{\omega})\right] \mathbf{p}(\tilde{\omega}) = -\mathbf{e}_{\text{inc}}(\tilde{\omega}), \tag{2.80}$$

where we have momentarily included all frequency arguments to emphasize that Eq. (2.80) is evaluated at the complex frequency $\tilde{\omega}$. Next, we will multiply on the left by $\mathbf{p}^\dagger/\tilde{\omega}$, and take the imaginary part of the entire equation, to arrive at

$$\mathbf{p}^\dagger \left\{ \operatorname{Im}\left[\frac{\mathbb{G}_0}{\tilde{\omega}} + \frac{\xi}{\tilde{\omega}} \right] \right\} \mathbf{p} = \operatorname{Im}\left[\left(\frac{\mathbf{e}_{\text{inc}}}{\tilde{\omega}} \right)^\dagger \mathbf{p} \right], \tag{2.81}$$

This equation can be regarded as a complex-valued extension of the real-valued, global conservation law of Eq. (2.25). In particular, the two terms on the left are both positive semidefinite, as can be proven by causality (cf. Sec. IX of the SM of Ref. [105]). To remove the shape dependence and focus on the material dependence, then we can drop the first term on the left-hand side of Eq. (2.81) and rewrite this equation as an inequality:

$$\mathbf{p}^\dagger \left[\operatorname{Im}\left(\frac{\xi}{\tilde{\omega}} \right) \right] \mathbf{p} \leq \operatorname{Im}\left[\left(\frac{\mathbf{e}_{\text{inc}}}{\tilde{\omega}} \right)^\dagger \mathbf{p} \right], \tag{2.82}$$

Equation (2.82) imposes a constraint on the strength of the complex-frequency polarization field that enters the near-field scattering amplitude $s(\tilde{\omega})$. The exact expression for the scattering amplitude is $s(\tilde{\omega}) = \frac{1}{\pi\tilde{\omega}} \operatorname{Tr} \mathbb{G}_0(\mathbf{x}, \mathbf{x}, \tilde{\omega})$. One can maximize the imaginary part of this amplitude subject to the constraint of Eq. (2.82) by exactly the procedure outlined in Sec. IX of the SM of Ref. [7]; doing so, one arrives at a simple result (remembering that we have dropped the sum-rule term):

$$\langle \rho \rangle \leq \frac{1}{\pi} \frac{|\chi(\tilde{\omega})|^2}{\operatorname{Im}[\tilde{\omega}\chi(\tilde{\omega})]} \mathbf{e}_{\text{inc}}^\dagger \mathbf{e}_{\text{inc}}. \tag{2.83}$$

As a reminder, the inner product of the incident field with itself is a volume integral of the square of the incident fields. The deep near field is dominated by the most rapidly decaying term in the incident fields; integrating only this contribution at the complex frequency gives $\mathbf{e}_{\text{inc}}^\dagger \mathbf{e}_{\text{inc}} = \frac{1}{16\pi d^3}$, where we have taken the arbitrary scattering body to fit in a half-space enclosure separated from the source by a distance d. Inserting this expression into the inequality, and normalizing by the free-space LDOS evaluated at $|\tilde{\omega}|$, we finally have a bandwidth-averaged bound [7]:

$$\frac{\langle \rho \rangle}{\rho_0(|\tilde{\omega}|)} \leq \frac{1}{8|k|^3 d^3} f(\omega), \tag{2.84}$$

where $f(\omega)$ is the bandwidth-averaged generalization of the material enhancement factor (discussed at real frequencies in Sect. 2.4.1.1):

$$f(\omega) = \frac{|\tilde{\omega}\chi|^2}{|\tilde{\omega}| \operatorname{Im}(\tilde{\omega}\chi)}. \tag{2.85}$$

The material enhancement factor of Eq. (2.85) is slightly simpler than that of Ref. [7], thanks to our use of the simpler integral-equation constraint of Eq. (2.81).

The bound of Eq. (2.84) is the key result: the bandwidth-averaged LDOS has an upper bound that is similar to that of the single-frequency LDOS, but reduced by the presence of a complex frequency. This reduction is significant for low-loss materials, for which Im χ might be quite small, in which case $\mathrm{Im}(\tilde{\omega}\chi) \approx (\Delta\omega)\chi$, wherein the bandwidth effectively provides the relevant loss. There is also an additional broadening due to dispersion, as χ is evaluated at the complex frequency $\tilde{\omega}$, at which Im χ will generally be larger. (There is *another* additional term in the more general version of the bound of Eq. (2.84) that exponentially decays with bandwidth, which we excluded for simplicity.) Hence, the bound of Eq. (2.84) has three properties that are quite theoretically pleasing. First, in the single-frequency limit, it asymptotically approaches the previously derived single-frequency bound. Second, in the all-frequency limit, it asymptotically approaches the previously derived sum rule. And, finally, in the nonzero- and finite-bandwidth regime, it intermediates between the two, with a smaller average response than the single-frequency bound, and a smaller total integrated response than the sum rule. This approach was extended to CDOS and NFRHT in Ref. [7], with similar features emerging. One interesting comparison point is to Ref. [178], which examined optimal materials for planar NFRHT designs. Unlike the power–bandwidth bounds, which increase with electron density and decrease with material loss, Ref. [178] found that the key material parameters in planar systems are simply the (ideally small) frequency at which surface polaritons are strongest, and the bandwidth over which they are strong. This finding has been experimentally corroborated [179], and it emerges theoretically in the more general NFRHT bounds of the next subsection.

Ref. [7] probed the feasibility of approaching the upper bounds in certain prototypical systems. Four key results were identified. First, for center frequencies close to the surface-plasmon frequencies of metals, planar systems supporting such plasmons are able to closely approach the bounds across a wide range of bandwidths. Second, double-cone (bowtie-antenna-like) antennas show a performance that can closely approach (nearly within 2X) their bounds across a wide range of bandwidths, for center frequencies coincident with their resonant frequencies. Third, these bounds were the first to enable systematic comparison of dielectric- and metal-based systems. Unlike the single-frequency case, the complex-frequency material enhancement factor does not diverge for lossless dielectrics (at nonzero bandwidth), which enables predictions of the center frequencies and bandwidths at which metals can be categorically superior to dielectrics, and vice versa. Finally, these bounds also enabled predictions of when 2D materials can be superior to bulk materials, and vice versa. The results highlight the power of fundamental limits more generally: they enable a high-level understanding of the landscape of a given physical design problem, identifying the material and architectural properties that really matter.

The "power–bandwidth" approach of Ref. [7] was recently generalized in Ref. [180]. Notice that the constraint of Eq. (2.81) is a global conservation law for real power; at the time that Ref. [7] was published, the reactive global conservation law, as well as the local-conservation-law approach, had not yet local-conservation-law approach had not yet been invented. Ref. [180] remedies this gap and shows that for dielectric scatterers, the use of additional conservation laws can significantly improve the resulting bounds. There is an interesting interplay between the quality factor of the sources and the bandwidth of interest, and there are useful semi-analytical bounds that can be derived from the global conservation laws applied to large-scale devices. Moreover, inverse-design structures are shown to come quite close to the improved complex-frequency, bounds.

2.4.3.2 Oscillator-Representation Bounds

An alternative to the complex-frequency approach to bandwidth averaging was very recently proposed in Ref. [176]. We will briefly summarize the (detailed) mathematical apparatus developed and highlight the key result for our purposes: a new, nearly tight bound for bandwidth-averaged NFRHT.

Before delving into scattering bodies, consider the bulk optical susceptibility of a material. It is known that the response of an isotropic passive material can be written as a linear combination of Drude–Lorentz oscillators:

$$\chi(\omega) = \sum_i \frac{\omega_p^2}{\omega_i^2 - \omega^2 - i\gamma\omega} c_i, \tag{2.86}$$

where ω_p is the "plasma frequency" of the material (related to its electron density [161, 181]), ω_i are the oscillator frequencies, γ are infinitesimal oscillator loss rates, and c_i are "oscillator strengths" that sum to unity, thanks to the sum rule of Eq. (2.50) discussed in Sect. 2.3.4. Often this representation is derived in single-electron quantum-material frameworks [161], but it applies more generally as a consequence of causality and passivity. (The technically rigorous mathematical statement uses the theory of Herglotz functions [182].) Any linear material's susceptibility must conform to the Drude–Lorentz linear combination of Eq. (2.86), perhaps not with a small number of oscillators (it is well known that effects such as inhomogeneous broadening lead to other lineshapes, such as the "Voigt" line-shape [183]), but with sufficiently many oscillators. It may seem counterintuitive to work with a representation that may need 1,000, or even 100,000 oscillators, instead of a different model with fewer parameters. From an optimization perspective, however, this is not correct. In the Drude–Lorentz representation of Eq. (2.86), the only degrees of freedom are the c_i coefficients, and the susceptibility is *linear* in these degrees of freedom. In many scenarios, large linear optimization problems are significantly easier to solve (sometimes even analytically) than large, nonlinear (and nonconvex) optimization problems.

Causality and passivity create three key ingredients that together lead to the Drude–Lorentz representation of Eq. (2.86): a Kramers–Kronig relation, a sum rule, and positivity of the imaginary part of the susceptibility. The exact sequence of transforming those ingredients to the Drude–Lorentz representation is detailed in Ref. [164]. One intuitive description is that the imaginary part of the susceptibility is a positive quantity and can be discretized into coefficients at many discrete frequencies along the real axis. Passivity implies that these coefficients are real, while the sum rule implies that their sum is constrained. Finally, the Kramers–Kronig relation guarantees that the imaginary parts of the susceptibilities are the *only* degrees of freedom; the real parts are entirely determined by the imaginary parts. Compiling the mathematical details of these steps leads to Eq. (2.86), which is a relation that many find intuitive, thanks largely to the fact that it can be derived in single-electron quantum mechanics.

The key idea of Ref. [176] is that there is a *wave-scattering operator* that exhibits nearly identical mathematical properties to material susceptibilities. This operator is the "$\mathbb{T}$" matrix. The $\mathbb{T}$ matrix is a scattering matrix that relates the polarization field induced in any scattering body to the incident fields impinging upon it [184]:

$$\mathbf{P}(\mathbf{x}, \omega) = \int_V \mathbb{T}(\mathbf{x}, \mathbf{x}', \omega)\mathbf{E}_{\text{inc}}(\mathbf{x}', \omega)\,\mathrm{d}\mathbf{x}', \tag{2.87}$$

or, in vector notation:

$$\mathbf{p} = \mathbb{T}\mathbf{e}_{\text{inc}}. \tag{2.88}$$

The $\mathbb{T}$ matrix is a *causal linear response function*, as the polarization field at $\mathbf{x}$ cannot be excited before the incident field exciting it reaches $\mathbf{x}'$. Just as causality implies a Kramers–Kronig relation for material susceptibilities, it was recognized in Ref. [176] that causality implies a Kramers–Kronig relation for $\mathbb{T}$ matrices. Sum rules come from the low- and high-frequency asymptotic behavior of Kramers–Kronig relations, and the $\mathbb{T}$ matrix satisfies a matrix-valued analog of the f-sum rule for material oscillator strengths. Finally, just as passivity implies that the imaginary parts of susceptibilities are positive, it similarly implies that the anti-Hermitian part of the $\mathbb{T}$ matrix is positive semidefinite. Together, these three ingredients imply a matrix-valued analog of Eq. (2.86) for any $\mathbb{T}$ matrix:

$$\mathbb{T}(\omega) = \sum_i \frac{\omega_p^2}{\omega_i^2 - \omega^2 - i\gamma\omega}\mathbb{T}_i, \tag{2.89}$$

where the Drude–Lorentz parameters are exactly the same as in Eq. (2.86) and the $\mathbb{T}_i$ are now matrix-valued coefficient degrees of freedom. The exact expression of Eq. (2.89) is for the case of reciprocal materials; in nonreciprocal terms, there is an extra term that makes the calculations more tedious but has no effect on most applications of interest. Analogous to the constraints on material oscillator strengths, passivity and the $\mathbb{T}$-matrix sum rule lead to constraints on the $\mathbb{T}_i$:

$$\sum_i \mathbb{T}_i = \mathbb{I}, \qquad \mathbb{T}_i \geq 0, \tag{2.90}$$

where $\mathbb{I}$ is the identity matrix. Equation (2.89), and its nonreciprocal analog, must hold for *any* linear electromagnetic scattering process. Even in scattering processes with complex interference phenomena, Fano resonances, etc., $\mathbb{T}(\omega)$ must exhibit lineshapes consistent with Eq. (2.89), which is shown in Ref. [176] to reveal surprising structure even in typical scattering problems.

Our interest in this chapter, however, is in fundamental limits, so we will focus on the utility of Eq. (2.89) to identify upper bounds in the application considered in Ref. [176], which is NFRHT. The approach in the paper requires a dozen or so mathematical steps explained in Sec. IX of the SM of Ref. [176]; the key is to transform the problem from one of thermal sources inside the hot body radiating power to the cold one to one of incoherent sources *between* the bodies radiating back to the emitter body. There are various other key steps, such as an appropriate renormalization of the point sources between the bodies. Ultimately, the culmination is the following: NFRHT is rewritten in terms of the total $\mathbb{T}$ matrix of the collective bodies, at which point the representation of Eq. (2.89) is inserted. Then, the entire frequency dependence of the problem is given by the collective products of the Drude–Lorentz oscillators and the Planck function, whose integrals can be determined analytically. Then one is left with a linear summation of given coefficients multiplying the unknown $\mathbb{T}_i$ degrees of freedom. The optimization over all possible $\mathbb{T}_i$, subject to the constraints of Eq. (2.90), has many unknowns, but can be done *analytically*, leading to a simple yet completely general bound on thermal HTC:

$$\text{HTC} \leq \beta \frac{T}{d^2}, \tag{2.91}$$

where T is the temperature, d is the separation, and $\beta \approx 0.11 k_B^2/\hbar$ is a numerical constant. Equation (2.91) is an unsurpassable limit that captures the key constraints imposed on every scattering $\mathbb{T}$ matrix. Strikingly, despite the relative simplicity of the approach, it offers the tightest bounds on NFRHT to date, only a factor of 5 larger than the best theoretical designs [178]. Previous approaches suggested strong material dependencies, with bounds that increased with electron density, whereas planar designs show the reverse trend. In this bound, use of a low-frequency sum rule in the $\mathbb{T}$-matrix representation leads to an electron-density-independent bound. Moreover, the optimization over $\mathbb{T}_i$ predicts precisely the same optimal peak transfer frequency as the best designs [176].

There are two sets of relaxations used to arrive at the bound of Eq. (2.91): first, beyond the representation theorem, no other Maxwell-equation constraints are imposed. Hence, the optimal $\mathbb{T}_i$ may not actually be physically realizable. Potentially, one could impose such constraints exactly by the local-conservation-law approach discussed above. Second, the heat transfer process is relaxed to the emission of the sources between the bodies into both the source and emitter, whereas the exact expression is the *difference* between the radiation into the emitter and receiver bodies. The latter relaxation leads to a linear dependence on $\mathbb{T}(\omega)$, as opposed to the quadratic dependence in the exact expression. It may be possible to optimize over the exact quadratic expression using manifold optimization techniques [176, 185, 186]. Tightening these relaxations may lead to a further tightening of the bound. Conversely, they may lead to the same bound, and improved design techniques [187] may identify structures that can achieve them.

2.4.4 Mode Volume

In this final section, we turn to the question of bounds on mode volume. Mode volume is a very different response function than any of those previously considered, as it is a property of an eigenfunction rather than a scattering quantity. There is no incident field in the definition of a mode volume, and hence the power-conservation and causality-based approaches of the previous sections are not immediately useful. In this section, we describe a method for bounding minimum mode volumes based on the optimization-theoretic notion of duality.

In optimization theory, the **dual** of an optimization problem is a second optimization problem, related to but distinct from the original, "primal" optimization problem [151]. The dual problem is formed by incorporating all constraints into the Lagrangian of the original optimization problem, introducing Lagrange multipliers as coefficients of the constraints, and optimizing out the primal variables, leaving only the Lagrange multipliers as degrees of freedom. An equivalent interpretation is that if one interprets a generic minimization optimization problem as the *minimax* of a Lagrangian, the dual problem is the *maximin* of the same Lagrangian. The dual program has two properties that can be quite useful for optimization and bounds: it is always a concave maximization problem (equivalent to a convex minimization problem, and therefore efficiently solvable by standard convex-optimization techniques), and its maximum is guaranteed to be a lower bound for the original, primal, minimization problem.

For many optimization problems, the dual cannot be expressed in a simple form; even among those problems for which it has a simple expression, it often has the trivial solution $-\infty$ as its maximum, giving a trivial lower bound. Ref. [188] showed that a very special class of electromagnetic design problems have a nontrivial, semi-analytical dual problem. In particular, for design problems in which the objective function to be minimized is the norm of a difference between the electric field $\mathbf{E}$ and some target field $\mathbf{E}_{\text{target}}$,

$$\mathcal{F} = \|\mathbf{E} - \mathbf{E}_{\text{target}}\|^2, \tag{2.92}$$

then one can impose the full Maxwell-equation constraints and identify a nontrivial, semi-analytical dual problem. One might suspect that objectives of the form of

Eq. (2.92) might be quite common: after all, a focusing metalens could have a target field that matches an Airy beam along a focal plane, a surface-pattern design intended to maximize spontaneous-emission enhancements could target the field at the location of the dipole, and so forth. But these cases do not work for the expression of Eq. (2.92): for a nontrivial dual problem, the field $\mathbf{E}_{\text{target}}$ must be specified *at every spatial point of the entire domain*. This includes, for example, the points within the scatterer, the points within any PML regions, etc. Knowing a target field at a single point, or on a focal plane, is not sufficient. And it is hard to think of any application in which we know the target field across the entire domain.

It turns out, however, that mode-volume minimization can be reformulated to target an objective specified over the entire domain. Mode volume, as specified in Eq. (2.22), is given by the integral of the field energy over all space divided by the field energy at a single point. Typically, the integral is treated as a normalization constant (taken to equal 1), and maximization of the field energy at a single point is the key objective. In Ref. [189], it was recognized that this convention could be reversed: the field energy at the point of interest can be fixed as a normalization constant, equal to 1, while minimizing the integral of the field energy can be the objective. Such an objective is exactly of the form of Eq. (2.92), with a target field of 0 everywhere! Physically, this makes intuitive sense: a minimum mode volume tries to minimize the field energy at every point, except for the "origin" of interest; everywhere else, it wants to drive the field as close to a target of 0 as possible.

Given this transformation, and a few others described in Ref. [189], one can use the formulation of Ref. [188] to specify a dual program for the mode-volume minimization problem. The solutions of this dual program can be formulated with the modeling language CVX [190] and solved with Gurobi [191], and those solutions represent fundamental lower bounds on the mode volume, given only a designable region and a refractive index of the material to be patterned.

First, the 2D TE case encapsulates scalar-wave physics: without vector fields, there are no field discontinuities across boundaries that can be responsible for large field amplitudes in "slot-mode" configurations [16, 18, 19]. There is also no near field for scalar waves, in the sense of large nonpropagating fields that culminate in a singularity at the location of a point source. In this case, the argument for a trivially small mode volume near a perfectly sharp tip *fails*: the lack of a singularity means that one cannot drive the field at the location of the source arbitrarily high. If there is no sharp-tip enhancement (as we will see), then dimensional arguments would require mode volume to scale with the square of the wavelength (in 2D), restoring the notion of a "diffraction-limited" mode volume. The only question, then, is the value of the coefficient of the squared wavelength. The duality-computed bounds confirm indeed that below some separation distance d, the mode-volume bounds asymptotically flatten out, to a small fraction of the square wavelength. This bound depends only on the available refractive index of the designable region and has been closely approached by inverse-designed structures [17, 189].

The 2D TM case is fundamentally different: sharp field discontinuities occur across material boundaries, and singularities in the near field of point sources imply the possibility for zero mode volume unless fabrication constraints, or similarly a nonzero source–scatterer separation distance, are enforced. In this case, the duality-based approach finds quite different scaling: the 2D TM mode-volume bounds scale as d^2, where d is the relevant source–scatterer distance (or sharp-tip radius of curvature), with no dependence on the wavelength. Intriguingly, this scaling is faster than the typical structure used for mode-volume minimization: a "bowtie antenna" [18, 19], whose optimal mode volume appears to scale only linearly with d (and hence linearly with wavelength, λ, as well). In Ref. [189], it is shown that inverse-designed structures appear to exhibit mode volumes that scale roughly as $d^{1.4}$, faster than the linear scaling of bowtie antennas but not quite as fast as the duality-based bound. At smaller length scales, these differences can be dramatic. For minimum feature sizes $d \approx 0.01\lambda$, the inverse-design curve falls about 5X below the bowtie-antenna curve, which itself is 40X above the mode-volume bound. Resolving this gap, either through identifying better designs or by identifying tighter bounds, could lead to significant reductions in mode volume through near-field engineering.

2.5 Summary and Looking Forward

Near-field optical response can require significant mathematical machinery, and the techniques to bound them even more so. We were careful above to give correct and sometimes nearly complete mathematical descriptions. Here, we can give a high-level summary of three of the prototypical response functions and application areas covered:

- LDOS, arguably the most important near-field response function, has single-frequency bounds that scale as $1/d^2$ and $|\chi(\omega)|^2/\operatorname{Im}\chi(\omega)$ [106]. This bound can be achieved at the surface-plasmon frequency of a given material; away from that frequency, inverse designs have shown good performance that can be relatively close to the bound, but generally it is also true that tighter bounds can be computed by using additional constraints. A sum rule is known for all-frequency LDOS [7, 174], which depends on the separation but *not* on the material; over finite bandwidth, bounds similar to the single-frequency expression can be found, albeit evaluated at the complex frequency. Again, these bounds are nearly achievable when the frequency range is centered around the surface-plasmon frequency of a material, but can be tightened in other scenarios (e.g., dielectric materials) [115]. The key open questions around LDOS are twofold: first, is there an analytical or semi-analytical bound that can be derived that is nearly achievable across all frequencies? And can one identify achievable bounds for only the *radiative* part of the LDOS, i.e., that fraction of power that is emitted to the far field?
- Near-field radiative heat transfer is one of the most technically challenging areas of near-field optics, both experimentally and theoretically, but an abundance of work makes it perhaps the area where we have the best understanding of what is possible. For planar bodies, there are simple and powerful transmission expres-

sions for NFRHT [57, 60], as well as an understanding of the optimal materials that lead to the largest response [178, 192, 193]. At a single frequency, semi-analytical bounds have been derived [171] that scale as $1/d^2$ with separation distance and *logarithmically* with $|\chi(\omega)|^2/\operatorname{Im}\chi$, both dependencies of which are exhibited by planar structures. Finally, when averaging against the Planck function to account for the thermal nature of the radiation, the recently developed oscillator theory of $\mathbb{T}$ matrices [176] enables a bound proportional only to $1/d^2$ and $k_B^2 T/\hbar$, with no material dependence. This bound can be approached within a factor of 5 by the best theoretical designs, showing a comprehensive understanding of what is possible in NFRHT, and the materials and structures needed to achieve that performance. One interesting open question is how this bound changes when one of the bodies must have a bandgap, as is required, for example, in thermophotovoltaics.
- Finally, mode volume is quite different from the other response functions considered above. It is a property of an eigenmode, instead of a scattered field, and hence some of the techniques based on power conservation do not lead to useful bounds in this case. The only approach we know of that leads to useful bounds relies on the *duality* technique of optimization theory. The most important question surrounding mode volume is how it scales with minimum feature size d. Ideally, it would scale as d^n, where n is the dimensionality of the system (either 2D or 3D), with no dependence on wavelength; this scaling would lead to the largest enhancements at highly subwavelength feature sizes. Certainly, such scaling is possible with plasmonic structures, but plasmonic structures are too lossy, and the concept of mode volume itself must be modified for plasmonic mode volume [25]. The question, then, is the optimal scaling for dielectric materials. Interestingly, the duality-based bounds of Ref. [189] suggest exactly d^n scaling. However, bowtie-antenna structures show d^{n-1} scaling, while inverse designs appear to show a scaling between these two. Hence, progress has been made on this crucial question, but it is still not fully resolved: what is the best possible scaling of mode volume with minimum feature size?

The theory of fundamental limits to near-field optical response is now sufficiently rich to be summarized in a book chapter, as we have done here. But the story is not complete: as we have seen in numerous examples, including the three above, there are still many response functions, material regimes, and frequency ranges at which there are gaps between the best known device structures and the best known bounds. Many of the bound techniques described herein have only been discovered in the past few years, and there are likely still significant strides to be made. The optical near field continues to offer a fertile playground for theoretical discovery, experimental demonstration, and new devices and technological applications.

2.6 Appendix: Complex Analysis for Sum Rules

Here we provide a brief summary of the basic rules of complex analysis, and how they are derived, emphasizing the key results relevant to sum rules. More expansive discussions of these ideas can be found in any good complex-analysis textbook.

First, we start with the definition of **complex differentiable**: a function f is complex differentiable if the limit

$$f'(z) = \lim_{h \to 0} \frac{f(z+h) - f(z)}{h} \tag{2.93}$$

exists for h along *any path* in the complex plane. The equality along any path is a very strong constraint and leads to the Cauchy–Riemann conditions on the derivatives of the real and imaginary parts of f. A function that is complex differentiable at every point on some domain Ω is **holomorphic** on Ω. A major theorem of complex analysis is that all such functions are also **complex analytic** (which means they have a convergent power series in a neighborhood of every point in Ω). From complex differentiability, it is a straight path to **Cauchy's integral theorem**: for f holomorphic on Ω, and a closed contour γ in Ω,

$$\oint_{\gamma} f(z)\,\mathrm{d}z = 0, \tag{2.94}$$

which can be proven by setting $f = u + iv$, $\mathrm{d}z = \mathrm{d}x + i\mathrm{d}y$, applying Green's/Stokes theorem, and using the Cauchy–Riemann conditions.

An important technique for integrals over open contours is **contour shifting**: if γ and $\tilde{\gamma}$ are contours with the same endpoints, then

$$\int_{\gamma} f(z)\,\mathrm{d}z = \int_{\tilde{\gamma}} f(z)\,\mathrm{d}z. \tag{2.95}$$

This follows directly from reversing the second contour, combining it with the first to make a closed contour, and applying Cauchy's integral theorem. Contour shifting is common in Casimir physics, for example, where the standard transformation is a "Wick rotation" from the positive real axis to the positive imaginary axis [194].

One can use contour shifting to prove an important integral formula. Consider the closed-contour integral $\oint_{\gamma} \frac{f(z)}{z-z_0}\,\mathrm{d}z$, where f is holomorphic on γ, but there is now a singularity in the integrand. For any arbitrary closed contour γ, one can follow the prescription of Fig. 2.5: first, make a tiny perforation in the contour and then use that perforation to shift to a modified contour that comprises two straight lines (whose integrals cancel by directionality) and a tiny circle at the origin. On the tiny circle, we can write $f(z) \approx f(z_0)$. On the circle, $z = z_0 + \varepsilon e^{i2\pi t}$, for t from 0 to 1, where ε is the radius of the circle on $\tilde{\gamma}$, such that

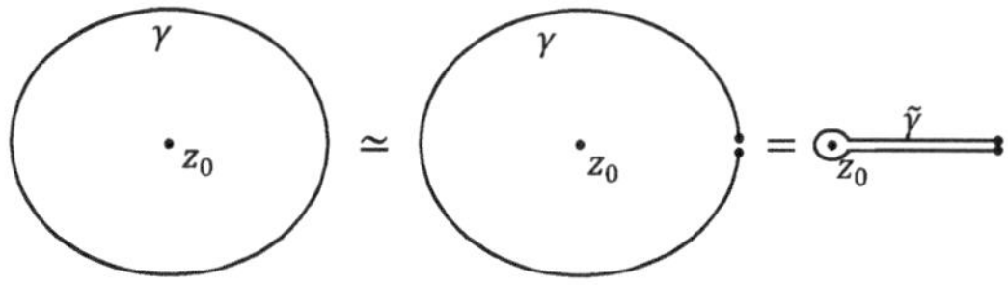

Fig. 2.5 Equivalent contours –the latter two by contour shifting– simplify the integration of *any* closed contour around a singularity (left) to that of a circle arbitrarily close to the singularity (right)

$$\begin{aligned}\oint_{\tilde{\gamma}} \frac{f(z)}{z-z_0} &\approx f(z_0) \oint_{\tilde{\gamma}} \frac{1}{z-z_0} \, \mathrm{d}z \\ &= f(z_0) \frac{1}{\varepsilon} \oint e^{-i2\pi t} \, \mathrm{d}\left(\varepsilon e^{i2\pi t}\right) \\ &= 2\pi i f(z_0). \end{aligned} \tag{2.96}$$

Equation (2.96) is **Cauchy's integral formula**.

One can take derivatives of Eq. (2.96) with respect to z_0 to yield an expression for the first derivative:

$$f'(z_0) = \frac{1}{2\pi i} \oint_{\gamma} \frac{f(z)}{(z-z_0)^2} \, \mathrm{d}z, \tag{2.97}$$

and more generally **Cauchy's differentiation formula**:

$$f^{(n-1)}(z_0) = \frac{(n-1)!}{2\pi i} \oint_{\gamma} \frac{f(z)}{(z-z_0)^n} \, \mathrm{d}z, \tag{2.98}$$

It is then one final step to get from Cauchy's differentiation formula to the residue theorem. Set the integrand in Eq. (2.98) to a function $g(z)$, which has a pole of order n at z_0. By a Laurent expansion, we can write *any* function with a pole of order n at z_0 in this form. Then we have the **residue theorem**:

$$\int_{\gamma} g(z) \, \mathrm{d}z = 2\pi i \sum_{\rho} \mathrm{Res}(f; z_0), \tag{2.99}$$

where the **residue** of f at z_0 is defined as

$$\mathrm{Res}(f; z_0) = \frac{1}{(n-1)!} \lim_{z \to z_0} \frac{d^{n-1}}{dz^{n-1}} \left[(z-z_0)^n f(z)\right]. \tag{2.100}$$

For $n = 1$, a simple pole, the residue is given by

$$\lim_{z \to z_0} \left[(z-z_0) f(z)\right]. \tag{2.101}$$

References

1. S.A. Maier, H.A. Atwater, Plasmonics: localization and guiding of electromagnetic energy in metal/dielectric structures. J. Appl. Phys. **98**(1), 10 (2005)
2. E. Betzig, R.J. Chichester, Single molecules observed by near-field scanning optical microscopy. Science **262**(5138), 1422–1425 (1993)
3. T. Taminiau, F. Stefani, F.B. Segerink, N. Van Hulst, Optical antennas direct single-molecule emission. Nat. Photonics **2**(4), 234–237 (2008)
4. M. Khajavikhan, A. Simic, M. Katz, J. Lee, B. Slutsky, A. Mizrahi, V. Lomakin, Y. Fainman, Thresholdless nanoscale coaxial lasers. Nature **482**(7384), 204–207 (2012)
5. R.F. Oulton, V.J. Sorger, T. Zentgraf, R.-M. Ma, C. Gladden, L. Dai, G. Bartal, X. Zhang, Plasmon lasers at deep subwavelength scale. Nature **461**(7264), 629–632 (2009)
6. I. Maksymov, M. Besbes, J.-P. Hugonin, J. Yang, A. Beveratos, I. Sagnes, I. Robert-Philip, P. Lalanne, Metal-coated nanocylinder cavity for broadband nonclassical light emission. Phys. Rev. Lett. **105**(18), 180502 (2010)
7. H. Shim, L. Fan, S.G. Johnson, O.D. Miller, Fundamental limits to near-field optical response over any bandwidth. Phys. Rev. X **9**, 011043 (2019)
8. K. Joulain, R. Carminati, J.-P. Mulet, J.-J. Greffet, Definition and measurement of the local density of electromagnetic states close to an interface. Phys. Rev. B: Condens. Matter Mater. Phys. **68**, 245405 (2003)
9. E. Yablonovitch, Inhibited spontaneous emission in solid-state physics and electronics. Phys. Rev. Lett. **58**(20), 2059 (1987)
10. E. Yablonovitch, Photonic crystals. J. Mod. Opt. **41**(2), 173–194 (1994)
11. J.D. Joannopoulos, S.G. Johnson, J.N. Winn, R.D. Meade, *Photonic Crystals: Molding the Flow of Light* (Princeton University Press, Princeton, 2011)
12. E.M. Purcell, Spontaneous emission probabilities at radio frequencies. Phys. Rev. **69**, 681 (1946)
13. M. Pelton, J. Vukovic, G.S. Solomon, A. Scherer, Y. Yamamoto, Three-dimensionally confined modes in micropost microcavities: quality factors and purcell factors. IEEE J. Quantum Electron. **38**(2), 170–177 (2002)
14. T. Kippenberg, S. Spillane, K. Vahala, Demonstration of ultra-high-q small mode volume toroid microcavities on a chip. Appl. Phys. Lett. **85**(25), 6113–6115 (2004)
15. N. Stoltz, M. Rakher, S. Strauf, A. Badolato, D. Lofgreen, P. Petroff, L. Coldren, D. Bouwmeester, High-quality factor optical microcavities using oxide apertured micropillars. Appl. Phys. Lett. **87**(3), 031105 (2005)
16. J.T. Robinson, C. Manolatou, L. Chen, M. Lipson, Ultrasmall mode volumes in dielectric optical microcavities. Phys. Rev. Lett. **95**, 143901 (2005)
17. X. Liang, S.G. Johnson, Formulation for scalable optimization of microcavities via the frequency-averaged local density of states. Opt. Express **21**, 30812–30841 (2013)
18. H. Choi, M. Heuck, D. Englund, Self-similar nanocavity design with ultrasmall mode volume for single-photon nonlinearities. Phys. Rev. Lett. **118**, 223605 (2017)
19. S. Hu, M. Khater, R. Salas-Montiel, E. Kratschmer, S. Engelmann, W.M. Green, S.M. Weiss, Experimental realization of deep-subwavelength confinement in dielectric optical resonators. Sci. Adv. **4**(8) (2018)
20. A. Taflove, A. Oskooi, S.G. Johnson, *Advances in FDTD Computational Electrodynamics: Photonics and Nanotechnology* (Artech house, Norwood, 2013)
21. E.N. Economou, *Green's Functions in Quantum Physics* (Springer Science & Business Media, Berlin, 2006)
22. F. Wijnands, J.B. Pendry, F.J. Garcia-Vidal, P.M. Bell, P.J. Roberts, L.M. Moreno, Green's functions for Maxwell's equations: application to spontaneous emission. Opt. Quantum Electron. **29**(2), 199–216 (1997)
23. Y. Xu, R.K. Lee, A. Yariv, Quantum analysis and the classical analysis of spontaneous emission in a microcavity. Phys. Rev. A **61**(3), 033807 (2000)

24. C. Sauvan, J.P. Hugonin, I.S. Maksymov, P. Lalanne, Theory of the spontaneous optical emission of nanosize photonic and plasmon resonators. Phys. Rev. Lett. **110**, 237401 (2013)
25. P. Lalanne, W. Yan, K. Vynck, C. Sauvan, J.-P. Hugonin, Light interaction with photonic and plasmonic resonances. Laser Photon. Rev. **12**, 1700113 (2018)
26. H. Zhang, O.D. Miller, Quasinormal coupled mode theory. arXiv:2010.08650 (2020)
27. D. Pellegrino, D. Balestri, N. Granchi, M. Ciardi, F. Intonti, F. Pagliano, A.Y. Silov, F.W. Otten, T. Wu, K. Vynck, et al., Non-lorentzian local density of states in coupled photonic crystal cavities probed by near-and far-field emission. Phys. Rev. Lett. **124**(12), 123902 (2020)
28. F.G. De Abajo, Optical excitations in electron microscopy. Rev. Mod. Phys. **82**(1), 209 (2010)
29. A. Cazé, R. Pierrat, R. Carminati, Spatial coherence in complex photonic and plasmonic systems. Phys. Rev. Lett. **110**, 063903 (2013)
30. R. Carminati, M. Gurioli, Purcell effect with extended sources: the role of the cross density of states. Opt. Express **30**, 16174–16183 (2022)
31. A. Canaguier-Durand, R. Pierrat, R. Carminati, Cross density of states and mode connectivity: probing wave localization in complex media. Phys. Rev. A **99**, 013835 (2019)
32. V. Krachmalnicoff, E. Castanié, Y. De Wilde, R. Carminati, Fluctuations of the local density of states probe localized surface plasmons on disordered metal films. Phys. Rev. Lett. **105**, 183901 (2010)
33. H.T. Dung, L. Knöll, D.-G. Welsch, Intermolecular energy transfer in the presence of dispersing and absorbing media. Phys. Rev. A **65**, 043813 (2002)
34. D. Martín-Cano, L. Martín-Moreno, F.J. García-Vidal, E. Moreno, Resonance energy transfer and superradiance mediated by plasmonic nanowaveguides. Nano Lett. **10**, 3129–3134 (2010)
35. J.A. Gonzaga-Galeana, J.R. Zurita-Sánchez, A revisitation of the Förster energy transfer near a metallic spherical nanoparticle: (1) efficiency enhancement or reduction? (2) the control of the Förster radius of the unbounded medium. (3) the impact of the local density of states. J. Chem. Phys. **139**(24), 244302 (2013)
36. J. Kästel, M. Fleischhauer, Suppression of spontaneous emission and superradiance over macroscopic distances in media with negative refraction. Phys. Rev. A **71**(1), 011804 (2005)
37. J. Kästel, M. Fleischhauer, Quantum electrodynamics in media with negative refraction. Laser Phys. **15**(1), 135–145 (2005)
38. D. Dzsotjan, A.S. Sørensen, M. Fleischhauer, Quantum emitters coupled to surface plasmons of a nanowire: a Green's function approach. Phys. Rev. B **82**(7), 075427 (2010)
39. D. Martín-Cano, A. González-Tudela, L. Martín-Moreno, F.J. García-Vidal, C. Tejedor, E. Moreno, Dissipation-driven generation of two-qubit entanglement mediated by plasmonic waveguides. Phys. Rev. B **84**, 235306 (2011)
40. A. Gonzalez-Tudela, D. Martin-Cano, E. Moreno, L. Martin-Moreno, C. Tejedor, F.J. Garcia-Vidal, Entanglement of two qubits mediated by one-dimensional plasmonic waveguides. Phys. Rev. Lett. **106**, 020501 (2011)
41. A. Otto, I. Mrozek, H. Grabhorn, W. Akemann, Surface-enhanced raman scattering. J. Phys. Condens. Matt. **4**(5), 1143 (1992)
42. S. Nie, S.R. Emory, Probing single molecules and single nanoparticles by surface-enhanced raman scattering. Science **275**(5303), 1102–1106 (1997)
43. K. Kneipp, Y. Wang, H. Kneipp, L.T. Perelman, I. Itzkan, R.R. Dasari, M.S. Feld, Single molecule detection using surface-enhanced raman scattering (sers). Phys. Rev. Lett. **78**(9), 1667 (1997)
44. K. Kneipp, H. Kneipp, I. Itzkan, R.R. Dasari, M.S. Feld, Surface-enhanced raman scattering and biophysics. J. Phys. Condens. Matt. **14**(18), R597 (2002)
45. D.A. Long, *Raman Spectroscopy* (McGraw-Hill, New York, 1977)
46. J. Michon, M. Benzaouia, W. Yao, O.D. Miller, S.G. Johnson, Limits to surface-enhanced raman scattering near arbitrary-shape scatterers. Opt. Express **27**, 35189–35202 (2019)
47. J.H. Lienhard IV, J.H. Lienhard V, *A Heat Transfer Textbook*, 4th ed. (Dover, Downers Grove, 2011)

48. D. Polder, M. Van Hove, Theory of radiative heat transfer between closely spaced bodies. Phys. Rev. B: Condens. Matter Mater. Phys. **4**(10), 3303–3314 (1971)
49. S.M. Rytov, Y.A. Kravtsov, V.I. Tatarskii, *Principles of Statistical Radiophysics* (Springer, New York, 1988)
50. S.-A. Biehs, M. Tschikin, R. Messina, P. Ben-Abdallah, Super-Planckian near-field thermal emission with phonon-polaritonic hyperbolic metamaterials. Appl. Phys. Lett. **102**(13), 131106 (2013)
51. O.D. Miller, S.G. Johnson, A.W. Rodriguez, Effectiveness of thin films in lieu of hyperbolic metamaterials in the near field. Phys. Rev. Lett. **112**, 157402 (2014)
52. S. Shen, A. Narayanaswamy, G. Chen, Surface phonon polaritons mediated energy transfer between nanoscale gaps. Nano Lett. **9**, 2909–2913 (2009)
53. E. Rousseau, A. Siria, G. Jourdan, S. Volz, F. Comin, J. Chevrier, J.-J. Greffet, Radiative heat transfer at the nanoscale. Nat. Photonics **3**(9), 514–517 (2009)
54. B. Song, Y. Ganjeh, S. Sadat, D. Thompson, A. Fiorino, V. Fernández-Hurtado, J. Feist, F.J. Garcia-Vidal, J.C. Cuevas, P. Reddy, et al., Enhancement of near-field radiative heat transfer using polar dielectric thin films. Nat. Nanotechnol. **10**(3), 253–258 (2015)
55. K. Kim, B. Song, V. Fernández-Hurtado, W. Lee, W. Jeong, L. Cui, D. Thompson, J. Feist, M.T.H. Reid, F.J. García-Vidal, J.C. Cuevas, E. Meyhofer, P. Reddy, Radiative heat transfer in the extreme near field. Nature **528**, 387–391 (2015)
56. J.J. Loomis, H.J. Maris, Theory of heat transfer by evanescent electromagnetic waves. Phys. Rev. B: Condens. Matter Mater. Phys. **50**(24), 18517–18524 (1994)
57. J.B. Pendry, Radiative exchange of heat between nanostructures. J. Phys. Condens. Matter **11**(35), 6621–6633 (1999)
58. K. Joulain, J.-P. Mulet, F. Marquier, R. Carminati, J.-J. Greffet, Surface electromagnetic waves thermally excited: radiative heat transfer, coherence properties and casimir forces revisited in the near field. Surf. Sci. Rep. **57**, 59–112 (2005)
59. P. Ben-Abdallah, K. Joulain, J. Drevillon, G. Domingues, Near-field heat transfer mediated by surface wave hybridization between two films. J. Appl. Phys. **106**(4), 44306 (2009)
60. S.-A. Biehs, E. Rousseau, J.-J. Greffet, Mesoscopic description of radiative heat transfer at the nanoscale. Phys. Rev. Lett. **105**, 234301 (2010)
61. M. Krüger, T. Emig, M. Kardar, Nonequilibrium electromagnetic fluctuations: heat transfer and interactions.' Phys. Rev. Lett. **106**, 210404 (2011)
62. J.P. Mulet, K. Joulain, R. Carminati, J.J. Greffet, Nanoscale radiative heat transfer between a small particle and a plane surface. Appl. Phys. Lett. **78**(19), 2931–2933 (2001)
63. A.W. Rodriguez, O. Ilic, P. Bermel, I. Celanovic, J.D. Joannopoulos, M. Soljačić, S.G. Johnson, Frequency-Selective Near-Field radiative heat transfer between photonic crystal slabs: a computational approach for arbitrary geometries and materials. Phys. Rev. Lett. **107**, 114302 (2011)
64. D. Dalvit, P. Milonni, D. Roberts, F. Da Rosa, *Casimir Physics*, vol. 834 (Springer, Berlin, 2011)
65. S.Y. Buhmann, *Dispersion Forces I: Macroscopic Quantum Electrodynamics and Ground-State Casimir, Casimir–Polder and van der Waals forces*, vol. 247 (Springer, Berlin, 2013)
66. A.W. Rodriguez, F. Capasso, S.G. Johnson, The casimir effect in microstructured geometries. Nat. Photonics **5**(4), 211–221 (2011)
67. P.S. Venkataram, S. Molesky, P. Chao, A.W. Rodriguez, Fundamental limits to attractive and repulsive casimir-polder forces. Phys. Rev. A **101**(5), 052115 (2020)
68. L. Zhu, S. Fan, Persistent directional current at equilibrium in nonreciprocal many-body near field electromagnetic heat transfer. Phys. Rev. Lett. **117**(13), 134303 (2016)
69. A. Raman, S. Fan, Photonic band structure of dispersive metamaterials formulated as a hermitian eigenvalue problem. Phys. Rev. Lett. **104**, 087401 (2010)
70. T. Wu, M. Gurioli, P. Lalanne, Nanoscale light confinement: the q's and v's. ACS Photonics **8**(6), 1522–1538 (2021)

71. C. Sauvan, T. Wu, R. Zarouf, E.A. Muljarov, P. Lalanne, Normalization, orthogonality, and completeness of quasinormal modes of open systems: the case of electromagnetism. Opt. Express **30**(5), 6846–6885 (2022)
72. T. Kato, *Perturbation Theory for Linear Operators*, vol. 132 (Springer Science & Business Media, Berlin, 2013)
73. D.C. Brody, E.-M. Graefe, Information geometry of complex hamiltonians and exceptional points. Entropy **15**(9), 3361–3378 (2013)
74. W. Shockley, H.J. Queisser, Detailed balance limit of efficiency of p-n junction solar cells. J. Appl. Phys. **32**, 510–519 (1961)
75. O.D. Miller, E. Yablonovitch, S.R. Kurtz, Strong internal and external luminescence as solar cells approach the Shockley-Queisser limit. IEEE J. Photovoltaics **2**, 303–311 (2012)
76. E. Yablonovitch, Statistical ray optics. J. Opt. Soc. Am. **72**(7), 899–907 (1982)
77. H.A. Wheeler, Fundamental limitations of small antennas. Proc. IRE **35**, 1479–1484 (1947)
78. L.J. Chu, Physical limitations of Omni-Directional antennas. J. Appl. Phys. **19**(12), 1163–1175 (1948)
79. R.F. Harrington, Effect of antenna size on gain, bandwidth, and efficiency. J. Res. Natl. Bur. Stand. Sect. D Radio Propag. **64D**, 1 (1960)
80. D.F. Sievenpiper, D.C. Dawson, M.M. Jacob, T. Kanar, S. Kim, J. Long, R.G. Quarfoth, Experimental validation of performance limits and design guidelines for small antennas. IEEE Trans. Antennas Propag. **60**, 8–19 (2012)
81. D.J. Bergman, Exactly solvable microscopic geometries and rigorous bounds for the complex dielectric constant of a two-component composite material. Phys. Rev. Lett. **44**(19), 1285–1287 (1980)
82. G.W. Milton, Bounds on the complex dielectric constant of a composite material. Appl. Phys. Lett. **37**(3), 300 (1980)
83. D.J. Bergman, Bounds for the complex dielectric constant of a two-component composite material. Phys. Rev. B **23**(6), 3058–3065 (1981)
84. G.W. Milton, Bounds on the complex permittivity of a two-component composite material. J. Appl. Phys. **52**(1981), 5286–5293 (1981)
85. G.W. Milton, Bounds on the transport and optical properties of a two-component composite material. J. Appl. Phys. **52**(8), 5294–5304 (1981)
86. C. Kern, O.D. Miller, G.W. Milton, Tight bounds on the effective complex permittivity of isotropic composites and related problems. Phys. Rev. Appl. **14**, 054068 (2020)
87. N. Fang, H. Lee, C. Sun, X. Zhang, Sub-diffraction-limited optical imaging with a silver superlens. Science **308**, 534–537 (2005)
88. R. Merlin, Radiationless electromagnetic interference: evanescent-field lenses and perfect focusing. Science **317**, 927–929 (2007)
89. C.W. McCutchen, Superresolution in microscopy and the Abbe resolution limit. J. Opt. Soc. Am. **57**(10), 1190–1192 (1967)
90. E.H.K. Stelzer, Beyond the diffraction limit? Nature **417**, 806–807 (2002)
91. N.I. Zheludev, What diffraction limit? Nat. Mat. **7**(6), 420–422 (2008)
92. D. Slepian, Prolate spheroidal wave functions, fourier analysis and uncertainty— I. Bell Syst. Tech. J. **40**(1), 43–63 (1961)
93. H.J. Landau, H.O. Pollak, Prolate spheroidal wave functions, fourier analysis and uncertainty—II. Bell Syst. Tech. J. **40**(1), 65–84 (1961)
94. P.J. Ferreira, A. Kempf, Superoscillations: faster than the nyquist rate. IEEE Trans. Sig. Proces. **54**(10), 3732–3740 (2006)
95. H. Shim, H. Chung, O.D. Miller, Maximal free-space concentration of electromagnetic waves. Phys. Rev. Appl. **14**(1), 014007 (2020)
96. C.E. Shannon, W. Weaver, *The Mathematical Theory of Communication* (University of Illinois Press, Urbana, 1949)
97. T.M. Cover, *Elements of Information Theory* (John Wiley & Sons, Hoboken, 1999)
98. W.D. Heiss, The physics of exceptional points. J. Phys. A: Math. Theor. **45**, 444016 (2012)
99. M.-A. Miri, A. Alù, Exceptional points in optics and photonics. Science **363**, eaar7709 (2019)

100. N. Bender, H. Yılmaz, Y. Bromberg, H. Cao, Creating and controlling complex light. APL Photonics **4**, 110806 (2019)
101. C.W. Hsu, B. Zhen, A.D. Stone, J.D. Joannopoulos, M. Soljačić, Bound states in the continuum. Nat. Rev. Mat. **1**, 1–13 (2016)
102. R.G. Newton, Optical theorem and beyond. Am. J. Phys. **44**(7), 639–642 (1976)
103. J.D. Jackson, *Classical Electrodynamics*, 3rd ed. (John Wiley & Sons, Hoboken, 1999)
104. D.R. Lytle, P.S. Carney, J.C. Schotland, E. Wolf, Generalized optical theorem for reflection, transmission, and extinction of power for electromagnetic fields. Phys. Rev. E **71**, 056610 (2005)
105. Z. Kuang, O.D. Miller, Computational bounds to light–matter interactions via local conservation laws. Phys. Rev. Lett. **125**, 263607 (2020)
106. O.D. Miller, A.G. Polimeridis, M.T. Homer Reid, C.W. Hsu, B.G. DeLacy, J.D. Joannopoulos, M. Soljacic, S.G. Johnson, Fundamental limits to optical response in absorptive systems. Opt. Express **24**, 3329–3364 (2016)
107. O.D. Miller, O. Ilic, T. Christensen, M.T.H. Reid, H.A. Atwater, J.D. Joannopoulos, M. Soljacic, S.G. Johnson, Limits to the optical response of graphene and two-dimensional materials. Nano Lett. **17**, 5408–5415 (2017)
108. Y. Ivanenko, M. Gustafsson, S. Nordebo, Optical theorems and physical bounds on absorption in lossy media. Opt. Express **27**, 34323–34342 (2019)
109. S. Nordebo, G. Kristensson, M. Mirmoosa, S. Tretyakov, Optimal plasmonic multipole resonances of a sphere in lossy media. Phys. Rev. B Condens. Matt. **99**, 054301 (2019)
110. Y. Yang, O.D. Miller, T. Christensen, J.D. Joannopoulos, M. Soljacic, Low-loss plasmonic dielectric nanoresonators. Nano Lett. **17**, 3238–3245 (2017)
111. O.D. Miller, S.G. Johnson, A.W. Rodriguez, Shape-independent limits to near-field radiative heat transfer. Phys. Rev. Lett. **115**, 204302 (2015)
112. Y. Yang, A. Massuda, C. Roques-Carmes, S.E. Kooi, T. Christensen, S.G. Johnson, J.D. Joannopoulos, O.D. Miller, I. Kaminer, M. Soljacic, Maximal spontaneous photon emission and energy loss from free electrons. Nat. Phys. **14**, 894–899 (2018)
113. O.D. Miller, K. Park, R.A. Vaia, Towards maximum optical efficiency of ensembles of colloidal nanorods. Opt. Express **30**, 25061–25077 (2022)
114. L. Tsang, J.A. Kong, K.-H. Ding, *Scattering of Electromagnetic Waves: Theories and Applications* (John Wiley & Sons, Inc., New York, 2000)
115. S. Molesky, P. Chao, A.W. Rodriguez, Hierarchical mean-field T operator bounds on electromagnetic scattering: upper bounds on near-field radiative purcell enhancement. Phys. Rev. Res. **2**, 043398 (2020)
116. Z. Kuang, L. Zhang, O.D. Miller, Maximal single-frequency electromagnetic response. Optica **7**, 1746–1757 (2020)
117. R.E. Hamam, A. Karalis, J.D. Joannopoulos, M. Soljačić, Coupled-mode theory for general free-space resonant scattering of waves. Phys. Rev. A **75**, 053801 (2007)
118. D.-H. Kwon, D.M. Pozar, Optimal characteristics of an arbitrary receive antenna. IEEE Trans. Antennas Propag. **57**, 3720–3727 (2009)
119. Z. Ruan, S. Fan, Design of subwavelength superscattering nanospheres. Appl. Phys. Lett. **98**(4), 043101 (2011)
120. I. Liberal, I. Ederra, R. Gonzalo, R. Ziolkowski, Upper bounds on scattering processes and Metamaterial-Inspired structures that reach them. IEEE Trans. Antennas Propag. **62**, 6344–6353 (2014)
121. I. Liberal, Y. Ra'di, R. Gonzalo, I. Ederra, S.A. Tretyakov, R.W. Ziolkowski, Least upper bounds of the powers extracted and scattered by bi-anisotropic particles. IEEE Trans. Antennas Propag. **62**, 4726–4735 (2014)
122. J.-P. Hugonin, M. Besbes, P. Ben-Abdallah, Fundamental limits for light absorption and scattering induced by cooperative electromagnetic interactions. Phys. Rev. B: Condens. Matter Mater. Phys. **91**, 180202 (2015)
123. A.E. Miroshnichenko, M.I. Tribelsky, Ultimate absorption in light scattering by a finite obstacle. Phys. Rev. Lett. **120**(3), 033902 (2018)

124. M. Gustafsson, M. Cismasu, Physical bounds and optimal currents on antennas. IEEE Trans. Antennas Propag. **60**(6), 2672–2681 (2012)
125. M. Gustafsson, K. Schab, L. Jelinek, M. Capek, Upper bounds on absorption and scattering. New J. Phys. **22**, 073013 (2020)
126. S. Molesky, P. Chao, W. Jin, A.W. Rodriguez, Global T operator bounds on electromagnetic scattering: upper bounds on far-field cross sections. Phys. Rev. Res. **2**, 033172 (2020)
127. M.I. Abdelrahman, F. Monticone, How thin and efficient can a metasurface reflector be? Universal bounds on reflection for any direction and polarization. arXiv:2208.05533 (2022)
128. L. Jelinek, M. Gustafsson, M. Capek, K. Schab, Fundamental bounds on the performance of monochromatic passive cloaks. Opt. Express **29**(15), 24068–24082 (2021)
129. K. Schab, L. Jelinek, M. Capek, M. Gustafsson, Upper bounds on focusing efficiency. Opt. Express **30**(25), 45705–45723 (2022)
130. W.C. Chew, M.S. Tong, B. Hu, Integral equation methods for electromagnetic and elastic waves. Synth. Lect. Comput. Electromagn. **3**, 1–241 (2008)
131. D.A. Miller, Spatial channels for communicating with waves between volumes. Opt. Lett. **23**(21), 1645–1647 (1998)
132. D.A. Miller, Communicating with waves between volumes: evaluating orthogonal spatial channels and limits on coupling strengths. Appl. Opt. **39**(11), 1681–1699 (2000)
133. R. Piestun, D.A. Miller, Electromagnetic degrees of freedom of an optical system. JOSA A **17**(5), 892–902 (2000)
134. D.A. Miller, Waves, modes, communications, and optics: a tutorial. Adv. Opt. Photonics **11**(3), 679–825 (2019)
135. L.N. Trefethen, D. Bau, *Numerical Linear Algebra* (Society for Industrial and Applied Mathematics, Philadelphia, 1997)
136. Z. Kuang, D.A.B. Miller, O.D. Miller, Bounds on the coupling strengths of communication channels and their information capacities. arXiv:2205.05150 (2022)
137. C. Ehrenborg, M. Gustafsson, Fundamental bounds on mimo antennas. IEEE Antennas Wireless Propag. Lett. **17**(1), 21–24 (2017)
138. C. Ehrenborg, M. Gustafsson, Physical bounds and radiation modes for mimo antennas. IEEE Trans. Antennas Propag. **68**(6), 4302–4311 (2020)
139. C. Ehrenborg, M. Gustafsson, M. Capek, Capacity bounds and degrees of freedom for mimo antennas constrained by q-factor. IEEE Trans. Antennas Propag. (2021)
140. Y. Gao, H. Vinck, and T. Kaiser, Massive mimo antenna selection: switching architectures, capacity bounds, and optimal antenna selection algorithms. IEEE Trans. Sig. Proces. **66**(5), 1346–1360 (2017)
141. S. Asaad, A.M. Rabiei, R.R. Müller, Massive mimo with antenna selection: fundamental limits and applications. IEEE Trans. Wireless Commun. **17**(12), 8502–8516 (2018)
142. S.G. Johnson, P. Bienstman, M.A. Skorobogatiy, M. Ibanescu, E. Lidorikis, J.D. Joannopoulos, Adiabatic theorem and continuous coupled-mode theory for efficient taper transitions in photonic crystals. Phys. Rev. E **66**, 66608 (2002)
143. S. Boyd, L. El Ghaoui, E. Feron, V. Balakrishnan, *Linear Matrix Inequalities in System and Control Theory* (SIAM, Philadelphia, 1994)
144. M.X. Goemans, D.P. Williamson, Improved approximation algorithms for maximum cut and satisfiability problems using semidefinite programming. J. ACM **42**(6), 1115–1145 (1995)
145. Z.Q. Luo, W.K. Ma, A. So, Y. Ye, S. Zhang, Semidefinite relaxation of quadratic optimization problems. IEEE Signal Process. Mag. **27**(3), 20–34 (2010)
146. S. Sojoudi, J. Lavaei, Physics of power networks makes hard optimization problems easy to solve, in *2012 IEEE Power and Energy Society General Meeting* (IEEE, Piscataway, 2012), pp. 1–8
147. E.J. Candes, T. Strohmer, V. Voroninski, Phaselift: exact and stable signal recovery from magnitude measurements via convex programming. Commun. Pure Appl. Math. **66**(8), 1241–1274 (2013)
148. R. Horstmeyer, R.Y. Chen, X. Ou, B. Ames, J.A. Tropp, C. Yang, Solving ptychography with a convex relaxation. New J. Phys. **17**(5), 053044 (2015)

149. M. Laurent, F. Rendl, Semidefinite programming and integer programming. Handbooks Oper. Res. Manag. Sci. **12**(C), 393–514 (2005)
150. L. Vandenberghe, S. Boyd, Semidefinite programming. SIAM Rev. **38**, 49–95 (1996)
151. S. Boyd, L. Vandenberghe, *Convex Optimization* (Cambridge University Press, Cambridge, 2004)
152. J. Park, S. Boyd, General Heuristics for Nonconvex Quadratically Constrained Quadratic Programming. arXiv:1703.07870 [math] (2017)
153. G. Angeris, J. Vučković, S. Boyd, Heuristic methods and performance bounds for photonic design. Opt. Express **29**(2), 2827–2854 (2021)
154. H. Shim, Z. Kuang, Z. Lin, O.D. Miller, Fundamental limits to multi-functional and tunable nanophotonic response. arXiv: 2112.10816 (2021)
155. H. Zhang, Z. Kuang, S. Puri, O.D. Miller, Conservation-law-based global bounds to quantum optimal control. Phys. Rev. Lett. **127**, 110506 (2021)
156. G. Angeris, T. Diamandis, J. Vučković, S. Boyd, Bounds on efficiency metrics in photonics. Preprint. arXiv:2204.05243 (2022)
157. G. Angeris, A note on generalizing power bounds for physical design. arXiv:2208.04411 (2022)
158. F.W. King, Sum rules for the optical constants. J. Math. Phys. **17**(8), 1509–1514 (1976)
159. V. Lucarini, J.J. Saarinen, K.-E. Peiponen, E. M. Vartiainen, *Kramers-Kronig Relations in Optical Materials Research* (Springer Science & Business Media, Berlin, 2005)
160. H.M. Nussenzveig, *Causality and Dispersion Relations* (Academic Press, New York, 1972)
161. E. Kaxiras, J.D. Joannopoulos, *Quantum Theory of Materials* (Cambridge University Press, Cambridge, 2019)
162. J. Skaar, K. Seip, Bounds for the refractive indices of metamaterials. J. Phys. D Appl. Phys. **39**, 1226–1229 (2006)
163. M. Gustafsson, D. Sjöberg, Sum rules and physical bounds on passive metamaterials. New J. Phys. **12**, 043046 (2010)
164. H. Shim, F. Monticone, O.D. Miller, Fundamental limits to the refractive index of transparent optical materials. Adv. Mat. **33**, 2103946 (2021)
165. C. Sohl, M. Gustafsson, G. Kristensson, Physical limitations on broadband scattering by heterogeneous obstacles. J. Phys. A: Math. Theor. **40**, 11165–11182 (2007)
166. R.G. Gordon, Three sum rules for total optical absorption cross sections. J. Chem. Phys. **38**(7), 1724 (1963)
167. E.M. Purcell, On the absorption and emission of light by interstellar grains. Astrophys. J. **158**, 433–440 (1969)
168. D.S. Jones, Scattering by inhomogeneous dielectric particles. Q. J. Mech. Appl. Mech. **38**(1), 135–155 (1985)
169. H. Shim, Z. Kuang, O.D. Miller, Optical materials for maximal nanophotonic response (Invited). Opt. Mat. Express **10**, 1561–1585 (2020)
170. S. Molesky, P.S. Venkataram, W. Jin, A.W. Rodriguez, Fundamental limits to radiative heat transfer: theory. Phys. Rev. B **101**(3), 035408 (2020)
171. P.S. Venkataram, S. Molesky, W. Jin, A.W. Rodriguez, Fundamental limits to radiative heat transfer: the limited role of nanostructuring in the near-field. Phys. Rev. Lett. **124**(1), 013904 (2020)
172. P. Ben-Abdallah, K. Joulain, Fundamental limits for noncontact transfers between two bodies. Phys. Rev. B: Condens. Matter Mater. Phys. **82**, 121419 (2010)
173. W. Jin, S. Molesky, Z. Lin, A.W. Rodriguez, Material scaling and frequency-selective enhancement of near-field radiative heat transfer for lossy metals in two dimensions via inverse design. Phys. Rev. B **99**(4), 041403 (2019)
174. S. Sanders, A. Manjavacas, Analysis of the limits of the local density of photonic states near nanostructures. ACS Photonics **5**, 2437–2445 (2018)
175. S.M. Barnett, R. Loudon, Sum rule for modified spontaneous emission rates. Phys. Rev. Lett. **77**(12), 2444–2446 (1996)

176. L. Zhang, F. Monticone, O.D. Miller, All electromagnetic scattering bodies are matrix-valued oscillators. arXiv: 2211.04457 (2022)
177. H. Hashemi, C.-W. Qiu, A.P. Mccauley, J.D. Joannopoulos, S.G. Johnson, Diameter-bandwidth product limitation of isolated-object cloaking. Phys. Rev. A **86**, 013804 (2012)
178. L. Zhang, O.D. Miller, Optimal materials for maximum large-area near-field radiative heat transfer. ACS Photonics **7**, 3116–3129 (2020)
179. R. Mittapally, J.W. Lim, L. Zhang, O.D. Miller, P. Reddy, E. Meyhofer, Probing the limits to near-field heat transfer enhancements in phonon-polaritonic materials. Nano Lett. **23**, 2187–2194 (2023)
180. P. Chao, R.K. Defo, S. Molesky, A. Rodriguez, Maximum electromagnetic local density of states via material structuring. Nanophotonics **12**, 549–557 (2022)
181. Z.-J. Yang, T.J. Antosiewicz, R. Verre, F.J. Garcia de Abajo, S.P. Apell, M. Kall, Ultimate limit of light extinction by nanophotonic structures. Nano Lett. **15**(11), 7633–7638 (2015)
182. A. Bernland, A. Luger, M. Gustafsson, Sum rules and constraints on passive systems. J. Phys. A Math. Theoret. **44**(14), 145205 (2011)
183. J.-M. Hartmann, C. Boulet, D. Robert, *Collisional Effects on Molecular Spectra: Laboratory Experiments and Models, Consequences for Applications* (Elsevier, Amsterdam, 2021)
184. R. Carminati, J.C. Schotland, *Principles of Scattering and Transport of Light* (Cambridge University Press, Cambridge, 2021)
185. P.-A. Absil, R. Mahony, R. Sepulchre, *Optimization Algorithms on Matrix Manifolds* (Princeton University Press, Princeton, 2009)
186. N. Boumal, B. Mishra, P.-A. Absil, R. Sepulchre, Manopt, a Matlab toolbox for optimization on manifolds. J. Mach. Learn. Res. **15**(42), 1455–1459 (2014)
187. S. Gertler, Z. Kuang, C. Christie, O.D. Miller, Many physical design problems are sparse QCQPs. arXiv:2303.17691 (2023)
188. G. Angeris, J. Vuckovic, S.P. Boyd, Computational bounds for photonic design. ACS Photonics **6**(5), 1232–1239 (2019)
189. Q. Zhao, L. Zhang, O.D. Miller, Minimum dielectric-resonator mode volumes. arXiv: 2008.13241 (2020)
190. M. Grant, S. Boyd, CVX: Matlab software for disciplined convex programming, version 2.1. http://cvxr.com/cvx (2014)
191. Gurobi Optimization, LLC, Gurobi optimizer reference manual (2019)
192. E. Rousseau, M. Laroche, J.-J. Greffet, Asymptotic expressions describing radiative heat transfer between polar materials from the far-field regime to the nanoscale regime. J. Appl. Phys. **111**(1), 014311 (2012)
193. M. Pascale, G.T. Papadakis, Tight bounds and the role of optical loss in polariton-mediated near-field heat transfer. Phys. Rev. Appl. **19**(3), 034013 (2023)
194. S.G. Johnson, Numerical methods for computing casimir interactions. In: Dalvit, D., Milonni, P., Roberts, D., da Rosa, F. (eds.) Casimir Physics. Lecture Notes in Physics, Springer, Berlin, Heidelberg. **834**, 175–218 (2011). https://doi.org/10.1007/978-3-642-20288-9_6

Chapter 3
Quasinormal Mode Theories and Applications in Classical and Quantum Nanophotonics

Juanjuan Ren, Sebastian Franke, and Stephen Hughes

Abstract This chapter presents theories and applications of optical quasinormal modes (QNMs), which can be used to solve a wide range of cavity problems in classical and quantum nanophotonics. Special emphasis is placed on obtaining intuitive, few-mode analytical expressions for the electromagnetic Green functions which connect to important figures of merit in cavity optics such as Purcell's formula. We give the basic background theory, starting from Maxwell's equations and classical mode expansion techniques, as well as normal modes, QNMs, and regularized QNMs, followed by a description of quantized QNMs, which form the foundation for developing rigorous quantum optical descriptions in nanophotonics and non-Hermitian resonant systems. We then show various instructive examples ranging from simple 1D cavities, which have analytical solutions, to plasmonic dimer modes, to complicated hybrid modes formed by coupled metal-dielectric systems, and QNM coupled-mode theory for coupled resonators.

3.1 Introduction

Mode theories are ubiquitous in optics and photonics. From transmission lines to fibers to cavities, an understanding of optical modes clarifies the underlying physics of light-matter coupling in many photonic materials and devices and forms a convenient starting point for understanding additional perturbations and couplings. Coupled-mode theory, as an example, has enjoyed tremendous success in explaining

J. Ren · S. Hughes (✉)
Department of Physics, Engineering Physics and Astronomy, Queen's University, Kingston, ON, Canada
e-mail: juanjuan.ren@queensu.ca; shughes@queensu.ca

S. Franke
Department of Physics, Engineering Physics and Astronomy, Queen's University, Kingston, ON, Canada

Technische Universität Berlin, Institut für Theoretische Physik, Nichtlineare Optik und Quantenelektronik, Berlin, Germany

R. Gordon (ed.), *Advances in Near-Field Optics*, Springer Series in Optical Sciences 244, https://doi.org/10.1007/978-3-031-34742-9_3

how two separate photonic systems couple and has a wide range of applications [1–6]. Optical modes are also a *requirement* in quantum optics when connecting to system-level quantization, which is essential when describing phenomena such as nonclassical light generation and cavity quantum electrodynamics (QED). This is particularly true in macroscopic cavity-QED, which ultimately must recover Maxwell's equations in the appropriate limit.

Nanophotonics offers additional challenges for understanding optical modes, partly because one usually requires a full 3D solution to the problems, and the vector field modes are much richer because of nanoscale patterning and control of the constituent materials. Historically, there have been claims that mode theories cannot even be expected to work for such media, e.g., because of material losses [7], but that is mainly due to a lack of a rigorous mode theory for such structures. Often one then adopts brute force numerical approaches such as finite-difference time-domain (FDTD) or finite-element methods, but these approaches can be tedious to use directly and often obscure the underling physics. Direct numerical solutions are also limited to the classical domain, yet the same numerical approaches can obtain the correct modes, which can then be used in very efficient and elegant ways, with analytical insight. Ultimately, numerical solutions are required for obtaining modes of complex nanostructures, but they can be obtained using established optical simulations that must be carried out anyway. However, these *open cavity modes* are not your typical modes that you learn as an undergraduate student, aka "normal modes," and they must be treated with appropriate care, in particular because losses (radiative and possibly nonradiative) are an inherent part of these mode solutions. This challenges the standard mode theories and also the usual quantization schemes for cavity modes.

In this chapter, we discuss open cavity modes in nanophotonics, which are already very diverse, and may include simple cavities, coupled cavities, plasmonic particles, dimers, and cavities coupled to waveguides, with an emphasis on solving important problems in classical and quantum nanophotonics. While there are many definitions of what constitutes a cavity mode in the literature (often with some ambiguity), we will concentrate on one unified picture, known as "quasinormal modes" (QNMs), which can be defined mathematically, and physically, as a mode solution to an open cavity problem with outgoing boundary conditions. The distinction from *normal modes* (NMs) will be clarified in the next theory section, but here we just remark that QNMs have complex eigenfrequencies, and thus they include losses (dissipation), while NMs have real eigenfrequencies. Since cavity NMs have no losses, they are not really a practical tool for studying cavity modes in realistic nanophotonic systems and often lead to ambiguous interpretations, especially for low-quality factor modes and coupled cavity modes. However, for single cavity modes with high-quality factors, they can be an extremely good approximation, since that is precisely the NM limit. Quasinormal modes not only fix certain limitations of NM theories applied to optics (such as Purcell's formula), but they also touch the heart of non-Hermitian physics, since if the radiation leaks out, energy is not conserved. Thus, any realistic cavity system, i.e., *with dissipation*, is a non-Hermitian problem from the beginning.

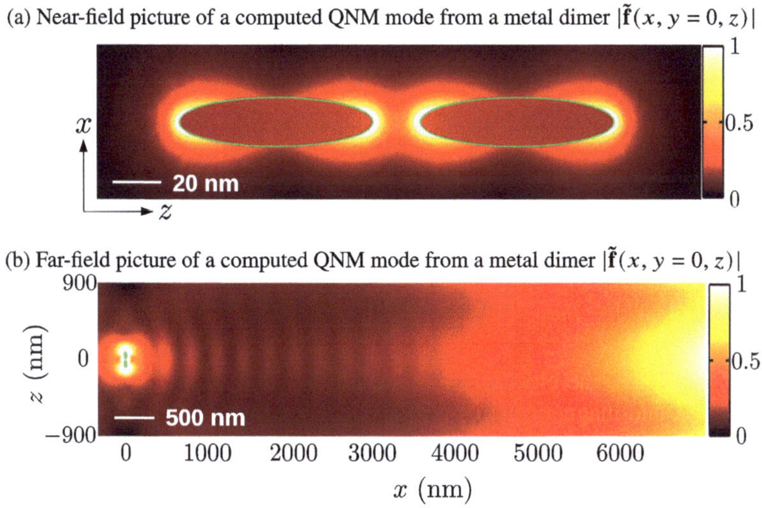

Fig. 3.1 Example calculation of a 3D QNM ($\tilde{\mathbf{f}}$, defined later, from the Helmholtz equation) obtained for a metal dimer using two ellipsoids in free space, whose dielectric constant is described through a Drude model, similar to gold but with more broadening. The center width and length of the ellipsoid are 20 nm and 80 nm, respectively, with a gap distance of 20 nm. We find a single QNM in the frequency regime of interest with $\hbar\tilde{\omega}_c = (2.066 - 0.148i)$ eV and a quality factor $Q_c \approx 7$. (**a**) Surface plot of the QNM profile in the near-field zone, in arbitrary units. (**b**) Surface plot of the same QNM profile, but now showing the far-field zone, where we clearly see the mode starting to spatially diverge. Although the mode spatially diverges, there is no divergence in time for the outgoing field profile, and the total field at these positions (far field) is always well behaved

Although we will discuss various types of optical QNMs in this chapter, we begin with a strikingly visual example: Fig. 3.1 shows the spatial profile of a QNM (labeled with $\tilde{\mathbf{f}}$) computed for a metal dimer structure with a 20 nm gap [8], which shows a typical localized plasmon mode formed in the near field, but in the far field, the mode starts to spatially diverge, which is a characteristic of cavity QNMs that exhibit finite losses. Remember that optical modes are source free solutions of Maxwell's equations, and although the mode is spatially diverging, it is not unphysical, and there is also no divergence in the time-dependent field.

3.2 Theory

In this section, we describe the main theoretical details and QNM formulas, starting from Maxwell's equations, and then we introduce both classical and quantized QNMs, as well as the main calculation and numerical methods; we also present a theory on coupled resonators using QNM coupled-mode theory, from both classical and quantum optics perspectives.

3.2.1 Maxwell's Equations, Helmholtz Equation, and Green Functions

The classical theory of light-matter interactions is governed by Maxwell's equations:

$$\begin{aligned} \nabla \cdot \mathbf{B}(\mathbf{r}, t) &= 0, & \nabla \times \mathbf{E}(\mathbf{r}, t) + \frac{\partial \mathbf{B}(\mathbf{r}, t)}{\partial t} &= 0, \\ \nabla \cdot \mathbf{D}(\mathbf{r}, t) &= \rho_{\mathrm{ext}}, & \nabla \times \mathbf{H}(\mathbf{r}, t) - \frac{\partial \mathbf{D}(\mathbf{r}, t)}{\partial t} &= \mathbf{J}_{\mathrm{ext}}, \end{aligned} \tag{3.1}$$

where $\mathbf{E}$ and $\mathbf{H}$ are the electric and magnetic fields, respectively, $\mathbf{D}$ and $\mathbf{B}$ are the displacement and magnetic induction fields, and ρ_{ext} and $\mathbf{J}_{\mathrm{ext}}$ are possible external charge and current densities. We also have the following constituent relationships relating to polarization, $\mathbf{P}$; magnetization, $\mathbf{M}$; and total current density, $\mathbf{J}$: $\mathbf{D} = \epsilon_0\mathbf{E} + \mathbf{P} = \epsilon_0\epsilon\mathbf{E}$, $\mathbf{H} = 1/\mu_0\mathbf{B} - \mathbf{M} = 1/(\mu_0\mu)\mathbf{B}$, and $\mathbf{J} = \partial\mathbf{P}/\partial t$. One can also define $\mathbf{J}$ in terms of a conductivity, but we prefer to work with complex dielectric constants for consistency with most approaches in optics; there is no loss in generality in doing this and it is a simple matter of choice.

Since we are interested in solving for the modes (i.e., source-free), we next assume there are no external sources, so $\rho_{\mathrm{ext}} = \mathbf{J}_{\mathrm{ext}} = 0$ and that the medium is isotropic and nonmagnetic.[1] Thus, equation set (3.1) becomes

$$\begin{aligned} \nabla \cdot \mathbf{H}(\mathbf{r}, t) &= 0, & \nabla \times \mathbf{E}(\mathbf{r}, t) + \mu_0\frac{\partial \mathbf{H}(\mathbf{r}, t)}{\partial t} &= 0, \\ \nabla \cdot \mathbf{D}(\mathbf{r}, t) &= 0, & \nabla \times \mathbf{H}(\mathbf{r}, t) - \frac{\partial \mathbf{D}(\mathbf{r}, t)}{\partial t} &= 0. \end{aligned} \tag{3.2}$$

Assuming harmonic fields given by $\mathbf{E}(\mathbf{r}, t) \equiv \mathbf{E}(\mathbf{r}, \omega)e^{-i\omega t}$ and $\mathbf{B}(\mathbf{r}, t) \equiv \mathbf{B}(\mathbf{r}, \omega)e^{-i\omega t}$, we obtain the (vector) Helmholtz equation:

$$\frac{1}{\epsilon(\mathbf{r}, \omega)}\nabla \times \nabla \times \mathbf{E}(\mathbf{r}, \omega) = \left(\frac{\omega}{c}\right)^2 \mathbf{E}(\mathbf{r}, \omega), \tag{3.3}$$

which can be used for calculating the fundamental optical modes, which will also depend on the chosen boundary conditions.

The inhomogeneous Helmholtz equation for an arbitrary polarization source, $\mathbf{P}_s(\mathbf{r}, \omega)$, or current source, $\mathbf{j}_s(\mathbf{r}, \omega) = -i\omega\mathbf{P}_s(\mathbf{r}, \omega)$, takes the following form:

$$\nabla \times \nabla \times \mathbf{E}(\mathbf{r}, \omega) - \left(\frac{\omega}{c}\right)^2 \epsilon(\mathbf{r}, \omega)\mathbf{E}(\mathbf{r}, \omega) = \left(\frac{\omega}{c}\right)^2 \frac{\mathbf{P}_s(\mathbf{r}, \omega)}{\epsilon_0}, \tag{3.4}$$

[1] An exception to this assumption will be made later when introducing a specific form for the mode normalization.

which can be used to define the (photonic) Green function:

$$\nabla \times \nabla \times \mathbf{G}(\mathbf{r}, \mathbf{r}', \omega) - \left(\frac{\omega}{c}\right)^2 \epsilon(\mathbf{r}, \omega)\mathbf{G}(\mathbf{r}, \mathbf{r}', \omega) = \left(\frac{\omega}{c}\right)^2 \mathbf{1}\delta(\mathbf{r} - \mathbf{r}'), \tag{3.5}$$

where the electric field solution is at $\mathbf{r}$ when a source electric field dipole is at $\mathbf{r}'$. The term $\mathbf{1}$ is the unit dyad. Keep in mind that $\mathbf{G}$ is also a dyad, or a 3×3 matrix, e.g., in a Cartesian coordinate system, we have $G_{\alpha\beta}$ with $\alpha, \beta = x, y, z$. The form of $\mathbf{G}$ will become clear below when we also obtain it analytically using mode expansions.

The Green function connects to many problems in classical and quantum optics [9–23]. A well-known example is that the photonic local density of states (LDOS) depends on $\text{Im}[\mathbf{G}(\mathbf{r}_0, \mathbf{r}_0, \omega)]$, and the same function at two-space points describes retarded dipole-dipole interactions and heat transfer. It also connects elegantly to the fluctuation-dissipation formula [24]. Indeed, it is probably difficult to find a modern book on nano-optics and not come across the Green function, e.g., see the excellent textbook by Novotny and Hecht [25]. The Green function is also useful for solving a wide range of problems in scattering theory, since it can be used to obtain the electric field response due to any arbitrary polarization source via

$$\mathbf{E}(\mathbf{r}, \omega) = \mathbf{E}_0(\mathbf{r}, \omega) + \frac{1}{\epsilon_0} \int \mathbf{G}(\mathbf{r}, \mathbf{r}', \omega) \cdot \mathbf{P}_s(\mathbf{r}', \omega) d\mathbf{r}', \tag{3.6}$$

where $\mathbf{E}_0(\mathbf{r}, \omega)$ is the homogeneous field solution. Inserting the second term from Eq. (3.6), i.e., the scattering part, into Eq. (3.4) results in Eq. (3.5), when using a point dipole source.

Below we list some useful properties (independent of coordinate system) of the photonic Green function [20, 26]:

$$\begin{aligned} &G_{ij}(\mathbf{r}, \mathbf{r}', \omega) = G_{ji}(\mathbf{r}', \mathbf{r}, \omega), \quad G^*_{ij}(\mathbf{r}, \mathbf{r}', \omega) = G_{ij}(\mathbf{r}, \mathbf{r}', -\omega), \\ &\int \text{Im}\left[\epsilon(\mathbf{r}'', \omega)\right] \mathbf{G}(\mathbf{r}, \mathbf{r}'', \omega) \cdot \mathbf{G}^*(\mathbf{r}'', \mathbf{r}', \omega) d\mathbf{r}'' = \text{Im}\left[\mathbf{G}(\mathbf{r}, \mathbf{r}', \omega)\right]. \end{aligned} \tag{3.7}$$

The total Green function can also be written in terms of the transverse and longitudinal solutions to Eq. (3.5), so that $\mathbf{G}(\mathbf{r}, \mathbf{r}') = \mathbf{G}^T(\mathbf{r}, \mathbf{r}') + \mathbf{G}^L(\mathbf{r}, \mathbf{r}')$.

3.2.2 *Dyson Equation for the Self-Consistent Green Function*

Let us consider the general expression for the displacement field, $\mathbf{D}(\mathbf{r}, \omega) = \epsilon_0\mathbf{E}(\mathbf{r}, \omega) + \mathbf{P}(\mathbf{r}, \omega) = \epsilon_0\epsilon(\mathbf{r}, \omega)\mathbf{E}(\mathbf{r}, \omega)$. If we use this polarization source as a *perturbation* to the permittivity ($\Delta\epsilon$), then $\epsilon \rightarrow \epsilon + \Delta\epsilon$, and

$$\mathbf{P}_s(\mathbf{r}, \omega) = \epsilon_0 \Delta\epsilon(\mathbf{r}, \omega)\mathbf{E}(\mathbf{r}, \omega), \tag{3.8}$$

so Eq. (3.6) becomes

$$\mathbf{E}(\mathbf{r}, \omega) = \mathbf{E}_0(\mathbf{r}, \omega) + \int \Delta\epsilon(\mathbf{r}', \omega)\mathbf{G}_0(\mathbf{r}, \mathbf{r}', \omega) \cdot \mathbf{E}(\mathbf{r}', \omega)\, d\mathbf{r}', \tag{3.9}$$

where $\mathbf{G}_0$ refers to the homogeneous solution without the polarization source. This solution can also be written *explicitly* in terms of the exact Green function [19]:

$$\mathbf{E}(\mathbf{r}, \omega) = \mathbf{E}_0(\mathbf{r}, \omega) + \int \Delta\epsilon(\mathbf{r}', \omega)\mathbf{G}(\mathbf{r}, \mathbf{r}', \omega) \cdot \mathbf{E}_0(\mathbf{r}', \omega)\, d\mathbf{r}', \tag{3.10}$$

where the exact $\mathbf{G}$ (i.e., including the scatterer) is obtained from the Dyson equation:

$$\mathbf{G}(\mathbf{r}, \mathbf{r}', \omega) = \mathbf{G}_0(\mathbf{r}, \mathbf{r}', \omega) + \int \Delta\epsilon(\mathbf{r}'', \omega)\mathbf{G}_0(\mathbf{r}, \mathbf{r}'', \omega) \cdot \mathbf{G}(\mathbf{r}'', \mathbf{r}', \omega)\, d\mathbf{r}''. \tag{3.11}$$

The Dyson equation is very useful for describing what happens when perturbations are added to a complex system in which the Green function is already known (including dipole emitters and quantum dots [27] and fabrication disorder [28]) and can be used to construct a solution outside the scattering volume in terms of the solution inside. This latter property is important for regularizing QNMs far away from the resonators [21], which is an essential aspect of QNM quantization [23].

3.2.3 Normal Modes, Completeness, and Green Function Expansions

Before discussing optical QNMs, it is useful to first give some background on optical NMs, as many of the techniques are the same, or at least very similar. Normal modes can be obtained when solving Eq. (3.3), subject to closed or periodic boundary conditions. For lossless systems, NMs represent a Hermitian system, and the eigenfrequencies are real (and continuous in general). The normalization of such modes, defined for the electric field, takes on a familiar form [16]:

$$\int_{\text{all space}} \epsilon(\mathbf{r})\mathbf{f}_i^*(\mathbf{r}) \cdot \mathbf{f}_j(\mathbf{r}) d\mathbf{r} = \delta_{i,j}, \tag{3.12}$$

where $\delta_{i,j}$ is the Kronecker delta function. These are the standard *textbook modes*. For waveguides, e.g., complicated photonic crystal waveguides, then "all space" for the integration limits can be replaced with a fundamental unit cell [29]. More generally, NMs for waveguide systems may also exist for complex (dispersive) permittivity and are normalized via

$$\int_{\text{all space}} \left. \frac{\partial \left[\omega^2 \epsilon(\mathbf{r}, \omega)\right]}{2\omega\partial\omega} \right|_{\omega=\omega_i} \mathbf{f}_i(\mathbf{r}) \cdot \mathbf{f}_j(\mathbf{r}) d\mathbf{r} = \delta_{i,j}, \tag{3.13}$$

which reduce to Eq. (3.12) if ϵ is purely real (and nondispersive, through the Kramers-Kronig relations) and if one replaces $\mathbf{f}_i(\mathbf{r}) \cdot \mathbf{f}_j(\mathbf{r}) \rightarrow \mathbf{f}_i^*(\mathbf{r}) \cdot \mathbf{f}_j(\mathbf{r})$ (power orthogonal). Actually, with material losses, one can also view these dissipative waveguide modes as a special case of QNMs, as we will see later, since the normalization will turn out to be of similar form, but without a diverging surface term. Moreover, for lossy waveguide systems, we have complex $\tilde{\beta}_\mu$ (wave vector) solutions, whose imaginary part represents spatial decay, e.g., through the Beer-Lambert law for lossy waveguide modes [30], which we are assuming are otherwise bound. Complex leaky modes can also be found, and they also relate explicitly to QNMs with diverging spatial fields [31], as discussed below, and require a more careful normalization. For complex waveguide modes, we also remark that one can define modes with either complex wave vectors, $\tilde{\beta}_\mu$, or complex $\tilde{\omega}_\mu$, which lead to quite different photonic band structures in general [32]. The complex $\tilde{\beta}_\mu$ is arguably the better picture for waveguide modes, or often one gets confused about stopped light features appearing in the band structure, when there is no stopped light in such lossy materials [33]. In such cases, for example, the group velocity is no longer a good/meaningful metric, and instead one should work with the energy velocity. In this chapter, we will work explicitly with open cavity modes.

For nondispersive media, the NMs satisfy a completeness relation [12]:

$$\epsilon(\mathbf{r}) \sum_\mu \mathbf{f}_\mu(\mathbf{r}) \mathbf{f}_\mu^*(\mathbf{r}') = \delta(\mathbf{r} - \mathbf{r}') \mathbf{1}, \tag{3.14}$$

where the sum includes both transverse and longitudinal modes and is only over the positive eigenmodes, $\mu > 0$, since NMs have a simple symmetry, with $\omega_\mu = -\omega_{-\mu}$ and $\mathbf{f}_{-\mu} = \mathbf{f}_\mu$. Note that the vector products are outer products, and return a dyad, so we could write $\mathbf{f}_\mu(\mathbf{r}) \otimes \mathbf{f}_\mu^*(\mathbf{r}')$, but we will assume this is implicit. For resonant cavities (or resonators[2]), we will be interested in the transverse modes with $\omega_\mu \neq 0$, but the longitudinal modes are important mathematically to ensure completeness.

A major advantage of using optical cavity modes is that they can be used as a basis to expand a solution for the Green function. This is extremely convenient when dealing with few mode systems such as optical cavities. Assuming an expansion of the form, $\mathbf{G}(\mathbf{r}, \mathbf{r}', \omega) = \sum_\mu A_\mu^{\rm NM}(\omega) \mathbf{f}_\mu(\mathbf{r}) \mathbf{f}_\mu^*(\mathbf{r}')$, and using the equations presented above, it is easy to derive [16, 20]

$$A^{\rm NM}(\omega) = \frac{\omega^2}{\omega_\mu^2 - \omega^2}, \tag{3.15}$$

and thus the *NM expansion for the Green function* is

[2] We define cavity or resonator as basically the same thing in this chapter, namely, a structure that allows resonant modes to form.

$$\mathbf{G}(\mathbf{r}, \mathbf{r}', \omega) = \sum_{\mu} \frac{\omega^2}{\omega_{\mu}^2 - \omega^2} \mathbf{f}_{\mu}(\mathbf{r}) \mathbf{f}_{\mu}^{*}(\mathbf{r}'), \tag{3.16}$$

which has a double pole per normal mode, $\omega_{\mu}^{\pm} = \pm\omega_{\mu}$. Also note that the vector products are conjugated.

3.2.4 Quasinormal Modes, Completeness, and Green Function Expansions

The QNMs are obtained from solutions of exactly the same Helmholtz equation for NMs, but now defined in complex frequency space:

$$\left\{ \nabla \times \nabla \times -\frac{\tilde{\omega}_{\mu}^2}{c^2} \epsilon(\mathbf{r}, \tilde{\omega}_{\mu}) \right\} \tilde{\mathbf{f}}_{\mu}(\mathbf{r}) = 0, \tag{3.17}$$

where ϵ defines the spatially dependent cavity, which can also be dispersive, and the solution is now solved with open boundary conditions, specifically with Silver-Müller radiations conditions [34, 35]:

$$\frac{\mathbf{r}}{|\mathbf{r}|} \times \nabla \times \tilde{\mathbf{f}}_{\mu}(\mathbf{r}) \rightarrow -i n_{\mathrm{B}} \tilde{k}_{\mu} \tilde{\mathbf{f}}_{\mu}(\mathbf{r}), \tag{3.18}$$

which are asymptotic relations for $|\mathbf{r}| \rightarrow \infty$ (n_{B} is the refractive index of background medium and $\tilde{k}_{\mu} = \tilde{\omega}_{\mu}/c$).

This *open* boundary condition renders the problem non-Hermitian, and in fact all open cavity systems produce non-Hermitian modes, with complex eigenfrequencies, $\tilde{\omega}_{\mu} = \omega_{\mu} - i\gamma_{\mu}$, where γ_{μ} is half the decay rate width. The quality factor of these modes is then given by $Q_{\mu} = \omega_{\mu}/(2\gamma_{\mu})$. Note that NMs are only recovered in the limit of no losses, i.e., when $Q_{\mu} \rightarrow \infty$. An interesting consequence of temporal energy loss is that the spatial profile of the QNMs diverges spatially, which is clear from the outgoing behavior $\tilde{\mathbf{f}}_{\mu}(r) \propto \exp(i\tilde{\omega}_{\mu} r/c) \propto \exp(\gamma_{\mu} r/c)$, outside the resonator.

The normalization of these QNMs is no longer given by Eq. (3.12). Following earlier notation of Refs. [36, 37], we denote the QNM norm by $\langle\langle \tilde{\mathbf{f}}_{\mu} | \tilde{\mathbf{f}}_{\mu} \rangle\rangle$. For dispersive media, one form of the norm is defined from [35, 38]:

$$\begin{aligned} \langle\langle \tilde{\mathbf{f}}_{\mu} | \tilde{\mathbf{f}}_{\mu} \rangle\rangle = \frac{1}{2\epsilon_0} \int_V \Bigg[& \epsilon_0 \left(\frac{\partial(\omega\epsilon(\mathbf{r}, \omega))}{\partial\omega} \Big|_{\tilde{\omega}_{\mu}} \right) \tilde{\mathbf{f}}_{\mu}(\mathbf{r}) \cdot \tilde{\mathbf{f}}_{\mu}(\mathbf{r}) \\ & - \mu_0 \left(\frac{\partial(\omega\mu(\mathbf{r}, \omega))}{\partial\omega} \Big|_{\tilde{\omega}_{\mu}} \right) \tilde{\mathbf{h}}_{\mu}(\mathbf{r}) \cdot \tilde{\mathbf{h}}_{\mu}(\mathbf{r}) \Bigg] \mathrm{d}\mathbf{r} \\ & - \frac{i}{2\epsilon_0 \tilde{\omega}_{\mu}} \int_{\partial V} \left[\left(r \frac{\partial \tilde{\mathbf{f}}_{\mu}(\mathbf{r})}{\partial r} \right) \times \tilde{\mathbf{h}}_{\mu}(\mathbf{r}) - \tilde{\mathbf{f}}_{\mu}(\mathbf{r}) \times \left(r \frac{\partial \tilde{\mathbf{h}}_{\mu}(\mathbf{r})}{\partial r} \right) \right] \cdot \hat{\mathbf{n}}\, \mathrm{d}A, \end{aligned} \tag{3.19}$$

which includes a volume integral over volume V and a surface integral over the outer surface of the volume, ∂V, where $\hat{\mathbf{n}}$ is the unit vector normal to and pointing outward from the surface and $\tilde{\mathbf{h}}_\mu(\mathbf{r})$ is the magnetic QNM with $\tilde{\mathbf{h}}_\mu(\mathbf{r}) = \frac{1}{i\tilde{\omega}_\mu\mu_0}\nabla \times \tilde{\mathbf{f}}_\mu(\mathbf{r})$.

An alternative approach uses a coordinate transform via the natural implementation of many numerical solutions that use a perfectly matched layer (PML)[3] region as part of the cavity structure. Thus, the surface integral is removed, and the QNM normalization using PML cavity modes is [39]

$$\begin{aligned}\langle\langle\tilde{\mathbf{f}}_\mu|\tilde{\mathbf{f}}_\mu\rangle\rangle = \frac{1}{2\epsilon_0}\int_{V-V_{\mathrm{PML}}}\Bigg[\epsilon_0\left(\frac{\partial(\omega\epsilon(\mathbf{r},\omega))}{\partial\omega}\Big|_{\tilde{\omega}_\mu}\right)\tilde{\mathbf{f}}_\mu(\mathbf{r})\cdot\tilde{\mathbf{f}}_\mu(\mathbf{r})\\ -\mu_0\left(\frac{\partial(\omega\mu(\mathbf{r},\omega))}{\partial\omega}\Big|_{\tilde{\omega}_\mu}\right)\tilde{\mathbf{h}}_\mu(\mathbf{r})\cdot\tilde{\mathbf{h}}_\mu(\mathbf{r})\Bigg]d\mathbf{r}\\ +\frac{1}{2\epsilon_0}\int_{V_{\mathrm{PML}}}\Bigg[\epsilon_0\left(\frac{\partial(\omega\epsilon_{\mathrm{PML}}(\mathbf{r},\omega))}{\partial\omega}\Big|_{\tilde{\omega}_\mu}\right)\tilde{\mathbf{f}}_\mu(\mathbf{r})\cdot\tilde{\mathbf{f}}_\mu(\mathbf{r})\\ -\mu_0\left(\frac{\partial(\omega\mu_{\mathrm{PML}}(\mathbf{r},\omega))}{\partial\omega}\Big|_{\tilde{\omega}_\mu}\right)\tilde{\mathbf{h}}_\mu(\mathbf{r})\cdot\tilde{\mathbf{h}}_\mu(\mathbf{r})\Bigg]d\mathbf{r},\end{aligned} \tag{3.20}$$

which has contributions from the PML as well. There are several choices for setting up PMLs to minimize reflections. For example, in a finite-element solver such as COMSOL (which we will use in this chapter), a coordinate transformation [40] can be applied, with the built-in stretched-coordinate PML. In the PML region, the coordinates are transferred from real space to the complex plane [41] to minimize boundary reflections; at the same time, $\epsilon_{\mathrm{PML}}(\mathbf{r},\omega)$ and $\mu_{\mathrm{PML}}(\mathbf{r},\omega)$ are usually set as the same values as the interior medium. For the nonmagnetic case with $n_{\mathrm{B}} = 1$, then $\epsilon_{\mathrm{PML}}(\mathbf{r},\omega) = 1$ and $\mu_{\mathrm{PML}}(\mathbf{r},\omega) = 1$ are used; in such a case, one has to use the complex spatial variables in the PML.

Other approximate forms of QNM normalization also exist, for handling the surface terms (regularization), which in practice are also very accurate and easy to check and use [35, 42–44].

Using these normalized QNMs, we define the QNM completeness relation as [45]

$$\epsilon(\mathbf{r})\sum_{\mu=\pm1,\pm2,\ldots}\frac{\tilde{\mathbf{f}}_\mu(\mathbf{r})\tilde{\mathbf{f}}_\mu(\mathbf{r}')}{2} = \delta(\mathbf{r}-\mathbf{r}')\mathbf{1}, \tag{3.21}$$

where $\epsilon(\mathbf{r})$ is given by $\epsilon_\infty(\mathbf{r}) \equiv \epsilon(\mathbf{r},\omega=\infty)$ [36]. Note that Eq. (3.21) has only been *proven* for the case of a dielectric sphere [45] and 1D resonators [36], and we will assume that this then applies to arbitrary shapes. Moreover, PML-normalized QNMs are known to form a suitable basis to achieve completeness [46].

[3] PML boundaries implement an absorbing/outgoing boundary condition numerically.

Similar to the NM Green function, we expand the Green function using QNMs:

$$\mathbf{G}(\mathbf{r}, \mathbf{r}', \omega) = \sum_{\mu} A_{\mu}(\omega)\tilde{\mathbf{f}}_{\mu}(\mathbf{r})\tilde{\mathbf{f}}_{\mu}(\mathbf{r}'), \tag{3.22}$$

where the QNM expansion coefficient is

$$A_{\mu}(\omega) = \frac{\omega}{2(\tilde{\omega}_{\mu} - \omega)}. \tag{3.23}$$

Unlike the NM expansion coefficient, $A^{\mathrm{NM}}(\omega)$ (Eq. (3.15)), the QNM expansion term only has one pole per mode. This is a consequence of the non-Hermitian nature of the problem: the terms $-\mu, \mu$ appearing in Eq. (3.22) cannot be combined into one term, since $\tilde{\mathbf{f}}_{\mu} \neq \tilde{\mathbf{f}}_{\mu}^{*}$. Interestingly, one can also use $A_{\mu}(\omega) = \frac{\omega^2}{2\tilde{\omega}_{\mu}(\tilde{\omega}_{\mu}-\omega)}$ (e.g., see [43, 45]), since the two are related through a sum relationship (so the total is the same). Note also that the vector field products in the Green function expansion no longer use the complex conjugate (cf. the normal mode form, Eq. (3.16)), and the QNM phase is an essential aspect of QNM theory, which manifests in unique spectral signatures such as Fano resonances between coupled QNMs [47, 48]. We will only consider the transverse QNMs, as those are the main contribution for the resonances of interest, and we will also assume that the QNMs are orthogonal to each other, which is not necessarily always the case with dispersive materials [49].

Exploiting the Green function expansion, one can invert such a relation and solve for the normalized QNMs of interest using complex frequencies. Following the techniques in Ref. [50], the QNM poles can be first found using a Padé approximation, which can exploit Maxwell solvers that can be solved with complex frequencies. Once the complex poles are found, the QNMs can be obtained in normalized form using a scattered field approach. This approach works as follows: for resonant structures near the resonance frequency, the total electric field separates as the scattered and background fields: $\mathbf{E}^{\mathrm{tot}} = \mathbf{E}^{\mathrm{scatt}} + \mathbf{E}^{\mathrm{BG}}$, where other contributions are negligible, and $\mathbf{E} \approx \mathbf{E}^{T}$ (transverse field). As an example, using a single QNM case, then the Green function is approximated as ($\mu = c$ is implicit)

$$\mathbf{G}(\mathbf{r}, \mathbf{r}', \tilde{\omega}_p) \approx A(\tilde{\omega}_p)\tilde{\mathbf{f}}(\mathbf{r})\tilde{\mathbf{f}}(\mathbf{r}'), \tag{3.24}$$

and using Eq. (3.6),

$$\mathbf{E}^{\mathrm{scatt}}(\mathbf{r}) = \mathbf{G}(\mathbf{r}, \mathbf{r}_0, \tilde{\omega}_p) \cdot \frac{\mathbf{d}(\tilde{\omega}_p)}{\epsilon_0}, \tag{3.25}$$

thus

$$\mathbf{d}(\tilde{\omega}_p) \cdot \mathbf{E}^{\mathrm{scatt}}(\mathbf{r}) = \frac{A(\tilde{\omega}_p)}{\epsilon_0}\mathbf{d}(\tilde{\omega}_p) \cdot \tilde{\mathbf{f}}(\mathbf{r})\tilde{\mathbf{f}}(\mathbf{r}_0) \cdot \mathbf{d}(\tilde{\omega}_p), \tag{3.26}$$

where $\mathbf{d}(\tilde{\omega}_p) = d(\tilde{\omega}_p)\,\hat{\mathbf{n}}_{\rm d}$ is the dipole source at $\mathbf{r}_0$ and the frequency $\tilde{\omega}_p$ close to $\tilde{\omega}_{\rm c}$. Typically, this has to be on the order of $\tilde{\omega}_p = (1 - 10^{-5})\tilde{\omega}_{\rm c}$ to $(1 - 10^{-8})\tilde{\omega}_{\rm c}$, depending on the quality factor Q of the dominant mode. At the dipole location, $\mathbf{r} = \mathbf{r}_0$, then $\mathbf{d}(\tilde{\omega}_p) \cdot \tilde{\mathbf{f}}(\mathbf{r}_0) = \sqrt{\frac{\epsilon_0 \mathbf{d}(\tilde{\omega}_p)\cdot\mathbf{E}^{\rm scatt}(\mathbf{r}_0)}{A(\tilde{\omega}_p)}}$, and we obtain the normalized QNM at all positions using Eq. (3.26):

$$\tilde{\mathbf{f}}(\mathbf{r}) = \sqrt{\frac{\epsilon_0}{A(\tilde{\omega}_p)\mathbf{d}(\tilde{\omega}_p) \cdot \mathbf{E}^{\rm scatt}(\mathbf{r}_0)}}\mathbf{E}^{\rm scatt}(\mathbf{r}). \tag{3.27}$$

The extension to two or more QNMs is straightforward.

Although this method is not a strict QNM normalization, it is reasonably simple, free of numerical integration errors, an extremely good approximation in many cases of interest, and easy to extend to obtain multiple QNMs. It is also easy to check its accuracy. Indeed, an analogous approach has also been used to compute multiple mechanical QNMs for optomechanical beams [51], with quantitative accuracy when compared against full numerical calculations. Moreover, for most practical examples of interest in nanophotonics, we find this dipole technique returns the same value as the norms defined with Eq. (3.19) and (3.20) (PML approach) to within a fraction of a percent (e.g., see [52]), but without the need to perform complicated integrations, which can lead to additional numerical errors as well, and typically very small spatial grids are required when using metal structures.

3.2.5 *Total Electromagnetic Fields and Regularized Quasinormal Modes*

It is important to note that individual QNMs, while divergent in space (assuming lossy media), contribute to a sum that is not divergent, as must be the case for the total fields. Also, there is no divergence for QNMs in time, unless they are obtained for a gain medium, where they can also be used and defined [52].

In the time domain, and within the resonator at position $\mathbf{r}$, the total electric field $\mathbf{E}(\mathbf{r}, t)$ has the form

$$\mathbf{E}(\mathbf{r}, t) = \sum_{\mu} a_{\mu}^{(0)}(t)\tilde{\mathbf{f}}_{\mu}^{(0)}(\mathbf{r}), \tag{3.28}$$

where $a_\mu(t) = e^{-i\tilde{\omega}_\mu t} a_\mu(t = 0)$ is the harmonic solution, obtained from the temporal part of the underlying wave equations.

For fields far outside the resonator, a few QNM expansions are no longer convenient to use as they diverge due to the radiation condition (the medium is unbounded), and as a consequence of the complex eigenfrequencies, with $\gamma_\mu > 0$

(lossy media). However, in such a case, one can use the Dyson equation, Eq. (3.11), to construct regularized QNMs [53]:

$$\tilde{\mathbf{F}}_\mu(\mathbf{R},\omega) = \int_{\text{res}} \Delta\epsilon_{\text{c}}(\mathbf{r},\omega)\mathbf{G}_{\text{B}}(\mathbf{R},\mathbf{r},\omega)\cdot\tilde{\mathbf{f}}_\mu(\mathbf{r})\,\text{d}\mathbf{r}, \tag{3.29}$$

where the integration is carried out only *inside* the resonator (Another form 'res'); and $\Delta\epsilon_{\text{c}}(\mathbf{r},\omega) = \epsilon_{\text{c}}(\mathbf{r},\omega) - \epsilon_{\text{B}}$ represent the permittivity difference of the resonator to the background medium with ϵ_{B}, and $\mathbf{G}_{\text{B}}(\mathbf{R},\mathbf{r},\omega)$ is the Green function for the background medium. We will use $\mathbf{R}$ to represent a point *outside* the resonator. Alternatively, these regularized fields can be constructed in the far field using near-field to far-field transformations [54]. Utilizing regularized fields ensures there is no enhanced spontaneous emission at all in the far field, as shown explicitly in Ref. [53]. Furthermore, we emphasize that this regularized QNM approach is based on fundamental Green identities and not just a phenomenological fix. We will discuss this in more detail in the next subsection with one-dimensional QNMs, since many of their properties can be obtained analytically.

Alternatively, using Green's identity, and the Helmholtz equation, we can derive another form of QNM regularization [8]:

$$\begin{aligned}\tilde{\mathbf{F}}'_\mu(\mathbf{R},\omega) =& \frac{c^2}{\omega^2}\int_{S'} \text{d}A_{\mathbf{s}}\mathbf{G}_{\text{B}}(\mathbf{R},\mathbf{s},\omega)\cdot[\hat{\mathbf{n}}\times\nabla\times\tilde{\mathbf{f}}_\mu(\mathbf{s})] \\ &- \frac{c^2}{\omega^2}\int_{S'} \text{d}A_{\mathbf{s}}[\hat{\mathbf{n}}\times\nabla\times\mathbf{G}_{\text{B}}(\mathbf{s},\mathbf{R},\omega)]^{\text{t}}\cdot\tilde{\mathbf{f}}_\mu(\mathbf{s}),\end{aligned} \tag{3.30}$$

where "t" refers to transpose and $\mathcal{S}'$ is a surface that surrounds the scattering volume, defined by $\Delta\epsilon(\mathbf{r},\omega)$. This form is identical to the results of the abovementioned near-field to far-field approach.

Since these regularized QNMs solve a scattering problem in real frequency space, their time-dependent solution naturally includes a retardation factor [55], e.g., $\Theta(t - n_{\text{B}}|\mathbf{R}-\mathbf{r}_0|/c)$, when propagating from point $\mathbf{r}_0$ to $\mathbf{R}$. For example, in the case of a 1D half open cavity, with boundary point $x = 0$ (closed) $x = L$ (open), we can start from the expression $E(x,t) = \sum_\mu a_\mu(t)\tilde{f}_\mu(x)$. This expression cannot be used in the far field $x \gg L$, because it diverges. Next, if we go into the frequency picture, $E(x,\omega) = \sum_\mu a_\mu(0)/(\omega-\tilde{\omega}_\mu)\tilde{f}_\mu(x)$; replacing $\tilde{f}_\mu(x)$ by $\tilde{F}_\mu(x,\omega) = \tilde{f}_\mu(L)e^{in_{\text{B}}\omega(x-L)/c}$, then one can see that the Fourier transform of $\tilde{F}_\mu(x,\omega)$ gives precisely $\Theta(t - n_{\text{B}}(x-L)/c)$. For the total field, however, one gets

$$E(x,t) = \sum_\mu a_\mu(0)\int_{-\infty}^{\infty}\text{d}\omega\frac{e^{i\omega(t-n_{\text{B}}(x-L)/c)}}{\omega-\tilde{\omega}_\mu}\tilde{f}_\mu(L), \tag{3.31}$$

whose evaluation depends on the $\gamma_\mu > 0$ (loss) or $\gamma_\mu < 0$ (gain) [52].

3.2.6 Analytical Example: One-Dimensional Quasinormal Modes for Dielectric Barriers

A useful way to appreciate many of the subtle and useful aspects of QNM theory is to begin with 1D examples, as many of their properties can be worked out analytically [43]. Although not so practical for connecting to real complex 3D problem in nanophotonics, they offer clear and transparent insights without worrying about overly complex numerical details and normalization issues. They also make the procedure and need for QNM regularization very clear as well, which is something that is typically overlooked in much of the QNM literature, and often misunderstood.

We consider a simple example of a single barrier, as shown in Fig. 3.2, which has a length L, defined between two points $x = a$ and $x = b$. The refractive index of the resonator is $n_{\rm R} = \sqrt{\epsilon_{\rm R}}$, which is embedded in a background medium with $n_{\rm B} = \sqrt{\epsilon_{\rm B}}$. The QNM eigenfrequencies are determined from [8]

$$n_{\rm R}\tilde{k}_\mu L = \mu\pi + \frac{i}{2}\ln\left[\frac{(n_{\rm R}-n_{\rm B})^2}{(n_{\rm R}+n_{\rm B})^2}\right], \tag{3.32}$$

where we made use of the outgoing boundary conditions:

$$\partial_x \tilde{f}_\mu(x)\big|_{x\nearrow a} = -i\tilde{\omega}_\mu n_{\rm B}/c\tilde{f}_\mu(a), \tag{3.33a}$$

$$\partial_x \tilde{f}_\mu(x)\big|_{x\searrow b} = i\tilde{\omega}_\mu n_{\rm B}/c\tilde{f}_\mu(b). \tag{3.33b}$$

Interestingly, the $\mu = 0$ QNM is purely imaginary and vanishes in the lossless limit. Within the cavity region, the QNM eigenfunctions are

$$\tilde{f}_\mu(x) = \frac{e^{in_{\rm R}\tilde{k}_\mu(x-x_0)} + (-1)^\mu e^{-in_{\rm R}\tilde{k}_\mu(x-x_0)}}{\sqrt{(-1)^\mu 2Ln_{\rm R}}}, \tag{3.34}$$

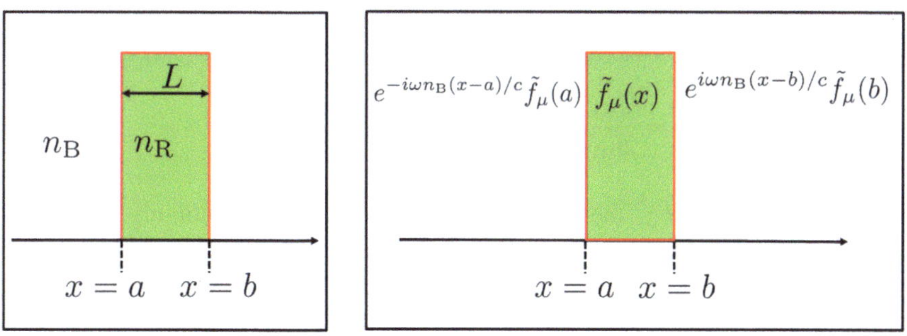

Fig. 3.2 Simple 1D barrier of length L, defined with refractive index $n_{\rm R}$, in a uniform background with $n_{\rm B}$. The right figure shows the form of the regularized QNM outside the barrier, which no longer spatially diverges

where the barrier center is taken to be at $x = x_0$. While the QNM solutions outside the barrier are

$$\tilde{f}_\mu(x > b) = e^{-i\tilde{\omega}_\mu n_\mathrm{B}/c(x-b)}\,\tilde{f}_\mu(b), \tag{3.35a}$$

$$\tilde{f}_\mu(x < a) = e^{-i\tilde{\omega}_\mu n_\mathrm{B}/c(a-x)}\,\tilde{f}_\mu(a). \tag{3.35b}$$

The 1D QNM norm has the simple form

$$\langle\langle \tilde{f}_\mu | \tilde{f}_\mu \rangle\rangle = \int_a^b \mathrm{d}x \epsilon(x) \left[\tilde{f}_\mu(x)\right]^2 + i\frac{n_\mathrm{B} c}{2\tilde{\omega}_\mu}\left\{\left[\tilde{f}_\mu(a)\right]^2 + \left[\tilde{f}_\mu(b)\right]^2\right\}, \tag{3.36}$$

with $a = x_0 - L/2$ and $b = x_0 + L/2$. For a constant permittivity $\epsilon(x) = n_\mathrm{R}^2$ in $[a, b]$, then Eq. (3.36) reduces to $\langle\langle \tilde{f}_\mu | \tilde{f}_\mu \rangle\rangle = 1$ when using Eq. (3.34). Also note that the latter (surface) term vanishes in the limit that $\gamma_\mu \to 0$.

We now show how to obtain the corresponding regularized QNMs. Using the barrier QNM example, and the surface-integral regularization of Eq. (3.30), we have

$$\tilde{F}'_\mu(x > b, \omega) = e^{in_\mathrm{B}k(x-b)}\,\tilde{f}_\mu(b), \tag{3.37a}$$

$$\tilde{F}'_\mu(x < a, \omega) = e^{-in_\mathrm{B}k(x-a)}\,\tilde{f}_\mu(a), \tag{3.37b}$$

where $k = \omega/c$. Physically, these solutions represent right-propagating and left-propagating plane waves with continuity conditions at the interfaces (boundaries): $\tilde{F}'_\mu(x, \omega)|_{x\nearrow a} = \tilde{f}_\mu(a)$ and $\tilde{F}'_\mu(x, \omega)|_{x\searrow b} = \tilde{f}_\mu(b)$; strikingly, these *regularized QNMs* are independent of frequency exactly at the boundary, and they become equal to the QNM at such positions.

Interestingly, the regularized QNMs using the Dyson solution, (3.29), yield a different solution:

$$\tilde{F}_\mu(x, \omega) = \frac{ik}{2n_\mathrm{B}} e^{in_\mathrm{B}kx} \int_a^b \mathrm{d}s\, \Delta\epsilon(s) e^{-in_\mathrm{B}ks}\,\tilde{f}_\mu(s), \tag{3.38}$$

for $x > b$ and

$$\tilde{F}_\mu(x, \omega) = \frac{ik}{2n_\mathrm{B}} e^{-in_\mathrm{B}kx} \int_a^b \mathrm{d}s\, \Delta\epsilon(s) e^{in_\mathrm{B}ks}\,\tilde{f}_\mu(s), \tag{3.39}$$

for $x < a$. Unlike $\tilde{F}'_\mu$, these solutions do not fulfill the continuity conditions. However, the total field, including all QNMs, is actually identical with both approaches [8].

Using the surface-integral regularization, one can reformulate the QNM Green function $G(x, x_s, \omega)$ for, e.g., $x_s \in [a, b]$ and $x > b$ as

$$G^{\text{reg}}_{\text{QNM}}(x, x_s, \omega) = \sum_\mu A_\mu(\omega)\tilde{f}_\mu(b)\tilde{f}_\mu(x_s)e^{in_\text{B}k(x-b)}. \tag{3.40}$$

In the one-dimensional case, it is also possible to derive an exact analytical solution of the single-barrier problem. The exact form is independent of a mode expansion and is obtained by solving a matrix problem that is formulated through continuity conditions at the boundary points and at $x = x_s$. For $x_s \in [a, b]$ and $x > b$, then

$$G_{\text{exact}}(x, x_s, \omega) = A(x_s, \omega)e^{in_\text{B}kx} = A(x_s, \omega)e^{in_\text{B}\omega x/c}, \tag{3.41}$$

where

$$A(x_s, \omega) = -\frac{i}{2n_\text{R}k}e^{-ik(n_\text{R}+n_\text{B})(b-a)/2}(1 + \Gamma_{\text{RB}})\frac{e^{ikn_\text{R}(b-a+x_s)} + \Gamma_{\text{RM}}e^{-ikn_\text{R}(b-a)}}{e^{ikn_\text{R}(b-a)} - \Gamma^2_{\text{RM}}e^{-ikn_\text{R}(b-a)}}, \tag{3.42}$$

with $\Gamma_{\text{RB}} = (n_\text{R} - n_\text{B})/(n_\text{R} + n_\text{B})$. We can immediately connect to the QNM solution via the denominator in Eq. (3.42). Indeed, the pole solution $e^{ikn_\text{R}(b-a)} - \Gamma^2_{\text{RB}}e^{-ikn_\text{R}(b-a)} = 0$ precisely determines the QNM eigenfrequencies of the single-barrier problem.

3.2.7 *Purcell's Formula for Enhanced Spontaneous Emission and a Generalized Effective Mode Volume*

Purcell's formula for enhanced spontaneous emission (SE) [56] is an essential figure of merit for characterizing light-matter interactions in terms of cavity mode parameters. Indeed, it is one of the standard metrics for explaining and describing enhanced light-matter interactions in optical cavities. It is also an important limit to recover in quantum optical theories, e.g., in a bad cavity limit (namely, when the coupling between an atom and the cavity is weak enough that the cavity mode can be adiabatically eliminated). Similar enhancements can also be used to characterize mechanical QNMs, [51, 57], using force displacements and the *elastic Purcell's formula*.

Purcell's formula, which describes the enhanced SE rate of a dipole or quantum emitter, $\Gamma(\mathbf{r}_0, \omega)$, depends on both space and frequency, and when normalized to the rate from a background homogeneous medium (with refractive index, n_B), $\Gamma_\text{B}(\mathbf{r}_0, \omega)$, is usually defined through

$$F_\text{P} \equiv \frac{3}{4\pi^2}\left(\frac{\lambda}{n_\text{B}}\right)^3\frac{Q}{V_{\text{eff}}}, \tag{3.43}$$

where V_{eff} is the effective mode volume; here we also assume the emitter is on resonance and at a field maximum position, with the same polarization as the cavity mode. However, this definition of "mode volume" is strictly in the context of NMs, which are power orthogonal, such that [42]

$$V_{\mathrm{NM},\mu} = \frac{\langle \mathbf{f}_\mu | \mathbf{f}_\mu \rangle}{\max\left[\epsilon(\mathbf{r})|\mathbf{f}_\mu(\mathbf{r})|^2\right]}, \tag{3.44}$$

where $\langle \mathbf{f}_\mu | \mathbf{f}_\mu \rangle$ is given by Eq. (3.12) or (3.13) for nondispersive or dispersive media, respectively. The problem with this definition is that for cavity modes with dissipation, the general result is ambiguous since the real mode profile of the open-system cavity diverges in space. In practice, this may not be noticeable for large quality factors, but it is problematic for low Q resonators, and becomes a larger problem for coupled cavity modes [47]. For NMs, the above effective mode volume represents a measure of spatial localization, but that is clearly not the case for QNMs. For coupled cavity problems, it causes additional problems (as we will show later), since it misses important modal phase effects.

For the rigorous cavity mode definition of effective mode volume for Purcell's formula (namely, with dissipation), we can use QNM theory, where $\tilde{V}_{\mathrm{QNM}}$ is obtained in the form of a "generalized mode volume" [42]:

$$\tilde{V}_{\mathrm{QNM},\mu}(\mathbf{r}_0) = \frac{\langle\langle \tilde{\mathbf{f}}_\mu | \tilde{\mathbf{f}}_\mu \rangle\rangle}{\epsilon(\mathbf{r}_0)\tilde{\mathbf{f}}_\mu^2(\mathbf{r}_0)}, \tag{3.45}$$

which is now a *complex and spatially dependent volume*. This expression is valid for emitters that are in a real dielectric medium (i.e., not lossy, so $\epsilon(\mathbf{r}_0)$ is purely real); otherwise, one has to address a local field problem (which can also be addressed also using QNM theory for inhomogeneous lossy cavities [58]). To connect to Purcell's formula, then one uses $V_{\mathrm{eff}}(\mathbf{r}_0) = \mathrm{Re}[\tilde{V}_{\mathrm{QNM}}(\mathbf{r}_0)]$, although this is not a true mode "volume" occupying some 3D space, but rather a measure of the inverse field squared which happens to have units of volume. This often leads to confusion, but one should really think of the QNM mode volume as simply being a measure of the local field amplitude squared (and at a point location). The main reason many researchers prefer to maintain the use of *mode volume* is likely out of convenience and respect for Purcell's formula, which can still work well for QNM systems, and can be quite intuitive. Indeed, in this regard, the original formula applies, and the QNMs are typically the ones that are usually computed numerically anyway,[4] e.g., using PMLs (which model open boundary conditions) in FDTD or finite-element solvers. Note also that the complex mode volume has been experimentally measured [59], so it can certainly be thought of an *observable*, much like complex dielectric constants, complex wave vectors, or complex frequencies; having these volumes complex is just a convenient way of explaining some of the important optical physics.

[4] Some works also compute approximate modes from a plane wave solution, i.e., not source-free, but this is manifestly *not* a mode [37].

Of course, it is natural to extend Purcell's original formula, but retain the elegance of projecting to cavity mode parameters. Ultimately, it is far more useful and general to write the dipole decay rate in terms of the projected LDOS: $\rho^{\rm L}(\mathbf{r}_0,\omega) \equiv \frac{6}{\pi\omega}\hat{\mathbf{n}}_{\rm d}\cdot{\rm Im}\,[\mathbf{G}(\mathbf{r}_0,\mathbf{r}_0,\omega)]\cdot\hat{\mathbf{n}}_{\rm d}$ [25], so that

$$\Gamma(\mathbf{r}_0,\omega) = \frac{\pi\omega|\mathbf{d}|^2}{3\hbar\epsilon_0}\rho^{\rm L}(\mathbf{r}_0,\omega) = \frac{2}{\hbar\epsilon_0}\mathbf{d}\cdot{\rm Im}\,[\mathbf{G}(\mathbf{r}_0,\mathbf{r}_0,\omega)]\cdot\mathbf{d}, \tag{3.46}$$

and the generalized Purcell factor is then

$$\begin{aligned} F_{\rm P}^{\rm cQNM}(\mathbf{r}_0,\omega) &= 1 + \frac{\mathbf{d}\cdot{\rm Im}\,[\mathbf{G}(\mathbf{r}_0,\mathbf{r}_0,\omega)]\cdot\mathbf{d}}{\mathbf{d}\cdot{\rm Im}\,[\mathbf{G}_{\rm B}(\mathbf{r}_0,\mathbf{r}_0,\omega)]\cdot\mathbf{d}}, \\ &= 1 + \frac{6\pi}{n_{\rm B}}\left(\frac{c}{\omega}\right)^3\hat{\mathbf{n}}_{\rm d}\cdot{\rm Im}\,[\mathbf{G}(\mathbf{r}_0,\mathbf{r}_0,\omega)]\cdot\hat{\mathbf{n}}_{\rm d}, \end{aligned} \tag{3.47}$$

where we assume real dipole moments ($\mathbf{d} = |\mathbf{d}|\hat{\mathbf{n}}_d = d\hat{\mathbf{n}}_d$, though this is not a model requirement), and we have used the analytical solution for the background homogeneous Green function in 3D space: ${\rm Im}[\mathbf{G}_{\rm B}] = \mathbf{1}n_{\rm B}\omega^3/(6\pi c^3)$; for a 2D TM (TE) case, ${\rm Im}[\mathbf{G}_{\rm B}] = \mathbf{1}\omega^2/(4c^2)$ (${\rm Im}[\mathbf{G}_{\rm B}] = \mathbf{1}\omega^2/(8c^2)$). The factor of 1 in Eq. (3.47) originates from background modes when the dipole is outside the cavity medium, which is derived from the Dyson equation [21]. For dipoles within the cavity structure, then this factor should be omitted, but for all our cases considered later, we will consider dipoles outside.

Importantly, since we have a QNM expansion for the Green function, we have the desired QNM solution for the Purcell factor in a fully analytical form, assuming of course that the QNMs are first computed. For example, in the case of a single QNM, $\mathbf{G} \approx \mathbf{G}_{\rm c}$, and one has

$$\Gamma_{\rm c}(\mathbf{r}_0,\omega) = \frac{2|\mathbf{d}|^2}{\hbar\epsilon_0}\hat{\mathbf{n}}_{\rm d}\cdot{\rm Im}\left[\frac{\omega\tilde{\mathbf{f}}_{\rm c}(\mathbf{r}_0)\tilde{\mathbf{f}}_{\rm c}(\mathbf{r}_0)}{2(\tilde{\omega}_{\rm c}-\omega)}\right]\cdot\hat{\mathbf{n}}_{\rm d}. \tag{3.48}$$

and thus

$$F_{\rm P}^{\rm cQNM}(\mathbf{r}_0,\omega) = 1 + \frac{\Gamma_{\rm c}(\mathbf{r}_0,\omega)}{\Gamma_{\rm B}(\mathbf{r}_0,\omega)} = 1 + \frac{6\pi}{n_{\rm B}}\left(\frac{c}{\omega}\right)^3\hat{\mathbf{n}}_{\rm d}\cdot{\rm Im}\left[\frac{\omega\tilde{\mathbf{f}}_{\rm c}(\mathbf{r}_0)\tilde{\mathbf{f}}_{\rm c}(\mathbf{r}_0)}{2(\tilde{\omega}_{\rm c}-\omega)}\right]\cdot\hat{\mathbf{n}}_{\rm d}. \tag{3.49}$$

Similarly, one can define photonic Lamb shifts from the corresponding real part of the QNM Green function.

Without using any modal approximations, these dipole decay rates can also be checked against full-dipole simulations using direct Maxwell's equation solvers, which is useful to check the accuracy of the QNM expansions. For example, one can compute the normalized power flow obtained from the numerical calculation:

$$F_{\mathrm{P}}^{\mathrm{num}}(\mathbf{r}_0,\omega) = \frac{\int_s \hat{\mathbf{n}}\cdot \mathbf{S}_{\mathrm{dipole,total}}(\mathbf{r};\omega)\mathrm{d}A}{\int_s \hat{\mathbf{n}}\cdot \mathbf{S}_{\mathrm{dipole,background}}(\mathbf{r};\omega)\mathrm{d}A}, \tag{3.50}$$

where $\mathbf{S}$ is the Poynting vector along the surface of a small sphere, s (typically 1 nm in radius) that is centered around the dipole, and $\hat{\mathbf{n}}$ is the normal unit vector pointing outward from the surface; "total" refers to the cavity structure and "background" is for the background medium only. Conveniently, such calculations can be performed when also obtaining the QNM. However, we stress that such calculations are very tedious, since they need to be computed at every single spatial position and frequency (if using a frequency-dependent solver), while the QNMs are already given as a function of space and defined over the QNM bandwidth of interest, so the QNM Purcell factor is analytic and can quickly be obtained for a wide range of spatial points and frequencies. For example, a QNM description can quickly obtain spatial map calculations of electron energy loss spectroscopy for plasmonic nanostructures [60], which is a notoriously difficult calculation, that requires the full two-space-point Green functions.

3.2.8 Nonradiative and Radiative Decay Rates and Beta Factors

An excitation dipole close to a finite-size lossy medium will result in radiative and nonradiative decay, and the ratio of these rates to the total decay rates can be defined through the radiative and nonradiative beta factors [61].[5] Thus, a dipole emitter with dipole moment $\mathbf{d}$ at $\mathbf{r}_0$ close to a lossy resonator will have two parts –radiative and nonradiative– that contribute to the total decay rate, Eq. (3.46). These can be obtained with full-dipole numerical simulations, or again through QNM projection, which improves both the intuition and efficiency; the separation is also formally required for QNM quantization.

3.2.8.1 Nonradiative Decay Rates

The nonradiative power emitted from a dipole that is dissipated to the lossy inhomogeneous cavity material can be obtained from

$$\Gamma_{\mathrm{vol}}^{\mathrm{nrad}}(\mathbf{r}_0,\omega) = \frac{2}{\hbar\omega}\int_{\mathrm{V}} \mathrm{Re}\left\{\mathbf{J}(\mathbf{r})\cdot\mathbf{E}^*(\mathbf{r})\right\}\mathrm{d}\mathbf{r}, \tag{3.51}$$

or using a surface integral,

[5] Note this is different to the wave vector solution for continuous modes, usually written with a subscript, e.g., β_k or $\tilde{\beta}_\mu$ as discussed earlier.

$$\Gamma_{\rm sur}^{\rm nrad}(\mathbf{r}_0,\omega) = -\frac{4}{\hbar\omega}\int_{\Sigma_{\rm lossy}} \hat{\mathbf{n}}\cdot\mathbf{S}(\mathbf{r},\omega)\,{\rm d}A, \tag{3.52}$$

where $\mathbf{J}(\mathbf{r},\omega) = -i\omega\epsilon_0(\epsilon-\epsilon_{\rm B})\mathbf{E}(\mathbf{r},\omega)$ and $\Sigma_{\rm lossy}$ is a closed surface only enclosing the lossy regime; $\mathbf{S}(\mathbf{r},\omega) = \frac{1}{2}{\rm Re}\{\mathbf{E}(\mathbf{r},\omega)\times\mathbf{H}^*(\mathbf{r},\omega)\}$ is the Poynting vector at this surface and unit vector $\hat{\mathbf{n}}$ is normal to $\Sigma_{\rm lossy}$, pointing outward.

These can be computed numerically:

$$\Gamma_{\rm vol,num}^{\rm nrad}(\mathbf{r}_0,\omega) = \frac{2}{\hbar\omega}\int_{\rm V} {\rm Re}\left\{\mathbf{J}_{\rm num}(\mathbf{r})\cdot\mathbf{E}_{\rm num}^*(\mathbf{r})\right\}{\rm d}\mathbf{r}, \tag{3.53}$$

$$\Gamma_{\rm sur,num}^{\rm nrad}(\mathbf{r}_0,\omega) = -\frac{4}{\hbar\omega}\int_{\Sigma_{\rm lossy}} \hat{\mathbf{n}}\cdot\mathbf{S}_{\rm num}(\mathbf{r},\omega)\,{\rm d}A, \tag{3.54}$$

or using the Green function solutions, $\mathbf{E}(\mathbf{r},\omega) = \mathbf{G}(\mathbf{r},\mathbf{r}_0,\omega)\cdot\frac{\mathbf{d}}{\epsilon_0}$, with $\mathbf{H}(\mathbf{r},\omega) = \frac{1}{i\omega\mu_0}\nabla\times\mathbf{E}(\mathbf{r},\omega)$. The volume integral form is more accurate if using a QNM expansion. For example, using the volume integral expression, we have

$$\begin{aligned}\Gamma_{\rm vol,QNM}^{\rm nrad}(\mathbf{r}_0,\omega) &= \frac{2}{\hbar\omega}\int_{\rm V_{lossy}} {\rm Re}\left\{\mathbf{J}(\mathbf{r})\cdot\mathbf{E}_{\rm QNM}^*(\mathbf{r})\right\}{\rm d}\mathbf{r},\\ &= \frac{2|\mathbf{d}|^2}{\hbar\epsilon_0}\int_{\rm V_{lossy}} \epsilon_I(\mathbf{r},\omega)\left|\mathbf{G}_{\rm QNM}(\mathbf{r},\mathbf{r}_0,\omega)\cdot\hat{\mathbf{n}}_{\rm d}\right|^2{\rm d}\mathbf{r},\end{aligned} \tag{3.55}$$

where $\mathbf{G}_{\rm QNM}(\mathbf{r},\mathbf{r}_0,\omega) = \sum_\mu A_\mu(\omega)\tilde{\mathbf{f}}_\mu(\mathbf{r})\tilde{\mathbf{f}}_\mu(\mathbf{r}_0)$ and $\epsilon_I = {\rm Im}(\epsilon)$. In the single mode limit, we obtain

$$\Gamma_{\rm c}^{\rm nrad}(\mathbf{r}_0,\omega) = \frac{2|\mathbf{d}|^2}{\hbar\epsilon_0}\left|A_{\rm c}(\omega)\left[\tilde{\mathbf{f}}_{\rm c}(\mathbf{r}_0)\cdot\hat{\mathbf{n}}_{\rm d}\right]\right|^2\int_{\rm V_{lossy}} \epsilon_{\rm I}(\mathbf{r},\omega)\left|\tilde{\mathbf{f}}_{\rm c}(\mathbf{r})\right|^2{\rm d}\mathbf{r}, \tag{3.56}$$

where all parameters are known analytically from the QNM solution.

3.2.8.2 Radiative Decay Rates

The radiative decay rate, from a dipole emitting radiation into the far field, is

$$\Gamma_{\rm sur}^{\rm rad}(\mathbf{r}_0,\omega) = \frac{4}{\hbar\omega}\int_{\Sigma_{\rm entire}} \hat{\mathbf{n}}\cdot\mathbf{S}(\mathbf{r},\omega)\,{\rm d}A, \tag{3.57}$$

which is carried out over a closed surface $\Sigma_{\rm entire}$ in the far field. As with the nonradiative decay, this can be computed numerically from

$$\Gamma^{\rm rad}_{\rm sur,num}(\mathbf{r}_0,\omega) = \frac{4}{\hbar\omega}\int_{\Sigma_{\rm entire}} \hat{\mathbf{n}}\cdot\mathbf{S}_{\rm num}(\mathbf{r},\omega)\,{\rm d}A. \tag{3.58}$$

Alternatively, we can again construct a solution in terms of the Greens functions and then do a QNM expansion to obtain a solution in terms of QNM quantities:

$$\begin{aligned}\Gamma^{\rm rad}_{\rm sur,QNM}(\mathbf{r}_0,\omega) \\ = \frac{4}{\hbar\omega}\int_{\Sigma_{\rm entire}} \hat{\mathbf{n}}\cdot\mathbf{S}(\mathbf{r},\omega)\,{\rm d}A = \frac{2}{\hbar\omega}\int_{\Sigma_{\rm entire}} \hat{\mathbf{n}}\cdot{\rm Re}\{\mathbf{E}(\mathbf{r},\omega)\times\mathbf{H}^*(\mathbf{r},\omega)\}\,{\rm d}A.\end{aligned} \tag{3.59}$$

If one chooses $\Sigma_{\rm entire}$ as a very far-field surface $\Sigma_{\rm far}$, then one can make the approximation $\mathbf{E}\times\mathbf{H}^* \rightarrow \hat{\mathbf{n}} n_{\rm B} c\epsilon_0|\mathbf{E}|^2 \rightarrow \hat{\mathbf{n}} n_{\rm B} c\epsilon_0|\mathbf{d}|^2|\mathbf{G}\cdot\mathbf{n}_d|^2/\epsilon_0^2$,

$$\begin{aligned}\Gamma^{\rm rad}_{\rm farsur,QNM}(\mathbf{r}_0,\omega) &= \frac{4}{\hbar\omega}\int_{\Sigma_{\rm far}} \hat{\mathbf{n}}\cdot\mathbf{S}(\mathbf{R}_{\rm far},\omega)\,{\rm d}A \\ &= \frac{2n_{\rm B}c|\mathbf{d}|^2}{\hbar\omega\epsilon_0}\int_{\Sigma_{\rm far}} \left|\mathbf{G}_{\rm QNM}(\mathbf{R}_{\rm far},\mathbf{r}_0,\omega)\cdot\hat{\mathbf{n}}_{\rm d}\right|^2\,{\rm d}A.\end{aligned} \tag{3.60}$$

Once again, we can reconstruct the Green function with QNMs, but now we need to use a regularized QNM function in the far field, so that

$$\mathbf{G}_{\rm QNM}(\mathbf{R}_{\rm far},\mathbf{r}_0,\omega) = \sum_\mu A_\mu(\omega)\tilde{\mathbf{F}}_\mu(\mathbf{R}_{\rm far},\omega)\tilde{\mathbf{f}}_\mu(\mathbf{r}_0), \tag{3.61}$$

which yields the general QNM form:

$$\Gamma^{\rm rad}_{\rm QNM}(\mathbf{r}_0,\omega) = \frac{2n_{\rm B}c|\mathbf{d}|^2}{\hbar\omega\epsilon_0}\int_{\Sigma_{\rm far}}\left|\sum_\mu A_\mu(\omega)\tilde{\mathbf{F}}_\mu(\mathbf{R}_{\rm far},\omega)\tilde{\mathbf{f}}_\mu(\mathbf{r}_0)\cdot\hat{\mathbf{n}}_{\rm d}\right|^2{\rm d}A \tag{3.62}$$

In the single mode limit,

$$\Gamma^{\rm rad}_{\rm c}(\mathbf{r}_0,\omega) = \frac{2n_{\rm B}c|\mathbf{d}|^2}{\hbar\omega\epsilon_0}\left|A_{\rm c}(\omega)\left[\tilde{\mathbf{f}}_{\rm c}(\mathbf{r}_0)\cdot\hat{\mathbf{n}}_{\rm d}\right]\right|^2\int_{\Sigma_{\rm far}}\left|\tilde{\mathbf{F}}_{\rm c}(\mathbf{R}_{\rm far},\omega)\right|^2{\rm d}A. \tag{3.63}$$

The regularized QNM (in the far field) can be computed and approximated in a number of ways, e.g., through the Dyson equation or from an efficient near-field to far-field transformation of the QNMs [54, 62]. Since these are also modal quantities, in practice we evaluate these expressions at the QNM pole frequencies, i.e., at $\omega = \omega_{\rm c}$.

3.2.8.3 Classical Radiative and Nonradiative Beta Factors

Using the QNM nonradiative and radiative decay rates, and within the bandwidth of the QNM of interest, we could define the QNM beta factors in the single mode limit. However, here we will just introduce a numerical estimate that we will also use later to test the quantum counterparts, which exploits quantized QNM parameters obtained from the classical solutions.

We can obtain the numerically exact beta factors using full-dipole simulations (within numerical error of course, such calculations are never exact), with (see "$\Gamma^{\rm rad}_{\rm sur,num}$" and "$\Gamma^{\rm nrad}_{\rm sur,num}$" in Eq. (3.58) and Eq. (3.54))

$$\beta^{\rm rad}_{\rm num}(\mathbf{r}_0,\omega) = \frac{\Gamma^{\rm rad}_{\rm sur,num}(\mathbf{r}_0,\omega)}{\Gamma_{\rm sur,num}(\mathbf{r}_0,\omega)}, \tag{3.64}$$

and

$$\beta^{\rm nrad}_{\rm num}(\mathbf{r}_0,\omega) = \frac{\Gamma^{\rm nrad}_{\rm sur,num}(\mathbf{r}_0,\omega)}{\Gamma_{\rm sur,num}(\mathbf{r}_0,\omega)} = \frac{-\int_{S_{\rm metal}} \hat{\mathbf{n}}\cdot\mathbf{S}_{\rm metal,total}(\mathbf{r}_{\rm metal},\omega){\rm d}A'}{\int_S \hat{\mathbf{n}}\cdot\mathbf{S}_{\rm dipole,total}(\mathbf{r},\omega){\rm d}A}. \tag{3.65}$$

Note that a disadvantage of this full-dipole approach is that it cannot project a modal quantity, which is not so insightful in the case of coupled QNMs. However, it is useful to verify the QNM solutions. Applications of using QNMs will be discussed later.

3.2.9 *Quantized Quasinormal Modes*

To rigorously quantize QNMs, we first summarize the general macroscopic Green function quantization approach for any lossy media [13, 63] and then use that as a starting point to connect to QNM quantization. In the dipole and rotating wave approximations, the Hamiltonian of a quantum emitter, modeled as a Fermionic two-level system (TLS), interacting with the medium-assisted photon field is

$$H = H_{\rm em} + H_{\rm a} + H_{\rm I}, \tag{3.66}$$

with

$$H_{\rm em} = \hbar\int_0^\infty {\rm d}\omega\,\omega\int {\rm d}\mathbf{r}\,\mathbf{b}^\dagger(\mathbf{r},\omega)\cdot\mathbf{b}(\mathbf{r},\omega), \tag{3.67a}$$

$$H_{\rm a} = \hbar\omega_0\sigma^+\sigma^-, \tag{3.67b}$$

$$H_{\rm I} = -\left[\sigma^+\int_0^\infty {\rm d}\omega\,\mathbf{d}_0\cdot\hat{\mathbf{E}}(\mathbf{r}_0,\omega) + {\rm H.a.}\right], \tag{3.67c}$$

where $H_{\rm em}$ is the energy of the lossy medium (and the spatial integration is performed with respect to all space $\mathbb{R}^3$), $\mathbf{b}^{(\dagger)}(\mathbf{r}, \omega)$ are the corresponding vector-valued bosonic annihilation (creation) operators, H.a. refers to Hermitian adjoint, and $\sigma^{\pm}$ are the Pauli operators. The terms $H_{\rm a}$ and $H_{\rm I}$ represent the TLS and emitter-field interactions, respectively, and ω_0, $\mathbf{r}_0$, and $\mathbf{d}_0$ are the transition frequency, spatial position, and dipole moment of the TLS (or atom), respectively.

The operator $\hat{\mathbf{E}}(\mathbf{r}, \omega)$ is the medium-assisted field operator, which solves the *quantized* Helmholtz equation:

$$\nabla \times \nabla \times \hat{\mathbf{E}}(\mathbf{r}, \omega) - \frac{\omega^2}{c^2}\epsilon(\mathbf{r}, \omega)\hat{\mathbf{E}}(\mathbf{r}, \omega) = i\omega\mu_0\hat{\mathbf{j}}_{\rm N}(\mathbf{r}, \omega), \tag{3.68}$$

where $\hat{\mathbf{j}}_{\rm N}(\mathbf{r}, \omega)$ is a quantum noise density, which accounts for the interaction of the free electromagnetic field with the lossy dielectric medium; this term is connected to the bosonic annihilation operators via [64]

$$\hat{\mathbf{j}}_{\rm N}(\mathbf{r}, \omega) = \omega\sqrt{\frac{\hbar\epsilon_0}{\pi}\epsilon_I(\mathbf{r}, \omega)}\mathbf{b}(\mathbf{r}, \omega), \tag{3.69}$$

which retains the fundamental commutation relations of the electromagnetic operators. Furthermore, $\epsilon_I(\mathbf{r}, \omega) = \mathrm{Im}[\epsilon(\mathbf{r}, \omega)] > 0$ is the imaginary part of the permittivity of the lossy medium. The formal solution of Eq. (3.68) is

$$\hat{\mathbf{E}}(\mathbf{r}, \omega) = \frac{i}{\omega\epsilon_0}\int \mathrm{d}\mathbf{s}\, \mathbf{G}(\mathbf{r}, \mathbf{s}, \omega) \cdot \hat{\mathbf{j}}_{\rm N}(\mathbf{s}, \omega), \tag{3.70}$$

where $\mathbf{G}(\mathbf{r}, \mathbf{r}', \omega)$ is exactly the same classical photon Green function introduced earlier (cf. Eq. (3.5)). In the time domain, the total electric field *operator* is

$$\hat{\mathbf{E}}(\mathbf{r}, t) = \int_0^\infty \mathrm{d}\omega\, \hat{\mathbf{E}}(\mathbf{r}, \omega, t) + \mathrm{H.a.}, \tag{3.71}$$

where ω is a mode component of the medium-photon fields. The time dependence of $\hat{\mathbf{E}}(\mathbf{r}, \omega, t)$ is obtained from the Heisenberg equation of motion with respect to the Hamiltonian (Eq. (3.67)), i.e., $\dot{\hat{\mathbf{E}}} = i/\hbar[\hat{\mathbf{E}}, H]$.

Using the noise current density $\hat{\mathbf{j}}_{\rm N}(\mathbf{r}, \omega)$, we can relate the electric field operator to the bosonic noise operators, $\mathbf{b}(\mathbf{r}, \omega)$, from

$$\hat{\mathbf{E}}(\mathbf{r}, \omega) = i\sqrt{\frac{\hbar}{\pi\epsilon_0}}\int \mathrm{d}\mathbf{s}\, \sqrt{\epsilon_I(\mathbf{s}, \omega)}\, \mathbf{G}(\mathbf{r}, \mathbf{s}, \omega) \cdot \mathbf{b}(\mathbf{s}, \omega). \tag{3.72}$$

We‘ note that this approach also includes the limit of a lossless dielectric [24], where the limits have to be performed very carefully, as explained further below. Next, we use the QNM Green function expansion, Eq. (3.22), to formulate the modal part of the electric field operator near or inside the scattering structure:

$$\hat{\mathbf{E}}_{\rm QNM}(\mathbf{r}) = i\sum_{\mu}\sqrt{\frac{\hbar\omega_\mu}{2\epsilon_0}}\,\tilde{\mathbf{f}}_\mu(\mathbf{r})\tilde{a}_\mu + \text{H.a.}, \tag{3.73}$$

where $\tilde{a}_\mu$ are the QNM operators, defined from the noise operators:

$$\tilde{a}_\mu = \lim_{\lambda\to\infty}\int_0^\infty d\omega \int_{V(\lambda)} d\mathbf{r}\,\tilde{\mathbf{L}}_\mu(\mathbf{r},\omega)\cdot\mathbf{b}(\mathbf{r},\omega), \tag{3.74}$$

and

$$\tilde{\mathbf{L}}_\mu(\mathbf{r},\omega) = \sqrt{\frac{2}{\pi\omega_\mu}}A_\mu(\omega)\sqrt{\epsilon_I^{(\alpha)}(\mathbf{r},\omega)}\,\tilde{\mathbf{f}}_\mu(\mathbf{r},\omega). \tag{3.75}$$

Conveniently, these expressions apply to all spatial positions, but note that $\tilde{\mathbf{f}}_\mu(\mathbf{r},\omega)$ is equal to the QNM $\tilde{\mathbf{f}}_\mu(\mathbf{r})$ for positions near or inside the cavity region (system part), and equal to the *regularized QNM* $\tilde{\mathbf{F}}_\mu(\mathbf{r},\omega)$, Eq. (3.29), for positions outside the cavity region (namely, the bath part in quantum optics).

We also highlight that the integration over all space, $\mathbb{R}^3$, is formally written as a limiting process over a sequence of volumes $V(\lambda)$, such that $V(\lambda)\to\mathbb{R}^3$ for $\lambda\to\infty$. To rigorously account for radiative losses within the macroscopic Green function quantization, we have introduced a sequence of permittivity functions $\epsilon^{(\alpha)}(\mathbf{r},\omega) = \epsilon(\mathbf{r},\omega)+\alpha\chi_{\rm L}(\omega)$, where $\alpha\chi_{\rm L}(\omega)$ is a spatially homogeneous Lorentz oscillator weighted by the parameters $\alpha\geq 0$ ($\epsilon_I^{(\alpha)}(\mathbf{r},\omega) = \text{Im}[\epsilon^{(\alpha)}(\mathbf{r},\omega)]$). We stress that the limits $\alpha\to 0$ and $\lambda\to\infty$ are not interchangeable; actually, the ordering of the limits with respect to λ and α is not only a requirement to obtain meaningful radiation processes but also to preserve the fundamental field commutation relations in the dielectric (nonabsorptive) limit [24].

The mode quantization scheme can be performed by first constructing proper annihilation and creation operators through a symmetrization transformation:

$$a_\mu = \sum_\eta\left[\mathbf{S}^{-1/2}\right]_{\mu\eta}\tilde{a}_\eta, \tag{3.76}$$

with $S_{\mu\eta} = [\tilde{a}_\mu, \tilde{a}_\eta^\dagger]$, where

$$S_{\mu\eta} = \lim_{\lambda\to\infty}\int_0^\infty d\omega\int_{V(\lambda)} d\mathbf{r}\,\tilde{\mathbf{L}}_\mu(\mathbf{r},\omega)\cdot\tilde{\mathbf{L}}_\eta^*(\mathbf{r},\omega), \tag{3.77}$$

is a dissipation-induced QNM overlap matrix, which yields a positive definite form [23, 24]. After applying the limit $\alpha\to 0$, $S_{\mu\eta}$ can be written as a sum $S_{\mu\eta} = S_{\mu\eta}^{\rm nrad} + S_{\mu\eta}^{\rm rad}$, where the *quantum S factors* are [23, 24]

$$S_{\mu\eta}^{\rm nrad} = \int_0^\infty {\rm d}\omega \frac{2A_\mu(\omega)A_\eta^*(\omega)}{\pi\sqrt{\omega_\mu\omega_\eta}} I_{\mu\eta}^{\rm nrad}(\omega), \tag{3.78}$$

$$S_{\mu\eta}^{\rm rad} = \int_0^\infty {\rm d}\omega \frac{2A_\mu(\omega)A_\eta^*(\omega)}{\pi\sqrt{\omega_\mu\omega_\eta}} \left[I_{\mu\eta}^{\rm rad}(\omega) + I_{\eta\mu}^{\rm rad*}(\omega)\right], \tag{3.79}$$

which represent the nonradiative and radiation contributions, respectively, with

$$I_{\mu\eta}^{\rm nrad}(\omega) = \int_{V_{\rm S}} {\rm d}\mathbf{s}\, \epsilon_I(\mathbf{s},\omega)\tilde{\mathbf{f}}_\mu(\mathbf{s})\cdot\tilde{\mathbf{f}}_\eta^*(\mathbf{s}), \tag{3.80}$$

$$I_{\mu\eta}^{\rm rad}(\omega) = \frac{1}{2\epsilon_0\omega}\oint_{\mathcal{S}} {\rm d}A_{\mathbf{s}} \left[\tilde{\mathbf{H}}_\mu(\mathbf{s},\omega)\times\hat{\mathbf{n}}_{\mathbf{s}}\right]\cdot\tilde{\mathbf{F}}_\eta^*(\mathbf{s},\omega), \tag{3.81}$$

where the unit vector $\hat{\mathbf{n}}_{\mathbf{s}}$ is normal to surface $\mathcal{S}$ (pointing outward) and the magnetic regularized QNMs is given as $\tilde{\mathbf{H}}_\mu(\mathbf{s},\omega) = \frac{1}{i\omega\mu_0}\nabla\times\tilde{\mathbf{F}}_\mu(\mathbf{s},\omega)$. While $S_{\mu\eta}^{\rm nrad}$ describes the absorption inside the scattering volume $V_{\rm S}$, $S_{\mu\eta}^{\rm rad}$ reflects the power flow of the regularized QNM fields through an outer surface $\mathcal{S}$. For practical calculations, it is also useful to apply a pole approximation to the above frequency integrals to obtain the symmetrization matrices through [54]

$$S_{{\rm p},\mu\eta}^{\rm nrad} = \frac{\sqrt{\omega_\mu\omega_\eta}}{i(\tilde{\omega}_\mu - \tilde{\omega}_\eta^*)} I_{\mu\eta}^{\rm nrad,p}, \tag{3.82}$$

$$S_{{\rm p},\mu\eta}^{\rm rad} = \frac{\sqrt{\omega_\mu\omega_\eta}}{i(\tilde{\omega}_\mu - \tilde{\omega}_\eta^*)} \left[I_{\mu\eta}^{\rm rad,p} + I_{\eta\mu}^{\rm rad,p*}\right], \tag{3.83}$$

with

$$I_{\mu\eta}^{\rm nrad,p} = \int_{V_{\rm S}} {\rm d}\mathbf{s}\, \sqrt{\epsilon_I(\mathbf{s},\omega_\mu)\epsilon_I(\mathbf{s},\omega_\eta)}\, \tilde{\mathbf{f}}_\mu(\mathbf{s})\cdot\tilde{\mathbf{f}}_\eta^*(\mathbf{s}), \tag{3.84}$$

and

$$I_{\mu\eta}^{\rm rad,p} = \frac{1}{2\epsilon_0\sqrt{\omega_\mu\omega_\eta}}\oint_{\mathcal{S}} {\rm d}A_{\mathbf{s}} \left[\tilde{\mathbf{H}}_\mu(\mathbf{s},\omega_\mu)\times\hat{\mathbf{n}}_{\mathbf{s}}\right]\cdot\tilde{\mathbf{F}}_\eta^*(\mathbf{s},\omega_\eta). \tag{3.85}$$

If one chooses a very far-field surface $\mathcal{S}_\infty$, then Silver-Müller radiations conditions can be used, $\hat{\mathbf{n}}_{\mathbf{s}}\times\tilde{\mathbf{H}}_\mu(\mathbf{s},\omega) \approx -n_{\rm B}c\epsilon_0\tilde{\mathbf{F}}_\mu(\mathbf{s},\omega)$, so that

$$S_{{\rm p},\mu\eta}^{\rm rad} = \frac{n_{\rm B}c}{i(\tilde{\omega}_\mu - \tilde{\omega}_\eta^*)} I_{\mu\eta}^{\rm sur}, \tag{3.86}$$

where

$$I_{\mu\eta}^{\rm sur} = \frac{1}{16\pi^2}\int_0^{2\pi}{\rm d}\varphi\int_0^\pi {\rm d}\vartheta\,\sin(\vartheta)\times\tilde{\mathbf{Z}}_\mu(\varphi,\vartheta,\omega_\mu)\cdot\tilde{\mathbf{Z}}_\eta^*(\varphi,\vartheta,\omega_\eta), \tag{3.87}$$

with

$$\tilde{\mathbf{Z}}_\mu(\varphi,\vartheta,\omega) = i\omega\mu_0 \oint_{S'} \mathrm{d}S' e^{-in_\mathrm{B}\omega\hat{\mathbf{R}}\cdot\mathbf{r}'/c}$$
$$\left[\tilde{\mathbf{J}}_\mu(\mathbf{r}') - \left(\tilde{\mathbf{J}}_\mu(\mathbf{r}')\cdot\hat{\mathbf{R}}\right)\hat{\mathbf{R}} - n_\mathrm{B}c\epsilon_0\hat{\mathbf{R}}\times\tilde{\mathbf{M}}_\mu(\mathbf{r}')\right]. \quad (3.88)$$

We use the radial basis vector $\hat{\mathbf{R}} = \hat{\mathbf{R}}(\varphi,\vartheta)$ in spherical coordinates, and $\tilde{\mathbf{J}}_\mu(\mathbf{r}')$ ($\tilde{\mathbf{M}}_\mu(\mathbf{r}')$) is the electric (magnetic) surface current on a near-field surface S′, close to the resonator [62], obtained from

$$\tilde{\mathbf{J}}_\mu(\mathbf{r}') = \hat{\mathbf{n}}' \times \tilde{\mathbf{h}}_\mu(\mathbf{r}'), \quad (3.89)$$

$$\tilde{\mathbf{M}}_\mu(\mathbf{r}') = -\hat{\mathbf{n}}' \times \tilde{\mathbf{f}}_\mu(\mathbf{r}'), \quad (3.90)$$

which use the magnetic QNMs $\tilde{\mathbf{h}}_\mu(\mathbf{r}') = \frac{1}{i\tilde{\omega}_\mu\mu_0}\nabla\times\tilde{\mathbf{f}}_\mu(\mathbf{r}')$ and $\hat{\mathbf{n}}'$ is a unit vector normal to the near-field surface S′, pointing outward.

We have achieved the desired goal of quantizing the QNMs for use in system-level quantum optics, as the symmetrized operators now fulfill the commutation relations $[a_\mu, a_\eta] = [a_\mu^\dagger, a_\eta^\dagger] = 0$ and $[a_\mu, a_\eta^\dagger] = \delta_{\mu\eta}$, for all μ,η and thus are proper annihilation and creation operators to construct QNM Fock states from the vacuum state $|\mathrm{vac}\rangle_\mathrm{em}$, which is defined via $\hat{b}_i(\mathbf{r},\omega)|\mathrm{vac}\rangle_\mathrm{em} = 0$. Importantly, this allows us to derive a (quantum) master equation, which we discuss next.

The construction of the QNM Fock space, along with the Hamiltonian, Eq. (3.67), can be used to derive a master equation for the density operator ρ in the combined QNM-TLS space [23, 65], which has the usual form

$$\partial_t\rho = -\frac{i}{\hbar}[H_\mathrm{S},\rho] + \mathcal{L}\rho, \quad (3.91)$$

where $H_\mathrm{S} = H_{\mathrm{QNM-a}} + H_\mathrm{a} + H_\mathrm{QNM}$ is the system Hamiltonian, where

$$H_{\mathrm{QNM-a}} = \hbar\sum_\mu \tilde{g}_\mu^\mathrm{s} a_\mu^\dagger\sigma^- + \mathrm{H.a.}, \quad (3.92)$$

is the TLS-QNM interaction Hamiltonian, and $\tilde{g}_\mu^\mathrm{s} = -i\sum_\eta[\mathbf{S}^{1/2}]_{\eta\mu}\sqrt{\omega_\eta/(2\hbar\epsilon_0)}\mathbf{d}_0\cdot\tilde{\mathbf{f}}_\eta(\mathbf{r}_0)$ the TLS-QNM coupling constant; H_a represents the TLS energy [Eq. (3.67b)]; and

$$H_\mathrm{QNM} = \hbar\sum_{\mu\eta}\chi_{\mu\eta}^+ a_\mu^\dagger a_\eta, \quad (3.93)$$

is the QNM Hamiltonian, including off-diagonal photon coupling terms through $\chi_{\mu\eta}^+ = [\chi_{\mu\eta} + \chi_{\eta\mu}^*]/2$ with $\chi_{\mu\eta} = \sum_\nu[\mathbf{S}^{-1/2}]_{\mu\nu}\tilde{\omega}_\nu[\mathbf{S}^{1/2}]_{\nu\eta}$.

Finally, the dissipator term, $\mathcal{L} = \mathcal{L}_{\text{em}} + \mathcal{L}_{\text{SE}}$ is the Lindblad super-operator, rigorously derived in a self-consistent way, which contains the QNM photon decay through

$$\mathcal{L}_{\text{em}}\rho = \sum_{\mu\eta} \chi^{-}_{\mu\eta} \left[2a_\eta \rho a^\dagger_\mu - a^\dagger_\mu a_\eta \rho - \rho a^\dagger_\mu a_\eta \right], \tag{3.94}$$

with $\chi^{-}_{\mu\eta} = i[\chi_{\mu\eta} - \chi^{*}_{\eta\mu}]/2$, and the atomic spontaneous emission through $\mathcal{L}_{\text{SE}}$ [23, 65]. In contrast to a multi-mode Jaynes-Cummings model, we now have dissipation-induced coupling between quantized QNMs, enforced through the quantization procedure. This turns out to have profound consequences on coupled QNM problems and leads to drastically different insight into QNM coupling models compared to a classical picture, e.g., there is no single QNM that produces a negative LDOS contribution, something that readily occurs with classical QNM expansions, since the QNM phase can be negative [47]. The approach is also able to give a first-principles picture of how Fano-like resonances appear between two discrete cavity modes [23]. In general, when multiple QNMs are involved, another (significant) advantage of the quantized QNM approach is that a mode approach can be used to solve for the input-output, by diagonalizing the Lindblad loss part. The appearing rates are shifted compared to γ_μ [65]. In contrast, this cannot be done using direct dipole simulations of the total beta factors or from knowledge of the classical QNMs.

3.2.10 *Bad Cavity Limit Solution from the Quantized Quasinormal Mode Master Equation*

Although one can now use the QNM master equation to explore unique regimes of multiphoton effects, as exemplified in Refs. [65, 66], it is useful to derive the quantum Purcell factor in a bad cavity limit, which allows us to compare with a purely classical solution. This is an important check for any quantum theory, as we expect classical-quantum correspondence in this limit (at least in the limit that the quantum emitter is treated as a harmonic oscillator). Physically, this just means that we consider a relatively small dipole moment, to ignore entanglement effects between the QNMs and the emitters, e.g., yielding vacuum Rabi oscillations in a *strong coupling regime*. Thus, we are effectively working in the weak coupling regime. We can do this in a way that uses no fitting parameters; everything comes from the QNM parameters, enabling a Maxwell-based approach to realistic quantum optics simulations, appropriate for studying multiphoton correlations between light and matter, including plasmonic modes.

Following Ref. [23], and the approach of Cirac [67] within the weak coupling limit, we obtain the TLS master equation for the reduced atomic density operator $\rho_{\text{a}} = \text{tr}_{\text{em}}\rho$, but now using the quantized QNM model:

$$\partial_t \rho_\mathrm{a} = -\frac{i}{\hbar}[H_\mathrm{a}, \rho_\mathrm{a}] + \frac{\Gamma_\mathrm{B}}{2}\mathcal{D}[\sigma^-]\rho_\mathrm{a} + \frac{\Gamma^\mathrm{qQNM}}{2}\mathcal{D}[\sigma^-]\rho_\mathrm{a}, \tag{3.95}$$

where $\mathcal{D}[A] = 2A\rho A^\dagger - A^\dagger A\rho - \rho A^\dagger A$, and the cavity-induced SE rate is

$$\Gamma^\mathrm{qQNM} = \sum_{\eta,\eta'} \tilde{g}_\eta S_{\eta\eta'} \tilde{g}^*_{\eta'} \frac{i(\omega_\eta - \omega_{\eta'}) + (\gamma_\eta + \gamma_{\eta'})}{(\Delta_{\eta a} - i\gamma_\eta)(\Delta_{\eta' a} + i\gamma_{\eta'})}, \tag{3.96}$$

with $\Delta_{\eta\mathrm{a}} = \omega_\eta - \omega_\mathrm{a}$ and Γ_B is the background SE rate defined earlier. Note that these rates are the SE rates at frequency ω_a; in a more general problem, such rates depend on the particular system Hamiltonian, where the rates have to be computed in a self-consistent way [21]. For simplicity, we also neglect photonic Lamb shifts, but these can easily be added into the theory, so this is not a model restriction, either classically or quantum mechanically.

Furthermore, a radiative and nonradiative SE rate can be obtained through the respective contributions of the symmetrization matrix $S_{\mu\eta}$, defined through Eqs. (3.81) and (3.80), respectively. One can then connect to the classical β-factors or equivalently the forms that are obtained from a (quantized) weak emitter-field coupling model using classical QNMs. It can be shown that they differ by two points: (i) The frequency dependence is different in both cases, as some parts in the quantum picture are evaluated at the QNM frequencies at the system level when defining the quantum S factors. (ii) The QNM approximation is done at different stages: while in the classical picture, the general Green function is kept until the very end, in the quantum picture, one has to introduce the mode expansion at the very beginning in order to define proper QNM Fock states.

The *quantum* Purcell factor is

$$F_\mathrm{P}^\mathrm{qQNM} = \frac{\Gamma^\mathrm{qQNM}}{\Gamma_\mathrm{B}} + 1. \tag{3.97}$$

In the limit of a single QNM, $\mu = \mathrm{c}$, we simply obtain the cavity-mediated SE rate:

$$\Gamma^\mathrm{qQNM} = \Gamma_\mathrm{c}^\mathrm{qQNM} = \frac{2S_\mathrm{cc}|\tilde{g}_\mathrm{c}|^2\gamma_\mathrm{c}}{(\omega_\mathrm{a} - \omega_\mathrm{c})^2 + \gamma_\mathrm{c}^2}, \tag{3.98}$$

and thus the single-QNM quantum Purcell factor is

$$F_\mathrm{P}^\mathrm{qQNM} = \frac{\Gamma_\mathrm{c}^\mathrm{qQNM}}{\Gamma_\mathrm{B}} + 1. \tag{3.99}$$

3.2.11 Coupled-Mode Theories Using Quasinormal Modes

Coupled-mode theory with cavity NMs has been tremendously successful [2, 6]. However, it is only rigorously valid in the limit of high Q resonators/modes, where the phase is also neglected [68]. Recently, coupled-mode theory using QNMs has been developed [52, 69–71], and it has been shown that standard coupled-mode theory breaks down, e.g., for studying exceptional points [52, 72–74], where two modes coalesce.

We will summarize the key results developed in [52, 75], specialized to coupled lossy cavities; however, we highlight that the theory also applies to coupled loss-gain amplifiers and gain resonators [52], which is another remarkable aspect of QNM theory. To quantify the form of the overlap integrals, we write the QNM norm as

$$\langle\langle \tilde{\mathbf{f}}_\mu | \hat{\epsilon}_\mu | \tilde{\mathbf{f}}_\mu \rangle\rangle \rightarrow \int_{\text{reg}} d\mathbf{r} \epsilon_\mu(\mathbf{r}) \tilde{\mathbf{f}}_\mu(\mathbf{r}) \cdot \tilde{\mathbf{f}}_\mu(\mathbf{r}) = 1, \tag{3.100}$$

and we will neglect dispersion for now; the results are easy enough to generalize to include dispersion if needed. The notation "reg" refers to a regularization procedure for the QNM norm, though the coupled-mode part of the overlap integrals will not be part of a numerical PML boundary condition.

Let us consider two cavities, cavity "1" and cavity "2," with a dielectric constant defined from $\epsilon_{1/2}(\mathbf{r})$, in a background medium with ϵ_B. Using operator notation for convenience, the dielectric constant difference operator is $\hat{V}_{1/2}$, which satisfies $\hat{\epsilon}_{1/2} = \hat{\epsilon}_\text{B} + \hat{V}_{1/2}$, and $\hat{\epsilon}_\text{t} = \hat{\epsilon}_\text{B} + \hat{V}_1 + \hat{V}_2$ is the total dielectric constant. We seek to derive a solution for the total system in terms of the bare (uncoupled) QNM quantities, whose eigenfunctions, eigenfrequencies, and coupling coefficients will be defined from $\tilde{\mathbf{f}}_{1/2}$, $\tilde{\omega}_{1/2}$, $\tilde{\kappa}_{12/21}$. To clarify the notation, the operator $\hat{\epsilon}_\text{t}$ is defined as $\langle \mathbf{r}|\hat{\epsilon}_\text{t}|\mathbf{r}'\rangle = \epsilon_\text{t}(\mathbf{r})\delta(\mathbf{r}-\mathbf{r}')$, and the electric field, following a projection onto space, is $\langle \mathbf{r}|\mathbf{E}\rangle = \mathbf{E}(\mathbf{r}, \omega)$.

So solve the general problem, we basically seek to obtain the total Green function: $\mathbf{G}(\mathbf{r}, \mathbf{r}') = \langle \mathbf{r}|\hat{\mathbf{G}}|\mathbf{r}'\rangle$, which is the Green function of the entire system (including both cavities). To simplify the analysis, we assume that each cavity yields one QNM of interest, and we assume that these modes are weakly coupled to each other, with

$$\langle \tilde{\mathbf{f}}_\alpha | \hat{\epsilon}_\text{t} | \tilde{\mathbf{f}}_\beta \rangle = \delta_{\alpha\beta}, \tag{3.101}$$

which can easily be checked numerically. Note that this weak coupling assumption is not a requirement for using coupled-mode theory, and relaxing this requirement simply results in more complicated overlap integrals (which are not needed for the problems we will exemplify). The QNM Green function is

$$\hat{\mathbf{G}} = \sum_{\alpha,\beta} \tilde{B}_{\alpha,\beta} |\tilde{\mathbf{f}}_\alpha\rangle \langle \tilde{\mathbf{f}}^*_\beta|, \tag{3.102}$$

whose solution can be written as a standard matrix problem:

$$\sum_{\alpha,\beta}[\tilde{\omega}_\alpha^2\langle\langle\tilde{\mathbf{f}}_i|\hat{\epsilon}_\alpha|\tilde{\mathbf{f}}_\alpha\rangle\rangle - \omega^2\langle\langle\tilde{\mathbf{f}}_i|\hat{\epsilon}_{\rm t}|\tilde{\mathbf{f}}_\alpha\rangle\rangle]\tilde{B}_{\alpha,\beta}\langle\langle\tilde{\mathbf{f}}_\beta|\hat{\epsilon}_{\rm t}|\tilde{\mathbf{f}}_j\rangle\rangle = \omega^2\langle\langle\tilde{\mathbf{f}}_i|\hat{\epsilon}_{\rm t}|\tilde{\mathbf{f}}_j\rangle\rangle. \quad (3.103)$$

We assume the solution is solved for cavity 1, and then we add cavity 2.[6] Using the bare QNMs as a basis, and the QNM completeness relation, we derive

$$\tilde{B}_{\alpha,\beta} = \frac{\omega/2}{(\omega-\tilde{\omega}_+)(\omega-\tilde{\omega}_-)}\begin{pmatrix}\tilde{\omega}_2-\omega & \tilde{\kappa}_{12}\\ \tilde{\kappa}_{21} & \tilde{\omega}_1-\omega\end{pmatrix}, \quad (3.104)$$

where

$$\tilde{\kappa}_{ij} = \frac{\tilde{\omega}_j}{2}\langle\langle\tilde{\mathbf{f}}_i|\hat{V}_i|\tilde{\mathbf{f}}_j\rangle\rangle, \quad (3.105)$$

with $(i, j = 1, 2)$ and

$$\langle\langle\tilde{\mathbf{f}}_i|\hat{V}_i|\tilde{\mathbf{f}}_j\rangle\rangle = \int_{\Omega_i}\mathbf{dr}[\epsilon_i(\mathbf{r})-\epsilon_{\rm B}]\tilde{\mathbf{f}}_i(\mathbf{r})\cdot\tilde{\mathbf{f}}_j(\mathbf{r}). \quad (3.106)$$

We stress this form uses the *unconjugated norm* in the QNM overlap integrals. Also note, we will assume that we can use the standard QNMs here as opposed to the regularized ones, since the coupling is really in the near-field regime, and we can also easily justify such an approximation.

An interesting manifestation of the QNM coupled-mode theory is that $\tilde{\kappa}_{12} \neq \tilde{\kappa}_{21}^*$ [52, 71], since we are dealing with open resonators whose modes do not satisfy Hermiticity. A consequence of this approach is that the criteria for exceptional point are much broader (and more difficult to define and find) [52].

Using Eq. (3.104) and Eq. (3.102), the Green function for the total coupled system is

$$\begin{aligned}\hat{\mathbf{G}} = &\frac{\frac{\omega}{2}(\tilde{\omega}_2-\omega)|\tilde{\mathbf{f}}_1\rangle\,\langle\tilde{\mathbf{f}}_1^*|}{(\omega-\tilde{\omega}_+)(\omega-\tilde{\omega}_-)} + \frac{\frac{\omega}{2}\tilde{\kappa}_{12}|\tilde{\mathbf{f}}_1\rangle\,\langle\tilde{\mathbf{f}}_2^*|}{(\omega-\tilde{\omega}_+)(\omega-\tilde{\omega}_-)}\\ &+\frac{\frac{\omega}{2}\tilde{\kappa}_{21}|\tilde{\mathbf{f}}_2\rangle\,\langle\tilde{\mathbf{f}}_1^*|}{(\omega-\tilde{\omega}_+)(\omega-\tilde{\omega}_-)} + \frac{\frac{\omega}{2}(\tilde{\omega}_1-\omega)|\tilde{\mathbf{f}}_2\rangle\,\langle\tilde{\mathbf{f}}_2^*|}{(\omega-\tilde{\omega}_+)(\omega-\tilde{\omega}_-)},\end{aligned} \quad (3.107)$$

where the two QNM pole frequencies are obtained explicitly:

$$\tilde{\omega}_\pm = \frac{\tilde{\omega}_1+\tilde{\omega}_2}{2} \pm \frac{\sqrt{4\tilde{\kappa}_{12}\tilde{\kappa}_{21}+(\tilde{\omega}_1-\tilde{\omega}_2)^2}}{2}. \quad (3.108)$$

[6] Of course, we could also start with cavity 2 and add in cavity 1.

It is interesting to see the effect this has on the criterion for finding exceptional points, which now occurs when

$$2\sqrt{\tilde{\kappa}_{21}\tilde{\kappa}_{12}}|_{\mathrm{EP}} \equiv K_1 + iK_2 = \pm i(\tilde{\omega}_1 - \tilde{\omega}_2), \tag{3.109}$$

which is highly nontrivial to achieve and in general will always return lossy QNMs for open cavity systems. Indeed, it seems impossible to find exceptional points if the two cavities have the same resonance, as is often assumed in the usual NM theories.

One can also diagonalize the basis to rewrite the coupled-mode-theory Green function as

$$\hat{\mathbf{G}}^{\mathrm{QNM}}_{\mathrm{CMT}} = \hat{\mathbf{G}}^{\mathrm{QNM}}_{+} + \hat{\mathbf{G}}^{\mathrm{QNM}}_{-} = \frac{\omega|\tilde{\mathbf{f}}^{+}\rangle\,\langle\tilde{\mathbf{f}}^{+*}|}{2(\tilde{\omega}_+ - \omega)} + \frac{\omega|\tilde{\mathbf{f}}^{-}\rangle\,\langle\tilde{\mathbf{f}}^{-*}|}{2(\tilde{\omega}_- - \omega)}, \tag{3.110}$$

where the hybrid QNMs are

$$|\tilde{\mathbf{f}}^{\pm}\rangle = \frac{-\tilde{\kappa}_{12}}{\sqrt{(\tilde{\omega}_\pm - \tilde{\omega}_1)^2 + \tilde{\kappa}_{12}^2}}|\tilde{\mathbf{f}}_1\rangle + \frac{\tilde{\omega}_\pm - \tilde{\omega}_1}{\sqrt{(\tilde{\omega}_\pm - \tilde{\omega}_1)^2 + \tilde{\kappa}_{12}^2}}|\tilde{\mathbf{f}}_2\rangle. \tag{3.111}$$

Only when one assumes $\tilde{\kappa} = \tilde{\kappa}_{12} \approx \tilde{\kappa}_{21}$, then this simplifies to

$$|\tilde{\mathbf{f}}^{\pm}\rangle \approx \frac{-\tilde{\kappa}}{\sqrt{(\tilde{\omega}_\pm - \tilde{\omega}_1)^2 + \tilde{\kappa}^2}}|\tilde{\mathbf{f}}_1\rangle + \frac{\tilde{\omega}_\pm - \tilde{\omega}_1}{\sqrt{(\tilde{\omega}_\pm - \tilde{\omega}_1)^2 + \tilde{\kappa}^2}}|\tilde{\mathbf{f}}_2\rangle. \tag{3.112}$$

3.2.11.1 Classical Purcell Factors Using Quasinormal Modes and Coupled-Mode Theory

Using the QNM coupled-mode theory result, the *classical Purcell factor* is simply

$$F_{\mathrm{P}}^{\mathrm{cQNM}}(\mathbf{r}_0, \omega) = 1 + \frac{\mathbf{d}\cdot\mathrm{Im}\{\mathbf{G}^{\mathrm{QNM}}_{\mathrm{CMT}}(\mathbf{r}_0, \mathbf{r}_0, \omega)\}\cdot\mathbf{d}}{\mathbf{d}\cdot\mathrm{Im}\{\mathbf{G}_{\mathrm{B}}(\mathbf{r}_0, \mathbf{r}_0, \omega)\}\cdot\mathbf{d}}, \tag{3.113}$$

or we can write in terms of the separate mode contributions from the "+" and "-" QNMs, with

$$F_{\mathrm{P},\pm}^{\mathrm{cQNM}}(\mathbf{r}_0, \omega) = \frac{\mathbf{d}\cdot\mathrm{Im}\{\mathbf{G}^{\mathrm{QNM}}_{\pm}(\mathbf{r}_0, \mathbf{r}_0, \omega)\}\cdot\mathbf{d}}{\mathbf{d}\cdot\mathrm{Im}\{\mathbf{G}_{\mathrm{B}}(\mathbf{r}_0, \mathbf{r}_0, \omega)\}\cdot\mathbf{d}}, \tag{3.114}$$

where $F_{\mathrm{P}}^{\mathrm{cQNM}} = 1 + F_{\mathrm{P},+}^{\mathrm{cQNM}} + F_{\mathrm{P},-}^{\mathrm{cQNM}}$. This is useful for clarifying the individual QNM contributions and for checking against non-coupled-mode-theory QNM results; we will not show this directly, but for detailed discussions on this point, see Refs. [52, 75].

The advantage of having an accurate coupled-mode theory is that we can explore a wide range of coupling regimes without having to recalculate the QNMs each time, e.g., as a function of resonator separation, and it helps to explain the underlying physics of QNM coupling. It also connects to current theories of mode coupling in quantum mechanics, typically only formulated as a coupled (quantized) NM problem (with heuristic dissipation terms).

3.2.11.2 Classical Purcell Factors Using Mode Expansions from Coupled-Mode Theory

To better appreciate the impact of using QNM results for coupled modes and how that affects the Purcell factor of a dipole emitter, it is useful to also consider the more standard NM results. This is an advantage also to connect to quantum field theories, as they typically use NMs, which we connect to below.

In a rotating wave approximation, the NM Green function has the form [12, 76]

$$\mathbf{G}^{\text{NM}}(\mathbf{r}, \mathbf{r}', \omega) \approx \sum_{\mu} \frac{\omega\, \mathbf{f}_{\mu}(\mathbf{r})\mathbf{f}^{*}_{\mu}(\mathbf{r}')}{2(\tilde{\omega}_{\mu} - \omega)}, \tag{3.115}$$

where we use QNM modes in the expansion, as these are the ones that are computed in practice. Moreover, we will assume that these modes are normalized as QNMs; otherwise, the NM normalization is ambiguous [42].[7]

Using the coupled-mode solutions, we obtain the NM Green function expansion result

$$\hat{\mathbf{G}}^{\text{NM}}_{\text{CMT}} = \frac{\omega|\tilde{\mathbf{f}}^{+}\rangle\, \langle\tilde{\mathbf{f}}^{+}|}{2(\tilde{\omega}_{+} - \omega)} + \frac{\omega|\tilde{\mathbf{f}}^{-}\rangle\, \langle\tilde{\mathbf{f}}^{-}|}{2(\tilde{\omega}_{-} - \omega)}, \tag{3.116}$$

and thus the following form for the NM classical Purcell factor:

$$F_{\text{P}}^{\text{cNM}}(\mathbf{r}_0, \omega) = 1 + \frac{\mathbf{d} \cdot \text{Im}\{\mathbf{G}^{\text{NM}}_{\text{CMT}}(\mathbf{r}_0, \mathbf{r}_0, \omega)\} \cdot \mathbf{d}}{\mathbf{d} \cdot \text{Im}\{\mathbf{G}_{\text{B}}(\mathbf{r}_0, \mathbf{r}_0, \omega)\} \cdot \mathbf{d}}. \tag{3.117}$$

The key difference, as remarked before, is the QNM phase. In the applications section of the chapter, we will show several examples that demonstrate the significant differences that this term can make, even for modes with very large quality factors.

---

[7] Though in practice, for high-Q resonators, the normalization is likely quite accurate for most problems [42].

3.2.11.3 Quantum Purcell Factors in the Bad Cavity Limit, Using Quasinormal Modes and Normal Modes Obtained from Coupled-Mode Theory

Next, we discuss the quantum QNM results and show how they connect to the coupled-mode theories. Since we have the hybrid modes, we can simply use the quantum approach given earlier to write down the quantum Purcell factor in a bad cavity limit,

Earlier, we introduced the quantum Purcell factor as $F_{\rm P}^{\rm qQNM} = \Gamma^{\rm qQNM}/\Gamma_{\rm B} + 1$ (Eq. (3.97)), where the quantum QNM decay rate can be separated into diagonal contribution and non-diagonal contributions [23]. For the lossy resonator structures we will show in the applications section (coupled whispering gallery modes from microdisk resonators), the nonradiative contribution to the quantum S factors completely dominates. Thus, we can write

$$\Gamma_{\rm diag}^{\rm qQNM} \approx \sum_{\mu} S_{\mu}^{\rm nrad} \frac{2|\tilde{g}_{\mu}|^2 \gamma_{\mu}}{\Delta_{\mu {\rm a}}^2 + \gamma_{\mu}^2}, \tag{3.118a}$$

$$\Gamma_{\rm ndiag}^{\rm qQNM} \approx \sum_{\mu,\eta \neq \mu} \tilde{g}_{\mu} S_{\mu\eta}^{\rm nrad} \tilde{g}_{\eta}^{*} K_{\mu\eta}. \tag{3.118b}$$

To make a connection to quantized NM theory, we first assume $S_{\mu\eta} = \delta_{\mu\eta}$, resulting in

$$\Gamma^{\rm qNM} = \sum_{\mu} \frac{2|\tilde{g}_{\mu}|^2 \gamma_{\mu}}{\Delta_{\mu {\rm a}}^2 + \gamma_{\mu}^2}, \tag{3.119}$$

which is the established result from the dissipative Jaynes-Cummings model in the bad cavity limit [77]. In such a model, there is no non-diagonal coupling, which we will show later is equivalent (or very similar) to the absence of the QNM phases in classical NM theory. The corresponding quantum NM Purcell factor is

$$F_{\rm P}^{\rm qNM} = \frac{\Gamma^{\rm qNM}}{\Gamma_{\rm B}} + 1. \tag{3.120}$$

3.3 Applications

3.3.1 One-Dimensional Quasinormal Modes and Green Functions for Dielectric Barriers

We begin the applications part by showing some one-dimensional QNMs and regularized QNMs, for a simple 1D barrier of length L. The theory is all analytic,

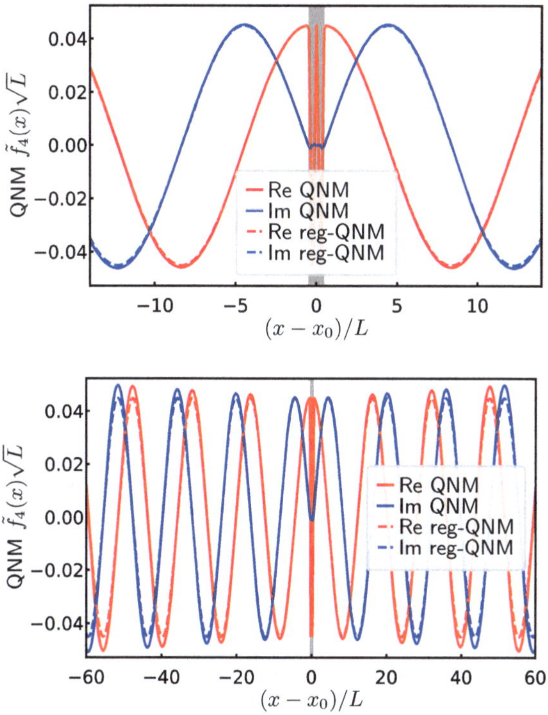

Fig. 3.3 Real and imaginary parts of the QNM (solid curves) and regularized QNM (dashed curves), with $\mu = 4$, for a single-barrier cavity with center x_0, length L, background refractive index $n_B = 1$, and slab index $n_R = 2\pi$. The gray dashed part marks the cavity region and we show two different spatial scales. The corresponding quality factor is $Q \approx 20$

For the (complex) QNMs, we use Eqs. (3.34) and (3.35) to compute the field profiles and use Eqs. (3.37a) and (3.37b) for the regularized QNMs. In the latter case, we apply a pole approximation at $k = \text{Re}[\tilde{k}_4]$, i.e., $\tilde{F}_\mu'^{(0)}(x, \omega_\mu)$ (with $\mu = 4$).

Figure 3.3 shows the results for both the real and imaginary parts of the QNM field and regularized QNM profile for two different distances. Both results are similar up to distances of around $x = 6L$, and then we begin to see the QNM diverge as a function of space, while the regularized QNM is well behaved, and essentially produces a sinusoidal pattern, when going into the far field.

To further demonstrate the use of QNM regularization on an observable, we next show how a single QNM expansion of the Green function compares with exact Green function calculations as a function of distance. Specifically, we compare the propagator, i.e., the imaginary part of the Green function, $\text{Im}[G(x, x_s, \omega)]$, obtained from the QNM expansion using $\tilde{f}_\mu(x)$ and $\tilde{f}_\mu(x_s)$ from Fig. 3.3 as input quantities, as well as the regularized form to an exact analytical solution of the single-barrier problem. We choose $x > b$ (outside the barrier) and $x_s = x_0$ (inside the barrier), so that the regularized form of the QNM Green function is precisely given by Eq. (3.40), while the analytical solution is given by Eq. (3.41).

In Fig. 3.4, we show the results for the three different approaches as a function of x from the cavity to the (right) background region, where we choose $\omega/c = k_4$, aligned with the real part of the QNM eigenfrequency. While both the QNM

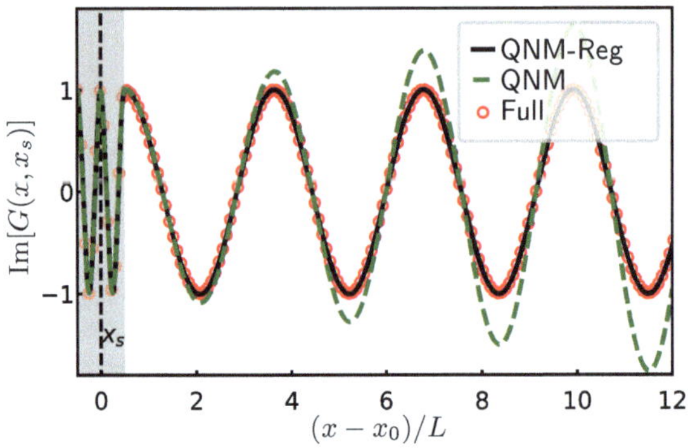

Fig. 3.4 Imaginary part of the Green function $G(x, x_s, \omega_4)$ versus x, obtained from the QNM expansion [using Eqs. (3.34)–(3.35)] and the regularized QNM approach (black solid) with a single QNM $\mu = 4$ and the exact analytical solution (red dots), for a single-barrier cavity (same parameters as Fig. 3.3). The gray dashed part marks the cavity region, containing the second spatial entry located at the center of the cavity ($x_s = x_0$). Note that propagators are weighted by $1/(\mathrm{Im}[G_{\mathrm{B}}]) = 2n_{\mathrm{B}}k$. In contrast to the (complex) QNM results, we see an excellent agreement between the exact solution [Eq. (3.41)] and the regularized QNM solution [Eq. (3.40)] at all spatial locations outside the cavity

and regularized QNM approach are in very good agreement with the exact solution inside the cavity region (gray area), from Eq. (3.41), the (complex) QNM solution [using Eq. (3.35)] fails to reproduce the behavior of the full solution because of the divergent character of $\tilde{f}_\mu(x)$. In contrast, *the regularized QNM approach, Eq.* (3.40), *is on top of the exact solution even for spatial positions far outside the cavity region*, which underlines the usefulness and accuracy of the QNM regularization technique.

3.3.2 Gold Dimer Quasinormal Modes for Localized Plasmons

In this subsection, we consider a 3D gold dimer (either with cylindrical nanorods as shown in Fig. 3.5a or with ellipsoids as shown in Fig. 3.6a) placed in a homogeneous medium with refractive index $n_{\mathrm{B}} = 1.0$ (free space). The dielectric constant of the gold-like material is assumed to be described by the Drude model:

$$\epsilon_{\mathrm{Drude}}(\omega) = 1 - \frac{\omega_{\mathrm{pl}}^2}{\omega^2 + i\omega\gamma_{\mathrm{pl}}}, \tag{3.121}$$

with $\hbar\omega_{\mathrm{pl}} = 8.2934$ eV ($\omega_{\mathrm{pl}} = 1.26 \times 10^{16}$ rad/s) and $\hbar\gamma_{\mathrm{pl}} = 0.0928$ eV ($\gamma_{\mathrm{pl}} = 1.41 \times 10^{14}$ rad/s).

For the dimer with cylindrical nanorods [23, 54], the radius and the length of the nanorod is $r_{\mathrm{cyli}} = 10$ nm and $L_{\mathrm{cyli}} = 80$ nm. The gap between the nanorods

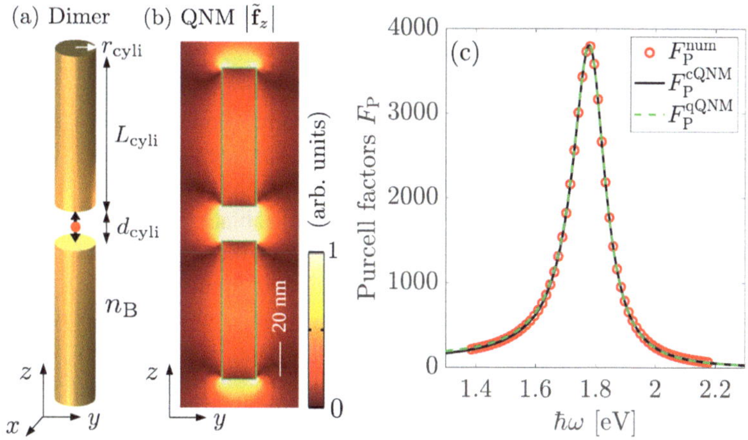

Fig. 3.5 (**a**) Schematic diagram of a cylindrical gold dimer with radius of $r_{\rm cyli} = 10\,\text{nm}$, length of $L_{\rm cyli} = 80\,\text{nm}$, and gap of $d_{\rm cyli} = 20\,\text{nm}$ placed at homogeneous medium with refractive index $n_{\rm B} = 1.0$. (**b**) Distribution of the dominant QNM $|\tilde{\mathbf{f}}_z|$ in the frequency regime of interest. (**c**) Purcell factors of a dipole placed at the gap center of a gold cylindrical dimer. Classical ($F_{\rm P}^{\rm cQNM}$, Eq. (3.49)) and quantum ($F_{\rm P}^{\rm qQNM}$, Eq. (3.99)) results agree with each other very well, and also agree well with the full-dipole solution ($F_{\rm P}^{\rm num}$, Eq. (3.50)). The quantum S factors and classical numerical beta factors are $S_{\rm p}^{\rm nrad} = 0.5830$ (Eq. (3.82)), $S_{\rm p}^{\rm rad} = 0.4197$ (Eq. (3.86)), $\beta_{\rm num}^{\rm nrad}(\omega_{\rm c}) = 0.5855$ (Eq. (3.65)), and $\beta_{\rm num}^{\rm rad}(\omega_{\rm c}) = 0.4057$ (Eq. (3.64))

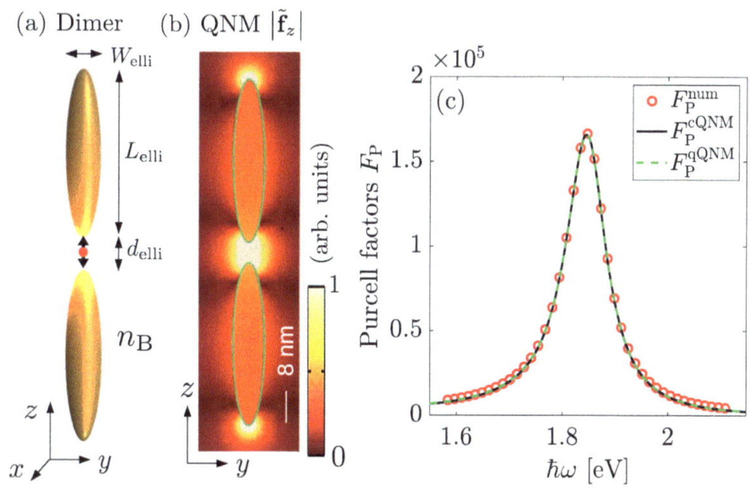

Fig. 3.6 (**a**) Schematic diagram of an ellipsoid gold dimer with center width of $W_{\rm elli} = 8\,\text{nm}$, length of $L_{\rm elli} = 40\,\text{nm}$, and gap of $d_{\rm elli} = 5\,\text{nm}$ placed at homogeneous medium with refractive index $n_{\rm B} = 1.0$. (**b**) Distribution of a dominant QNMs $|\tilde{\mathbf{f}}_z|$ in the frequency regime of interest. (**c**) Purcell factors of a dipole placed at the gap center of the dimer. Both classical ($F_{\rm P}^{\rm cQNM}$, Eq. (3.49)) and quantum ($F_{\rm P}^{\rm qQNM}$, Eq. (3.99)) results agree well with the full-dipole solution ($F_{\rm P}^{\rm num}$, Eq. (3.50)). The quantum S factors and classical numerical beta factors are now dominated by nonradiative decay: $S_{\rm p}^{\rm nrad} = 0.9544$ (Eq. (3.82)), $S_{\rm p}^{\rm rad} = 0.0443$ (Eq. (3.86)), $\beta_{\rm num}^{\rm nrad}(\omega_{\rm c}) = 0.9568$ (Eq. (3.65)), and $\beta_{\rm num}^{\rm rad}(\omega_{\rm c}) = 0.0441$ (Eq. (3.64))

is $d_{\text{cyli}} = 20\,\text{nm}$. We find a single QNM dominating in the frequency regime of interest, with a complex QNM eigenfrequency of $\hbar\tilde{\omega}_{\text{cyli}} = (1.780 - 0.068i)\,\text{eV}$, yielding a quality factor of $Q_{\text{cyli}} = 13$. The corresponding mode profile (the dominant z-component) is shown in Fig. 3.5b. One can see that the mode is mostly concentrated in the gap region. With a $z-$polarized dipole at the gap center, the corresponding classical Purcell factors $F_{\text{P}}^{\text{cQNM}}$ (solid black curve, Eq. (3.49)) are shown in Fig. 3.5c, which agree extremely well (indeed they are indistinguishable in values) with the full-dipole numerical solutions $F_{\text{P}}^{\text{num}}$ (red circles, Eq. (3.50)), which verifies the accuracy of the QNM and the validity of the single mode approximation.

Next, we also compute the quantum Purcell factor which we stress uses an entirely different approach to the classical results (involving nonlocal fields). We obtain $S_{\text{p}}^{\text{nrad}} = 0.5830$ (Eq. (3.82)), and $S_{\text{p}}^{\text{rad}} = 0.4197$ (Eq. (3.86)), which are very close to the on-resonance classical beta factors $\beta_{\text{num}}^{\text{nrad}}(\omega_{\text{c}}) = 0.5855$ (Eq. (3.65)), and $\beta_{\text{num}}^{\text{rad}}(\omega_{\text{c}}) = 0.4057$ (Eq. (3.64)) from the full numerical dipole solution. Thus, we obtained the total $S = S_{\text{p}}^{\text{nrad}} + S_{\text{p}}^{\text{rad}} = 1.0027 \approx 1$. The total quantum Purcell factors $F_{\text{P}}^{\text{qQNM}}$ (dashed green curve, Eq. (3.99)) are shown in Fig. 3.5c, which again agree extremely well with the full-dipole solutions (red circles). Indeed, this agreement is rather remarkable given the quite different numerical calculations involved, which is a testament to the power of the entire classical and quantum QNM approaches that we present; we stress again that neither solution uses any fitting parameters.

To demonstrate the robustness of the theory for different designs, let us also consider a different dimer design with ellipsoid nanorods [65] shown in Fig. 3.6a, where the length and the width of the nanorod are $L_{\text{elli}} = 40\,\text{nm}$ and $W_{\text{elli}} = 8\,\text{nm}$. The gap is $d_{\text{elli}} = 5\,\text{nm}$. A single QNM is found at $\hbar\tilde{\omega}_{\text{elli}} = (1.847 - 0.047i)\,\text{eV}$ with quality factor of $Q_{\text{elli}} = 19.5$. The mode profile is shown in Fig. 3.6b. The classical Purcell factors $F_{\text{P}}^{\text{cQNM}}$ (solid black curve, Eq. (3.49)) of a $z-$polarized dipole at the gap center are agreeing well with the full-dipole solutions $F_{\text{P}}^{\text{num}}$ (red circles, Eq. (3.50)), as shown in Fig. 3.6c.

In addition, for the quantum QNM results, we obtained $S_{\text{p}}^{\text{nrad}} = 0.9544$ (Eq. (3.82)), and $S_{\text{p}}^{\text{rad}} = 0.0443$ (Eq. (3.86)), which are very close to the on-resonance classical beta factors $\beta_{\text{num}}^{\text{nrad}}(\omega_{\text{c}}) = 0.9568$ (Eq. (3.65)), and $\beta_{\text{num}}^{\text{rad}}(\omega_{\text{c}}) = 0.0441$ (Eq. (3.64)) from the full-dipole solution. In addition, we determined that the total $S = S_{\text{p}}^{\text{nrad}} + S_{\text{p}}^{\text{rad}} = 0.9987 \approx 1$, as expected for an isolated single QNM. From these quantum S factor numbers, the nonradiative part accounts for the vast majority of the total contribution (i.e., most of the energy are dissipated within the metal). The final quantum Purcell factors $F_{\text{P}}^{\text{qQNM}}$ obtained through Eq. (3.99) are shown in Fig. 3.6c (dashed green curve), agreeing very well with the full numerical dipole solutions (red circles).

It should be noted that a far-field limit, combined with efficient near-field to far-field transformations [54] (Eq. (3.86)), is employed when calculating the radiative S parameters. For both dimer examples shown above, the near-field surface is selected as a cuboid enclosing the dimer and is 50 nm away from the dimer. The grid size is set as 0.5 nm on this near-field surface. The angle grid in Eq. (3.87) is set as $\pi/20$

($\vartheta = 0 : \frac{\pi}{20} : \pi$, and $\varphi = 0 : \frac{\pi}{20} : 2\pi$), which is proved to be small enough to get numerically convergent values. As for the volume integration related with nonradiative S parameters (Eq. (3.82)), this is performed within COMSOL directly. Further details of this approach are discussed in Ref. [54].

3.3.3 Hybrid Metal-Dielectric Quasinormal Modes for Metal Dimers on Photonic Crystal Cavity beams

Next, we consider a complex hybrid structure, where a gold ellipsoid dimer is placed on the top of a photonic crystal (PC) cavity [54, 65, 68], as shown in Fig. 3.7. Similar coupled dielectric-antenna designs have been studied in Ref. [78], using a high-quality factor whispering gallery mode, which show highly unusual lineshapes in the combined spectral response.

The width and the height of the PC beam are $W_{\rm PC} = 376$ nm and $h_{\rm PC} = 200$ nm. The finite length (y-direction) of the PC beam is $L_{\rm PC} = 8500$ nm. From the center of the beam, there are 14 air holes (a_1 to a_{14}) distributed in the positive y-direction, while in the negative y-direction, another 14 air holes are distributed symmetrically to the center of the beam. The distance between the two center air holes (two a_1) is

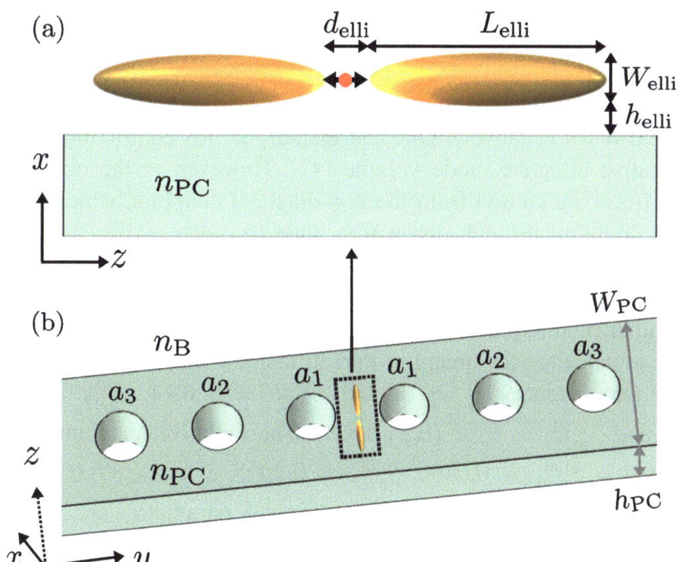

Fig. 3.7 Schematic diagram of the hybrid metal-dielectric cavity structure consisting of a gold ellipsoid dimer and a photonic crystal beam cavity. The minimum distance between the dimer surface and the PC is $h_{\rm elli} = 5$ nm. The gap of the dimer is $d_{\rm elli} = 5$ nm. The width and the length of the ellipsoid nanorod are $W_{\rm elli} = 8$ nm and $L_{\rm elli} = 40$ nm

126 nm. The radius of the air holes for a_1 to a_7 is increased linearly from 68 nm to 86 nm. The center-to-center distance for air holes (a_1 to a_2,..., a_6 to a_7) is increased linearly from 264 nm to 299 nm. Starting from a_8 to a_{14}, the radius and the center-to-center distance are fixed at 86 nm and 306 nm.

For the gold dimer, we use the same gold ellipsoid dimer shown in the last section, second example, i.e., with parameters $W_{\rm elli} = 8$ nm, $L_{\rm elli} = 40$ nm, and $d_{\rm elli} = 5$ nm. In addition, the dimer is placed on the top of the center of the PC beam. The smallest surface-to-surface distance between them is $h_{\rm elli} = 5$ nm.

The refractive index of the PC beam is $n_{\rm PC} = 2.04$, and the gold dimer is described by the Drude model. The are two dominant QNMs found in the frequency regime of interest. The first one is with eigenfrequency of $\hbar\tilde{\omega}_1 = \hbar\tilde{\omega}_{\rm plas-like} = (1.7945 - 0.0482i)$ eV and quality factor of $Q_1 = 18.6$, which is called plasmon-like mode since it shows similar properties as the plasmon mode of the isolated gold dimer, including the eigenfrequency, the low-quality factor, and the strong localized field in the dimer region (see Fig. 3.8a–c, z-component, $|\tilde{\mathbf{f}}_{1z}|$). The second one is called as PC-like mode, with eigenfrequency $\hbar\tilde{\omega}_1 = \hbar\tilde{\omega}_{\rm PC-like} = (1.6062 - 0.0001i)$ eV of and quality factor of $Q_2 = 6123$. As shown in Fig. 3.8d–f (z-component, $|\tilde{\mathbf{f}}_{2z}|$), the PC-like QNM has a significant field distribution in the PC beam, mainly in the dimer region.

Considering a z-polarized dipole at the gap center of the gold dimer, the corresponding classical Purcell factors ($F_{\rm P}^{\rm cQNM}$, Eq. (3.47), dashed black curve) are shown in Fig. 3.9a, b, which show a Fano-like lineshape and agree well with the full numerical dipole solutions ($F_{\rm P}^{\rm num}$, Eq. (3.50), red circles). In addition, we also give the separate contribution from the two dominant QNMs (see solid green (blue) curve for plasmon-like (PC-like) QNM contribution. The specific lineshape of the separate contributions is well explained by the QNM phases, where one of them contributes negatively. One can then argue this contribution as stemming from a negative effective mode volume [47]. However, in the quantum picture, the phase effects are caused from the non-diagonal coupling, which is a quantum mechanical coupling through dissipation. One can also easily change the QNM basis of the classical solution, to avoid having to argue from the viewpoint of negative contributions, while in the quantum picture, the mode basis is forced through fundamental requirements of the bosonic field commutation relations.

In order to calculate the quantum Purcell factors, we obtained the nonradiative quantum S parameters $S_{\rm p,11}^{\rm nrad} = 0.9230$, $S_{\rm p,22}^{\rm nrad} = 0.8074$, $S_{\rm p,12}^{\rm nrad} = -0.0004 - -0.0223i$, and $S_{\rm p,21}^{\rm nrad} = S_{\rm p,12}^{\rm nrad*}$(Eq. (3.82)). The radiative quantum S parameters are calculated as $S_{\rm p,11}^{\rm rad} = 0.0406$, $S_{\rm p,22}^{\rm rad} = 0.2055$, $S_{\rm p,12}^{\rm rad} = -0.0006 - 0.0001i$, and $S_{\rm p,21}^{\rm rad} = S_{\rm p,12}^{\rm rad*}$(Eq. (3.86)), so that the total S parameters are $S_{11} = 0.9636$, $S_{22} = 1.0129$, $S_{12} = -0.0010 - 0.0224i$, and $S_{21} = S_{12}^*$. As shown in Fig. 3.9c, d, the final quantum Purcell factors ($F_{\rm P}^{\rm qQNM}$, Eq. (3.97), solid black curve, related to above S parameters) show very good agreement with full numerical dipole solutions (red circles, $F_{\rm P}^{\rm num}$, Eq. (3.50)).

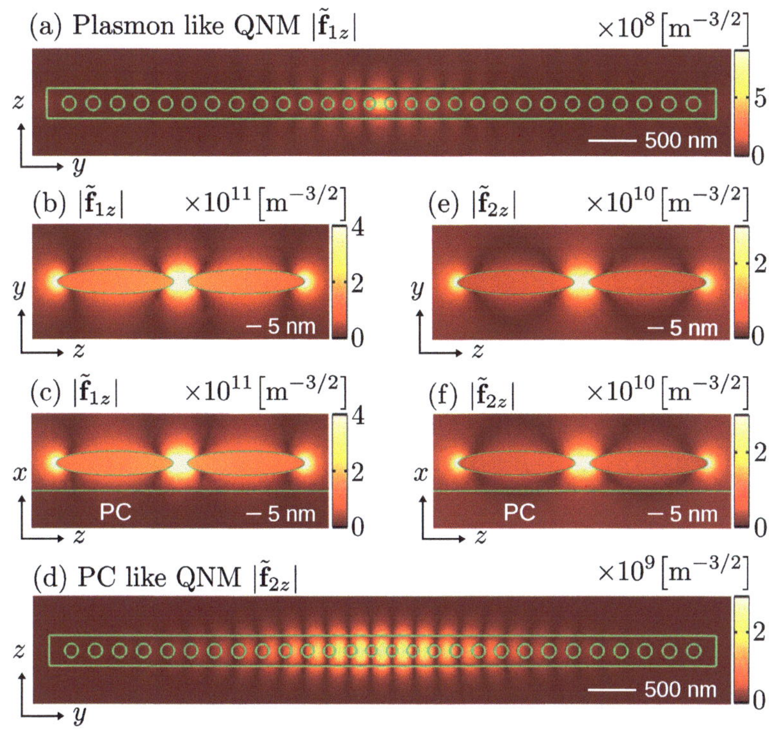

Fig. 3.8 Spatial profile of the QNM for the hybrid structure formed in Fig. 3.7. The are two dominant QNMs in the frequency regime of interest: (**a–c**) Plasmon-like QNMs $|\tilde{\mathbf{f}}_{1z}|$ (absolute value of the dominant $z-$ component), and (**d–f**) PC-like QNMs $|\tilde{\mathbf{f}}_{2z}|$ at different spatial region. Note that for better display, different color bars are used. For the plasmon-like mode with low-quality factor ($Q_1 = 18.6$), the mode mainly lives near the dimer. For the PC-like mode with a high-quality factor ($Q_2 = 6123$), the mode lives on both PC and mostly the dimer. All calculations are fully 3D

Note again that when calculating the radiative S parameters, we are using far-field limits combined with near-field to far-field transformations (Eq. (3.86)), which significantly reduces the computational run time of the numerical calculations [54]. The near-field surface for the plasmon-like mode is selected as a cuboid (six surfaces in total) enclosing the dimer only, which is 4 nm away from the dimer surface in the negative-x direction and is 50 nm away from the dimer surface from the other five directions (including $+x$, $\pm y$, and $\pm z$ direction). This is reasonable as we can see that the mode profile of the plasmon-like mode is highly concentrated on the dimer region. As for the PC-like mode, we choose a cuboid, enclosing both the dimer and the PC beam, which is 50 nm away from each direction. The grid size on the near-field surface is 0.5 nm. The angle grid used in Eq. (3.87) is again selected as $\pi/20$. For the volume integration related with nonradiative S parameters, we use Eq. (3.82), as we did for the metal dimer examples.

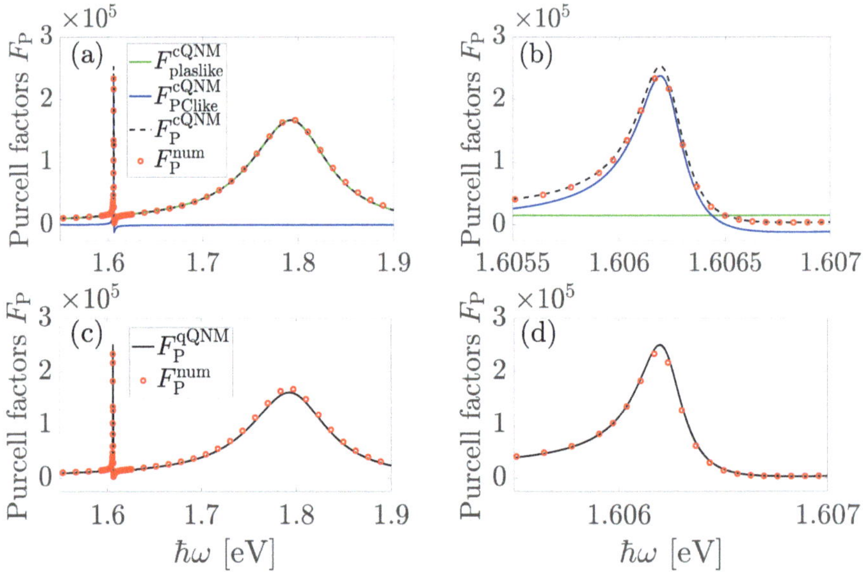

Fig. 3.9 Classical Purcell factors (F_P^{cQNM}, Eq. (3.47), dashed black curve) and quantum Purcell factors (F_P^{qQNM}, Eq. (3.97), solid black curve) at the dimer gap center. Both of them show very good agreement with the full numerical dipole solutions (F_P^{num}, Eq. (3.50), red circles)

3.3.4 Coupled-Cavity Quasinormal Modes

In this final subsection, we consider two whispering gallery mode (WGM) resonators, using two microdisks [75], as shown in Fig. 3.10a. For these simulations, we use a 2D model as the disks are relatively large (micron scale). The diameter of both microdisks is $D = 10\,\mu\text{m}$. The gap distance d_{gap} is varied from 600 nm to 1200 nm. The oscillating dipole (polarized out of the (xy) plane, i.e., a line current) can be located in the gap, which is d_L (d_R) away from the left (right) microdisk. Below, we will consider either $d_L = 10\,\text{nm}$ or $d_R = 10\,\text{nm}$. Both disks are lossy with a refractive index $n_L = 2 + 10^{-5}i$ and $n_R = 2 + 10^{-4}i$. The background medium is free space, i.e., $n_B = 1.0$.

To understand the coupled system, first we must understand the bare QNMs properties for the individual microdisks, which will be used as input for the coupled QNM theory. It is well known that microdisk resonators support WGMs. Here, in 2D case (xy plane), we focus on the TM mode $(\tilde{\mathbf{h}}_x, \tilde{\mathbf{h}}_y, \tilde{\mathbf{f}}_z)$, where the electric field QNMs only have a z-component and the magnetic-field QNMs $\tilde{\mathbf{h}}$ are in the xy-plane. The indices to describe the WGM modes are radial mode number q and azimuthal mode number m, which are describing the number of mode intensity nodes along the radial direction within the resonator, and the number of wavelengths in the azimuthal direction along the equator, separately. Also note, with the same mode number q, m, two degenerate WGMs will propagate along opposite directions [36, 79–81]; they

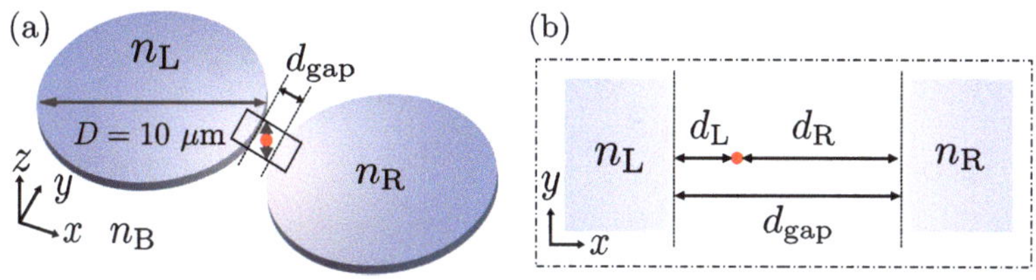

Fig. 3.10 (**a**) 3D schematic diagram of coupled loss-loss microdisks with gap distance d_{gap}. In the simulations, we use a 2D model. Both disks have the same disk diameter, $D = 10\,\mu m$. The refractive index of the left (right) microdisk is $n_L = 2 + 10^{-5}i$ ($n_R = 2 + 10^{-4}i$). (**b**) Zoom in of the gap region. The z−polarized dipole (red dot) is modeled within the gap, which is d_L (d_R) away from the left (right) microdisk ($d_L + d_R = d_{gap}$)

are either counterclockwise (CCW) $E_{ccw}(\mathbf{r}, \phi) = E(\mathbf{r})e^{(im\phi)}$, or clockwise (CW) $E_{cw}(\mathbf{r}, \phi) = E(\mathbf{r})e^{(-im\phi)}$ (the increase of angle ϕ is along CCW direction). One can obtain two degenerate standing modes by linearly superposing CCW and CW modes [79–81]. Note here we only excite one of the standing modes by using a TM dipole (z-polarized) located along the x-axis; see also Ref. [52].

To be specific, a TM mode with $q = 1$ and $m = 37$ is selected because it has a relatively high-quality factor (see below) and its resonance (at the telecommunication band) is far away from other modes, so it is safe to make a single mode approximation.

Using the inverse Green function approach described earlier, the QNM eigenfrequency for the left microdick only is obtained at $\hbar\tilde{\omega}_L = \hbar\omega_L - i\hbar\gamma_L = (0.8337 - 0.00000412i)$ eV, which yields a quality factor $Q_L \sim 10^5$. While for the right microdisk only (more lossy than the left one), then the QNM eigenfrequency is $\hbar\tilde{\omega}_R = \hbar\omega_R - i\hbar\gamma_R \approx \hbar\omega_L - i10\hbar\gamma_L$, which yields a quality factor $Q_R \sim 10^4$. The corresponding QNM profiles ($\text{Re}[\tilde{\mathbf{f}}_z^L]$ and $\text{Re}[\tilde{\mathbf{f}}_z^R]$) are shown in Fig. 3.12a, b, without any coupling (i.e., the QNMs are located mainly on their own resonators).

Using the above two bare (uncoupled) QNMs as input, next we apply the coupled QNM theory (CQT) to get the hybridized QNMs for the loss-loss resonator system. Firstly, for any specific gap distance between two microdisks, the coupling coefficients $\tilde{\kappa}_{LR/RL}$ are calculated through Eq. (3.105). We highlight again that, in general, the couplings are nonreciprocal, i.e., $\tilde{\kappa}_{RL} \neq \tilde{\kappa}_{LR}^*$. Then the analytical eigenfrequencies $\tilde{\omega}_\pm$ for the coupled system are computed via Eq. (3.108), as shown in Fig. 3.11, which agree extremely well with the numerical simulations (black circles, approximate eigenfrequency solver in COMSOL, which is accurate for resonator modes with very high-quality factors). The coupled QNMs $\tilde{\mathbf{f}}^\pm$ are then given through Eq. (3.111). As an example, we show the coupled QNM profile with $d_{gap} = 800$ nm in Fig. 3.12c, d, where now the coupled modes are located over both resonators.

Considering a z-polarized dipole at $d_L = 10$ nm (close to the left resonator), the Purcell factors with various gap distance are given in Fig. 3.13a, c. Firstly, with the analytic CQT, Eq. (3.113) give the Purcell factors F_P^{cQNM} from classical

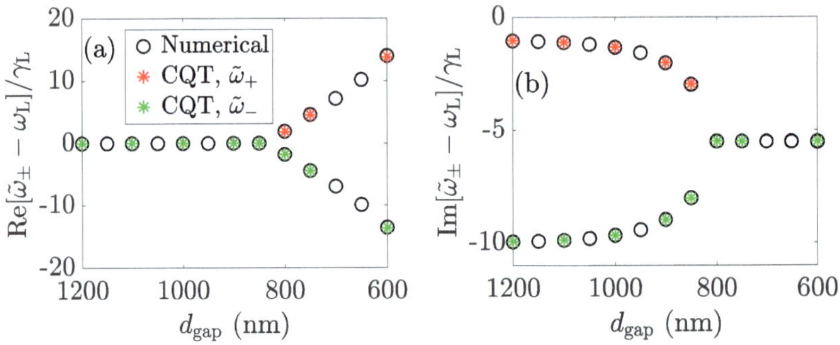

Fig. 3.11 QNM eigenfrequencies for the coupled loss-loss resonators from the analytic coupled QNM theory (CQT, $\tilde{\omega}_\pm$, Eq. (3.108)) and direct numerical solutions (using an approximate eigenfrequency solver in COMSOL). Excellent agreement is obtained with no fitting parameters. Adapted from [75] with permission from ACS Photonics

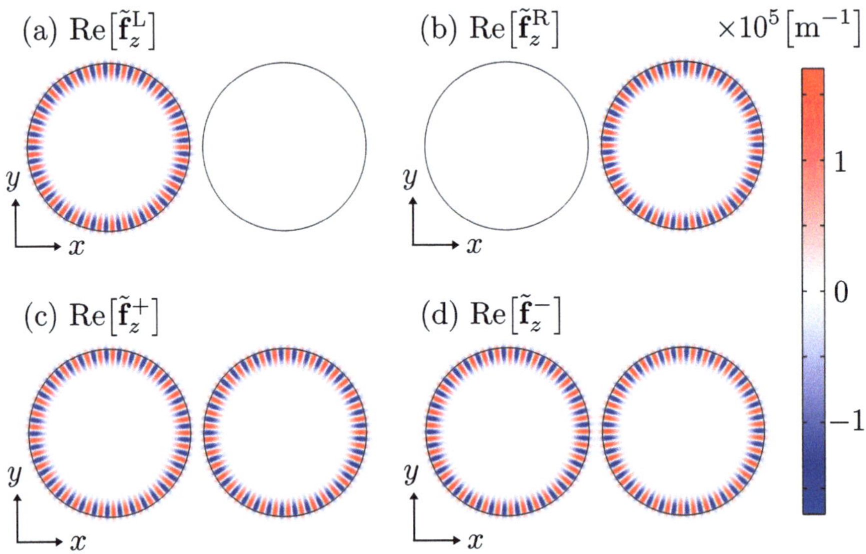

Fig. 3.12 QNM profiles for: (**a**, **b**) individual QNM (Re$[\tilde{\mathbf{f}}_z^\mathrm{L}]$ for the left microdisk alone, Re$[\tilde{\mathbf{f}}_z^\mathrm{R}]$ for right microdisk alone), and (**c**, **d**) coupled QNMs for the combined loss-loss resonator system (hybrid QNMs (Re$[\tilde{\mathbf{f}}_z^\pm]$, using the CQT result, Eq. (3.111)). The gap between the microdisks is $d = 800\,\mathrm{nm}$. Note the QNM profile difference near the gap between the '+' and '−' hybrid modes

QNMs (solid black curve, total contribution from two coupled QNMs $\tilde{\mathbf{f}}^\pm$), which agrees very well with full numerical dipole solutions, $F_\mathrm{P}^\mathrm{num}$ (red circles, Eq. (3.50)). However, when using coupled normal-mode theory, the Purcell factors $F_\mathrm{P}^\mathrm{cNM}$ (solid magenta curve, Eq. (3.117)) are showing a clear departure from the full numerical dipole solutions, because the NM approach does not account for the significant role of the complex mode phase, which is the QNM phase.

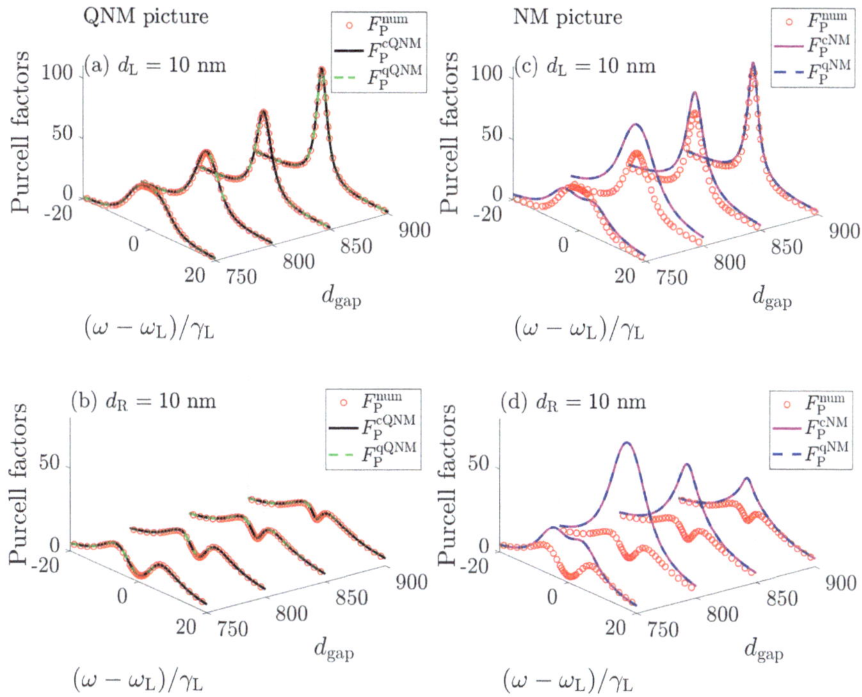

Fig. 3.13 Purcell factors from classical QNMs (F_P^{cQNM}, Eq. (3.113), solid black curve), quantized QNMs (F_P^{qQNM}, Eq. (3.97), dashed green curve), classical NM (F_P^{cNM}, Eq. (3.117), solid magenta curve), and dissipative Jaynes-Cummings model (F_P^{qNM}, Eq. (3.120), dashed blue curve) for various gap sizes, compared with full-dipole solutions (F_P^{num}, Eq. (3.50), red circles) when dipole placed at (a, c) 10 nm from the left resonator and (b,d) 10 nm from the right resonator. Adapted from [75] with permission from ACS Photonics

To obtain the *quantum Purcell factors*, using the rigorous quantized QNM theory (in the bad cavity limit), we first calculate the quantum S parameters for the coupled systems with the help of CQT. Note that the nonradiative part of S easily dominates, so we can make the approximation $S \sim S^{nrad}$, and these values are shown in Table 3.1 for different gap distances. Then the quantum Purcell factors (F_P^{qQNM}, Eq. (3.97), dashed green curve) from the quantized QNMs are shown in Fig. 3.13a, which are seen to be in excellent agreement with full numerical dipole solutions and the classical QNMs results.

Notably, in the quantized QNM theory, if one neglects the non-diagonal terms ($S_{+-} = S_{-+} = 0$), and assumes $S_{++} = S_{--} = 1$, then Eq. (3.120) gives the Purcell factors (F_P^{qNM}, dashed blue) from the general dissipative Jaynes-Cummings model, as shown in Fig. 3.13c. One will find that it differs greatly from the full-

Table 3.1 Quantum S parameters $S \sim S^{\text{nrad}}$ (Eq. (3.82)) for the coupled loss-loss resonators. Adapted from [75] with permission from ACS Photonics

Gap distance	S_{++}	S_{+-}	S_{-+}	S_{--}
$d_{\text{gap}} = 750\,\text{nm}$	1.4137	$0.0003 + 0.9999i$	$0.0003 - 0.9999i$	1.4145
$d_{\text{gap}} = 800\,\text{nm}$	2.6439	$0.0015 + 2.4483i$	$0.0015 - 2.4483i$	2.6453
$d_{\text{gap}} = 850\,\text{nm}$	1.7768	$-0.00003 - 1.4696i$	$-0.00003 + 1.4696i$	1.7782
$d_{\text{gap}} = 900\,\text{nm}$	1.2929	$0.0001 + 0.8200i$	$0.0001 - 0.8200i$	1.2935

dipole solutions, but surprisingly it matches perfectly with the above classical NM model ($F_{\text{P}}^{\text{cNM}}$, Eq. (3.117)). Their equivalence is fully discussed in Ref. [75], using the various photon Green function contributions.

When the dipole is placed at $d_{\text{R}} = 10$ nm (i.e., close to the right microdisk), the corresponding Purcell factors are shown in Fig. 3.13b, d. Once again, we find that the Purcell factors from both classical ($F_{\text{P}}^{\text{cQNM}}$, Eq. (3.113), solid black curve) and quantum QNMs ($F_{\text{P}}^{\text{qQNM}}$, Eq. (3.97), dashed green curve) show excellent agreement with each other and with the full numerical dipole solutions $F_{\text{P}}^{\text{num}}$ (red circles, Eq. (3.50)), which verifies the extremely high accuracy of the classical/quantized QNMs and the validity of the QNM coupled-mode theory. In contrast, in the normal mode picture, both classical ($F_{\text{P}}^{\text{cNM}}$, Eq. (3.117), solid magenta curve) and quantum ($F_{\text{P}}^{\text{qNM}}$, Eq. (3.120), dashed blue curve) results show a clear departure from the full-dipole solution as before.

Although the NM theory can fail significantly here, we highlight that QNM theory of coupled modes can be used to develop more accurate NM theories, both at the classical and quantum levels [75]. In such approaches, the QNM complex eigenfrequencies are used instead of the real NM eigenfrequencies, to the commonly used lossless coupled-mode theory, and for the quantum approach this is complimented by a Jaynes-Cummings model with an intercavity coupling in the bare resonator regime. In this way, improved NM model can recover the interference effects in the Purcell factor calculations, which can be a good approximation for high Q resonators. Related work in this regard, in Refs. [82, 83], quantizes the electric field using lossless NMs (i.e., in the standard way), for coupled cavities, and then projects onto the real lossy cavity modes by a non-Hermitian projection operator.

3.4 Conclusions and Future Prospects

We have presented a selection of QNM theories and applications, showing how different nanophotonic cavity structures can be quantitatively described in terms of the underlying QNMs. We have also discussed how QNM theories differ from the usual NM theories, and we have shown the impact of the QNM phase for forming hybrid dielectric-plasmon modes, exhibiting Fano-like resonances, and for coupled high-Q microdisk resonators, showing a dramatic breakdown of NM theory. We also showed how QNM theory revises Purcell's formula by using a generalized and well-defined effective mode volume. Moreover, we discussed how to quantize QNMs and developed a powerful QNM master equation which can be used to explore *genuine quantum optical* effects in nanophotonics [65, 66]. As a simple example, we showed how the bad cavity limits fully recover all the classical QNM regimes and Purcell factor results presented in this chapter, even for highly unusual lineshapes. The quantum description gives a fundamental picture of how QNMs quantum mechanically couple through dissipation and avoids many of the conceptual problem with classical mode theories including negative effective mode volumes.

Applications of QNMs expand into many other areas of optical physics. For example, QNMs can also be used to explore gain cavity modes and coupled gain-loss systems near exceptional points [52], already extending current theories on the standard LDOS Purcell factor (i.e., Eq. (3.46) fails), which is no longer applicable [84]. Quantized gain QNMs can also be used to explore new regimes in quantum optics [85], and can study nontrivial and unexpected phenomena often with unusual interpretations, including chiral power flow from linear dipoles near index-modulated ring resonators [74], where once again NM theory does not work, but QNMs yield a quantitatively good explanation of the light-matter interactions [86]. The concept of hybrid QNMs has also been used to model metal resonators coupled to 2D semiconductors, e.g., QNMs offer a first-principles explanation of strong coupling between metal resonators and direct gap excitons in transition metal dichalcogenide (TMDC) monolayers [87, 88], without recourse to guessing an ill-defined single coupling parameter for heuristic coupled NMs, which yields little insight to the underlying mechanism of strong coupling in such systems. In contrast, a QNM theory explains strong coupling from the concept of hybridized modes and explains how such features evolve, e.g., as a function of gap size and temperatures. It is clear that QNMs have profound uses in optical cavity physics, and we envision widespread use and continued developments, in both classical and quantum optical domains. Indeed, quantized QNMs offer a possible solution to quantitatively model system-bath interactions in the ultrastrong coupling regime [89, 90].

Finally, we highlighted that QNMs also apply to other modes of relevance in nanophotonics, not just optical modes, including mechanical QNMs of optomechanical beams, which can be used to rigorously define an elastic Purcell factor, e.g., with force displacements [51, 57].

Acknowledgments We acknowledge Queen's University and the Natural Sciences and Engineering Research Council of Canada for the financial support and CMC Microsystems for the provision of COMSOL Multiphysics to facilitate this research. We also acknowledge support from the Alexander von Humboldt Foundation through a Humboldt Research Award. We thank Reuven Gordon, Philip Kristensen, Chris Gustin, Chelsea Carlson, Marten Richter, and Andreas Knorr for their useful comments and collaborations related to some of the work presented in this chapter.

References

1. A. Yariv, Coupled-mode theory for guided-wave optics. IEEE J. Quant. Electron. **9**(9), 919–933 (1973)
2. H.A. Haus, W. Huang, Coupled-mode theory. Proc. IEEE **79**(10), 1505–1518 (1991)
3. W.-P. Huang, Coupled-mode theory for optical waveguides: an overview. J. Opt. Soc. Am. A **11**(3), 963 (1994)
4. A.W. Snyder, Coupled-mode theory for optical fibers. J. Opt. Soc. Am. **62**(11), 1267 (1972)
5. J. Love. A.W. Snyder, *Optical Waveguide Theory*, 2nd ed. (Chapman and Hall, Boca Raton, 1983)
6. S. Fan, W. Suh, J.D. Joannopoulos, Temporal coupled-mode theory for the Fano resonance in optical resonators. J. Opt. Soc. Am. A **20**(3), 569 (2003)
7. A.F. Koenderink, On the use of Purcell factors for plasmon antennas. Opt. Lett. **35**(24), 4208–4210 (2010)
8. S. Franke, J. Ren, S. Hughes, Impact of mode regularization for quasinormal mode perturbation theories, e-print: arXiv:2204.03982 (2022)
9. J.E. Sipe, New Green-function formalism for surface optics. J. Opt. Soc. Am. B **4**(4), 481 (1987)
10. K. Joulain, J.-P. Mulet, F. Marquier, R. Carminati, J.-J. Greffet, Surface electromagnetic waves thermally excited: radiative heat transfer, coherence properties and Casimir forces revisited in the near field. Surf. Sci. Rep. **57**(3–4), 59–112 (2005)
11. W.D. Newman, C.L. Cortes, Z. Jacob, Enhanced and directional single-photon emission in hyperbolic metamaterials. J. Opt. Soc. Am. B **30**(4), 766 (2013)
12. M. Wubs, L.G. Suttorp, A. Lagendijk, Multiple-scattering approach to interatomic interactions and superradiance in inhomogeneous dielectrics. Phys. Rev. A **70**(5), 053823 (2004)
13. H.T. Dung, L. Knöll, D.-G. Welsch, Three-dimensional quantization of the electromagnetic field in dispersive and absorbing inhomogeneous dielectrics. Phys. Rev. A **57**, 3931–3942 (1998)
14. S. Hughes, Coupled-cavity QED using planar photonic crystals. Phys. Rev. Lett. **98**, 083603 (2007)
15. G.S. Agarwal, Quantum electrodynamics in the presence of dielectrics and conductors. I. Electromagnetic-field response functions and black-body fluctuations in finite geometries. Phys. Rev. A **11**, 230–242 (1975)
16. K. Sakoda, *Optical Properties of Photonic Crystals* (Springer Science & Business Media, Berlin, 2005)
17. P. Anger, P. Bharadwaj, L. Novotny, Enhancement and quenching of single-molecule fluorescence. Phys. Rev. Lett. **96**, 113002 (2006)
18. K. Joulain, R. Carminati, J.-P. Mulet, J.-J. Greffet, Definition and measurement of the local density of electromagnetic states close to an interface. Phys. Rev. B **68**, 245405 (2003)
19. O.J.F. Martin, N.B. Piller, Electromagnetic scattering in polarizable backgrounds. Phys. Rev. E **58**, 3909–3915 (1998)
20. C.P. Van Vlack, Dyadic Green Functions and their Applications. PhD thesis, Queen's University (2012)
21. R.-C. Ge, C. Van Vlack, P. Yao, J.F. Young, S. Hughes, Accessing quantum nanoplasmonics in a hybrid quantum dot–metal nanosystem: Mollow triplet of a quantum dot near a metal nanoparticle. Phys. Rev. B **87**, 205425 (2013)
22. O.D. Miller, A.G. Polimeridis, M.T. Homer Reid, C.W. Hsu, B.G. DeLacy, J.D. Joannopoulos, M. Soljačić, S.G. Johnson, Fundamental limits to optical response in absorptive systems. Opt. Express **24**(4), 3329 (2016)
23. S. Franke, S. Hughes, M.K. Dezfouli, P.T. Kristensen, K. Busch, A. Knorr, M. Richter, Quantization of quasinormal modes for open cavities and plasmonic cavity quantum electrodynamics. Phys. Rev. Lett. **122**, 213901 (2019)

24. S. Franke, J. Ren, S. Hughes, M. Richter, Fluctuation-dissipation theorem and fundamental photon commutation relations in lossy nanostructures using quasinormal modes. Phys. Rev. Res. **2**, 033332 (2020)
25. L. Novotny, B. Hecht, *Principles of Nano-Optics* (Cambridge University Press, Cambridge, 2006)
26. L. Knöll, S. Scheel, D.-G. Welsch, QED in dispersing and absorbing media (2000), arXiv:quant-ph/0006121
27. C. Carlson, D. Dalacu, C. Gustin, S. Haffouz, X. Wu, J. Lapointe, R.L. Williams, P.J. Poole, S. Hughes, Theory and experiments of coherent photon coupling in semiconductor nanowire waveguides with quantum dot molecules. Phys. Rev. B **99**, 085311 (2019)
28. M. Patterson, S. Hughes, S. Combrié, N.-V.-Quynh Tran, A. De Rossi, R. Gabet, Y. Jaouën, Disorder-induced coherent scattering in slow-light photonic crystal waveguides. Phys. Rev. Lett. **102**, 253903 (2009)
29. V.S.C. Manga Rao, S. Hughes, Single quantum-dot Purcell factor and β factor in a photonic crystal waveguide. Phys. Rev. B **75**(20), 205437 (2007)
30. H.M.K. Wong, M.K. Dezfouli, L. Sun, S. Hughes, A.S. Helmy, Nanoscale plasmonic slot waveguides for enhanced Raman spectroscopy. Phys. Rev. B **98**, 085124 (2018)
31. R. Sammut, A.W. Snyder, Leaky modes on a dielectric waveguide: orthogonality and excitation. Appl. Opt. **15**(4), 1040 (1976)
32. P. Yao, C. Van Vlack, A. Reza, M. Patterson, M.M. Dignam, S. Hughes, Ultrahigh purcell factors and lamb shifts in slow-light metamaterial waveguides. Phys. Rev. B **80**, 195106 (2009)
33. A. Reza, M.M. Dignam, S. Hughes, Can light be stopped in realistic metamaterials? Nature **455**(7216), E10–E11 (2008)
34. P.A. Martin, *Multiple Scattering: Interaction of Time-Harmonic Waves with N Obstacles*. Number 107 (Cambridge University Press, Cambridge, 2006)
35. P.T. Kristensen, R.-C. Ge, S. Hughes, Normalization of quasinormal modes in leaky optical cavities and plasmonic resonators. Phys. Rev. A **92**(5), 053810 (2015)
36. P.T. Leung, S.Y. Liu, K. Young, Completeness and orthogonality of quasinormal modes in leaky optical cavities. Phys. Rev. A **49**(4), 3057–3067 (1994)
37. P.T. Kristensen, S. Hughes, Modes and mode volumes of leaky optical cavities and plasmonic nanoresonators. ACS Photonics **1**(1), 2–10 (2014)
38. E.A. Muljarov, W. Langbein, R. Zimmermann, Brillouin-Wigner perturbation theory in open electromagnetic systems. EPL (Europhys. Lett.) **92**(5), 50010 (2010)
39. C. Sauvan, J.P. Hugonin, I.S. Maksymov, P. Lalanne, Theory of the spontaneous optical emission of nanosize photonic and plasmon resonators. Phys. Rev. Lett. **110**, 237401 (2013)
40. W. Yan, R. Faggiani, P. Lalanne, Rigorous modal analysis of plasmonic nanoresonators. Phys. Rev. B **97**(20), 205422 (2018)
41. J.-P. Berenger, A perfectly matched layer for the absorption of electromagnetic waves. J. Comput. Phys. **114**(2), 185–200 (1994)
42. P.T. Kristensen, C. Van Vlack, S. Hughes, Generalized effective mode volume for leaky optical cavities. Opt. Lett. **37**(10), 1649–1651 (2012)
43. P.T. Kristensen, K. Herrmann, F. Intravaia, K. Busch, Modeling electromagnetic resonators using quasinormal modes. Adv. Opt. Photonics **12**(3), 612 (2020)
44. B. Stout, R. Colom, N. Bonod, R.C. McPhedran, Spectral expansions of open and dispersive optical systems: Gaussian regularization and convergence. New J. Phys. **23**(8), 083004 (2021)
45. K.M. Lee, P.T. Leung, K.M. Pang, Dyadic formulation of morphology-dependent resonances. I. Completeness relation. JOSA B **16**(9), 1409–1417 (1999)
46. B. Vial, F. Zolla, A. Nicolet, M. Commandré, Quasimodal expansion of electromagnetic fields in open two-dimensional structures. Phys. Rev. A **89**, 023829 (2014)
47. M.K. Dezfouli, R. Gordon, S. Hughes, Modal theory of modified spontaneous emission of a quantum emitter in a hybrid plasmonic photonic-crystal cavity system. Phys. Rev. A **95**, 013846 (2017)

48. J.R. de Lasson, P.T. Kristensen, J. Mørk, N. Gregersen, Semianalytical quasi-normal mode theory for the local density of states in coupled photonic crystal cavity–waveguide structures. Opt. Lett. **40**(24), 5790 (2015)
49. M. Perrin, Eigen-energy effects and non-orthogonality in the quasi-normal mode expansion of Maxwell equations. Opt. Express **24**(24), 27137–27151 (2016)
50. Q. Bai, M. Perrin, C. Sauvan, J.-P. Hugonin, P. Lalanne, Efficient and intuitive method for the analysis of light scattering by a resonant nanostructure. Opt. Express **21**(22), 27371–27382 (2013)
51. A.-W. El-Sayed, S. Hughes, Quasinormal-mode theory of elastic Purcell factors and Fano resonances of optomechanical beams. Phys. Rev. Res. **2**, 043290 (2020)
52. J. Ren, S. Franke, S. Hughes, Quasinormal modes, local density of states, and classical purcell factors for coupled loss-gain resonators. Phys. Rev. X **11**, 041020 (2021)
53. R.-C. Ge, P.T. Kristensen, J.F. Young, S. Hughes, Quasinormal mode approach to modelling light-emission and propagation in nanoplasmonics. New J. Phys. **16**(11), 113048 (2014)
54. J. Ren, S. Franke, A. Knorr, M. Richter, S. Hughes, Near-field to far-field transformations of optical quasinormal modes and efficient calculation of quantized quasinormal modes for open cavities and plasmonic resonators. Phys. Rev. B **101**(20), 205402 (2020)
55. M. Ismail Abdelrahman, B. Gralak, Completeness and divergence-free behavior of the quasi-normal modes using causality principle. OSA Continuum **1**(2), 340–348 (2018)
56. E.M. Purcell, Spontaneous emission probabilities at radio frequencies. Phys. Rev. **69**, 681 (1946)
57. M.K. Schmidt, L.G. Helt, C.G. Poulton, M.J. Steel, Elastic Purcell effect. Phys. Rev. Lett. **121**, 064301 (2018)
58. R.-C. Ge, J.F. Young, S. Hughes, Quasi-normal mode approach to the local-field problem in quantum optics. Optica **2**(3), 246 (2015)
59. K.G. Cognée, W. Yan, F. La China, D. Balestri, F. Intonti, M. Gurioli, A.F. Koenderink, P. Lalanne, Mapping complex mode volumes with cavity perturbation theory. Optica **6**(3), 269–273 (2019)
60. R.-C. Ge, S. Hughes, Quasinormal mode theory and modelling of electron energy loss spectroscopy for plasmonic nanostructures. J. Opt. **18**(5), 054002 (2016)
61. R.-C. Ge, S. Hughes, Design of an efficient single photon source from a metallic nanorod dimer: A quasi-normal mode finite-difference time-domain approach. Opt. Lett. **39**(14), 4235–4238 (2014)
62. M.J. Barth, R.R. McLeod, R.W. Ziolkowski, A near and far-field projection algorithm for finite-difference time-domain codes. J. Electromagn. Waves Appl. **6**(1–4), 5–18 (1992)
63. W. Vogel, D.-G. Welsch, *Quantum Optics* (John Wiley & Sons, Hoboken, 2006)
64. S. Scheel, L. Knöll, D.-G. Welsch, QED commutation relations for inhomogeneous kramers-kronig dielectrics. Phys. Rev. A **58**(1), 700 (1998)
65. S. Franke, M. Richter, J. Ren, A. Knorr, S. Hughes, Quantized quasinormal-mode description of nonlinear cavity-QED effects from coupled resonators with a Fano-like resonance. Phys. Rev. Res. **2**, 033456 (2020)
66. S. Hughes, S. Franke, C. Gustin, M.K. Dezfouli, A. Knorr, M. Richter, Theory and limits of on-demand single-photon sources using plasmonic resonators: a quantized quasinormal mode approach. ACS Photonics **6**(8), 2168–2180 (2019)
67. J.I. Cirac, Interaction of a two-level atom with a cavity mode in the bad-cavity limit. Phys. Rev. A **46**, 4354–4362 (1992)
68. P.T. Kristensen, J.R. de Lasson, M. Heuck, N. Gregersen, J. Mørk, On the theory of coupled modes in optical cavity-waveguide structures. J. Lightwave Technol. **35**(19), 4247–4259 (2017)
69. B. Vial, Y. Hao, A coupling model for quasi-normal modes of photonic resonators. J. Opt. **18**(11), 115004 (2016)
70. K.G. Cognée, Hybridization of open photonic resonators. PhD dissertation, University of Amsterdam (2020)
71. C. Tao, J. Zhu, Y. Zhong, H. Liu, Coupling theory of quasinormal modes for lossy and dispersive plasmonic nanoresonators. Phys. Rev. B **102**(4), 045430 (2020)

72. M.-A. Miri, A. Alù, Exceptional points in optics and photonics. Science **363**(6422), eaar7709 (2019)
73. A. Pick, B. Zhen, O.D. Miller, C.W. Hsu, F. Hernandez, A.W. Rodriguez, M. Soljačić, S.G. Johnson, General theory of spontaneous emission near exceptional points. Opt. Express **25**(11), 12325 (2017)
74. H.-Z. Chen, T. Liu, H.-Y. Luan, R.-J. Liu, X.-Y. Wang, X.-F. Zhu, Y.-B. Li, Z.-M. Gu, S.-J. Liang, H. Gao, L. Lu, L. Ge, S. Zhang, J. Zhu, R.-M. Ma, Revealing the missing dimension at an exceptional point. Nat. Phys. **16**(5), 571–578 (2020)
75. J. Ren, S. Franke, S. Hughes, Connecting classical and quantum mode theories for coupled lossy cavity resonators using quasinormal modes. ACS Photonics **9**(1), 138–155 (2022)
76. P. Yao, S. Hughes, Macroscopic entanglement and violation of Bell's inequalities between two spatially separated quantum dots in a planar photonic crystal system. Opt. Express **17**(14), 11505–11514 (2009)
77. G.S. Agarwal, *Quantum Optics* (Cambridge University Press, Cambridge, 2013)
78. H.M. Doeleman, E. Verhagen, A. Femius Koenderink, Antenna–cavity hybrids: matching polar opposites for Purcell enhancements at any linewidth. ACS Photonics **3**(10), 1943–1951 (2016)
79. A. Mazzei, S. Götzinger, L. de S. Menezes, G. Zumofen, O. Benson, V. Sandoghdar, Controlled coupling of counterpropagating whispering-gallery modes by a single rayleigh scatterer: a classical problem in a quantum optical light. Phys. Rev. Lett. **99**(17), 173603 (2007)
80. I. Teraoka, S. Arnold, Resonance shifts of counterpropagating whispering-gallery modes: degenerate perturbation theory and application to resonator sensors with axial symmetry. J. Opt. Soc. Am. B **26**(7), 1321 (2009)
81. K.G. Cognée, H.M. Doeleman, P. Lalanne, A.F. Koenderink, Cooperative interactions between nano-antennas in a high-Q cavity for unidirectional light sources. Light Sci. Appl. **8**(1), 115 (2019)
82. D.P. Fussell, M.M. Dignam, Quasimode-projection approach to quantum-dot–photon interactions in photonic-crystal-slab coupled-cavity systems. Phys. Rev. A **77**, 053805 (2008)
83. C. Vendromin, M.M. Dignam, Nonlinear optical generation of entangled squeezed states in lossy nonorthogonal quasimodes: an analytic solution, e-print: arXiv:2210.06521 (2022)
84. S. Franke, J. Ren, M. Richter, A. Knorr, S. Hughes, Fermi's golden rule for spontaneous emission in absorptive and amplifying media. Phys. Rev. Lett. **127**, 013602 (2021)
85. S. Franke, J. Ren, S. Hughes, Quantized quasinormal-mode theory of coupled lossy and amplifying resonators. Phys. Rev. A **105**, 023702 (2022)
86. J. Ren, S. Franke, S. Hughes, Quasinormal mode theory of chiral power flow from linearly polarized dipole emitters coupled to index-modulated microring resonators close to an exceptional point. ACS Photonics **9**(4), 1315–1326 (2022)
87. C. Carlson, R. Salzwedel, M. Selig, A. Knorr, S. Hughes, Strong coupling regime and hybrid quasinormal modes from a single plasmonic resonator coupled to a transition metal dichalcogenide monolayer. Phys. Rev. B **104**, 125424 (2021)
88. E.V. Denning, M. Wubs, N. Stenger, J. Mørk, P.T. Kristensen, Quantum theory of two-dimensional materials coupled to electromagnetic resonators. Phys. Rev. B **105**, 085306 (2022)
89. W. Salmon, C. Gustin, A. Settineri, O. Di Stefano, D. Zueco, S. Savasta, F. Nori, S. Hughes, Gauge-independent emission spectra and quantum correlations in the ultrastrong coupling regime of open system cavity-QED. Nanophotonics **11**(8), 1573–1590 (2022)
90. C. Gustin, S. Franke, S. Hughes, Gauge-invariant theory of truncated quantum light-matter interactions in arbitrary media. Phys. Rev. A **107**, 013722 (2023)

Chapter 4
Probing the Optical Near-Field

Renaud Bachelot and Ludovic Douillard

Abstract Since the description of diffraction from the seventeenth century and the development of optical microscopy that has followed, many approaches have been developed for breaking the diffraction limit. Following the first proposition of near-field optical microscopy made in the 1920s, the first experimental demonstrations started in the early 1980s through scanning near-field optical microcopy, within the context of the swift development of scanning probe microscopies.

Since, different alternative approaches and concepts have emerged for probing the optical near-field with a sub-wavelength resolution.

The chapter is divided into four main sections. In Sect. 4.2, important theoretical principles will be reminded. They will allow the reader to acquire a general background in near-field optics. Section 4.3 describes how it is possible to probe the near-field with physical optical nano-antenna. In particular, different approaches of scanning near-field optical microscopy will be discussed. Section 4.4 is dedicated to the way free electrons can be used for probing the near-field. Section 4.5 deals with the use of nanoscale photochemistry for probing the optical near-field.

These three approaches present respective features and assets that will be illustrated by examples of achievements from the literature.

4.1 Introduction

About 30 years ago, a new topic bridging optics and nanotechnology was introduced: nano-optics [45, 147, 158, 159]. This topic deals with light-matter interaction at the nanoscale, that is to say at a scale which is much smaller than the light wavelength. At this scale, the role of evanescent waves is crucial and the optical

R. Bachelot (✉)
Light, nanomaterials, nanotechnologies (L2n) Laboratory, CNRS EMR 7004, University of Technology of Troyes (UTT), Troyes Cedex, France
e-mail: renaud.bachelot@utt.fr

L. Douillard
Université Paris-Saclay, CEA, CNRS, SPEC, Gif sur Yvette, France

R. Gordon (ed.), *Advances in Near-Field Optics*, Springer Series in Optical Sciences 244, https://doi.org/10.1007/978-3-031-34742-9_4

properties of materials are generally different from what they are at the microscale, opening plenty of new avenues and applications to be explored. Relying on both topics (photonics + nanotechnologies), nano-optics and nanophotonics are areas of rapid growth. It is conducted by several institutes and laboratories all over the world. Nowadays, nano-optics sessions are programmed in almost all of the major conferences dedicated to material science, physics, physical-chemistry, nanotechnology, and optics. We are likely to live a pivotal time of the nano-optics history when the applications will be developed while scientific and technological challenges will continue to be addressed. Within this exciting context, it appears crucial to have available efficient and reliable tools for probing the optical near-field with a nanometer optical resolution, that is to say, with a precision which is much smaller than the light wavelength. In order to illustrate this point, let us cite an extract from the report *Nanosciences and Nanotechnologies*, published by the Royal Society and Royal Academy of Engineering in July 2014: "However, the industrialists believe that a nanotechnology breakthrough has occurred in the tools used to observe and measure properties and processes at the nanoscale level." This affirmation is particularly true in nano-optics.

Over the past four centuries, optical microscopy has been considerably developed, making micronic optical observation a routine [126]. In particular, the correction of chromatic and geometrical aberrations in objective lenses has allowed one to achieve the theoretical resolution. This diffraction-limited resolution theory was addressed by Ernst Abbe in 1873 and later assessed by Lord Rayleigh in 1896 to quantify the measure of separation between two Airy patterns that is necessary to distinguish them as separate entities (Eq. 4.1):

$$R = 0.610\frac{\lambda}{\mathrm{NA}} \tag{4.1}$$

where R is the radius of the diffraction spot related to the circular objective pupilla (first zero of the Bessel function describing the Airy pattern, i.e., the dark disk ([22], chapter 8.5)), λ is the light wavelength in vacuum, and NA is the numerical aperture of the objective, defined as $nsin\theta$, where n is the index of refraction of the medium in which the lens is working (1.00 for air, 1.33 for pure water, and, typically, 1.52 for immersion oil) and θ is the maximal half-angle of the cone of light that can enter or exit the objective lens. Nowadays, NA of 1.48 is usual, permitting 200 nm resolution in the visible (see, e.g., high resolution fluorescence imaging in Hedde and Nienhaus [80]). However, the optical images obtained by modern microscopy and spectroscopy are fundamentally limited in the sense that the spatial resolution remains diffraction-limited since it is proportional to the wavelength.

This chapter deals with approaches and tools that have been developed and used for breaking the diffraction limit in order to access to nanoscale optical information, i.e., with a precision much better than the wavelength of the light. Recent powerful techniques are out of the scope of this chapter, such as *stimulated emission depletion* (STED), *stochastic optical reconstruction microscopy* (STORM), and *photoactivated localization microscopy* (PALM) [52], which have led to Nobel prizes in Chemistry in 2014. One of the laureates was E. Betzig, who had intensively worked on near-field microscopy for breaking the diffraction limit since the 1980s [19]. In other words, he has been fully aware of the different concepts expounded in the chapter.

The chapter is divided into four main sections. In Sect. 4.2, important theoretical principles will be reminded. They will allow the reader (i) to understand what is meant by "near-field" and "evanescent waves," and (ii) to consider the following sections with a valuable background. Section 4.3 describes how it is possible to probe the near-field with physical optical nano-antenna. In particular, different approaches of scanning near-field optical microscopy will be discussed. Section 4.4 is dedicated to the way free electrons can be used for probing the near-field. Section 4.5 deals with the use of nanoscale photochemistry for probing the optical near-field.

4.2 Principles

4.2.1 Huygens Fresnel Formalism

To describe the concept of diffraction limit, it is worth using the Huygens-Fresnel (HF) principle, which describes the way light interacts with a surface (see illustration in Fig. 4.1).

When incident light shines an object, the surface of the latter generates, through oscillation of dipoles (electrons), spherical wavelets that act as secondary sources, named "Huygens wavelets," because this physical picture was proposed by Christiaan Huygens in 1678. As a result, point-like sources emit spherical waves from

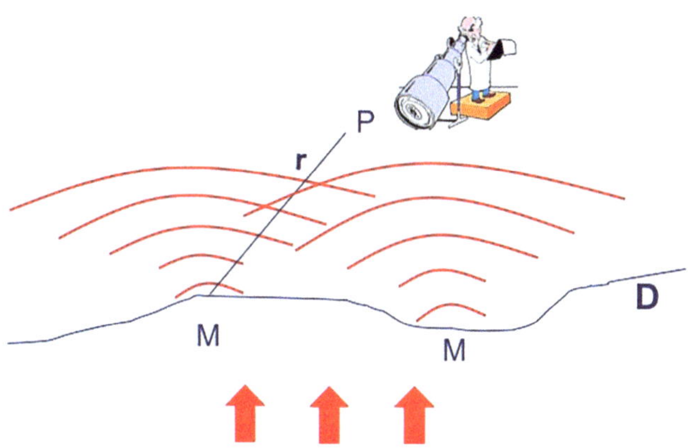

Fig. 4.1 Huygens Fresnel principle. Huygens wavelets are generated at every illuminated point M of surface D, and interfere with each other

any point M of the illuminated domain D in Fig. 4.1. Light propagates before being detected at P so that diffraction plays a spatial *averaging* role: all the wavelets from the sample quickly mix together coherently. In other words, these wavelets interfere with each other (Fresnel contribution in 1818), leading to a spatial averaging in the far-field. This effect is at the origin of the diffraction limit: by far-field observation, it is not possible to get full information from a single dipole. In particular, it is not possible to get any information about the evanescent field of the dipole, and only the propagating components of the dipole radiation can be detected in the far-field (see Sect. 4.2.2). As a result, it is not possible to distinguish two adjacent dipoles that are separated by a distance smaller than R, defined in Eq. (4.1). This wavelet mixing takes place very quickly, over the first micron from the surface. It is thus significant even with light source of low coherence length. The Huygens-Fresnel principle can be described by a scalar Eq. (4.2):

$$\Psi(P) = \int_D \Psi_o(M) Q \frac{\exp(ikr)}{r} dS \tag{4.2}$$

where $\Psi(P)$ is the resulting complex field at P (the observation point), M is a point generating a wavelet from elementary surface element dS, $k = 2\pi/\lambda$ is the wave number, and $r = MP$, the viewing distance. Ψ_0 is a weighting function that is 0 out of domain D, and Q is an inclination factor depending on the angle between the incident beam and MP direction. The Huygens-Fresnel principle leads to Rayleigh Sommerfeld and Fresnel Kirchoff Integral equations and resulting well-known approximation of Fresnel and Fraunhoffer, building up the theory of diffraction [22].

Huygens-Fresnel principle can be compared to the more sophisticated Green formalism to describe light field that is far-field scattered from a surface. The Green tensor $\overleftrightarrow{G}$ is the solution of the wave equation with a point-like source, as a second term (Eq. 4.3):

$$\nabla \times \nabla \times \overleftrightarrow{G}\left(\boldsymbol{r}, \boldsymbol{r}', \omega\right) - \epsilon(r)\frac{\omega^2}{c^2}\overleftrightarrow{G}\left(\boldsymbol{r}, \boldsymbol{r}', \omega\right) = \delta\left(\boldsymbol{r} - \boldsymbol{r}'\right)\overleftrightarrow{I} \tag{4.3}$$

where $\overleftrightarrow{I}$ is the unit tensor and δ is the Dirac function. This formalism allows one to express the field $\boldsymbol{E}$ at position $\boldsymbol{r}$, resulting from the dipole source $\boldsymbol{p}$ at position $\boldsymbol{r}'$ (see Fig. 4.2 and Eq. 4.4).

$$\boldsymbol{E}\left(\boldsymbol{r}, \omega\right) = \mu_0 \omega^2 \overleftrightarrow{G}\left(\boldsymbol{r}, \boldsymbol{r}', \omega\right) \boldsymbol{p}\left(\boldsymbol{r}'\right) \tag{4.4}$$

For a distribution of dipole sources on a surface, the total resulting field at $\boldsymbol{r}$ can be expressed by an integral (Eq. 4.5) that reminds us of Eq. 4.2:

$$\boldsymbol{E}\left(\boldsymbol{r}, \omega\right) = i\omega\mu_0 \int \overleftrightarrow{G}\left(\boldsymbol{r}, \boldsymbol{r}', \omega\right) \boldsymbol{j}\left(\boldsymbol{r}', \omega\right) d\boldsymbol{r}' \tag{4.5}$$

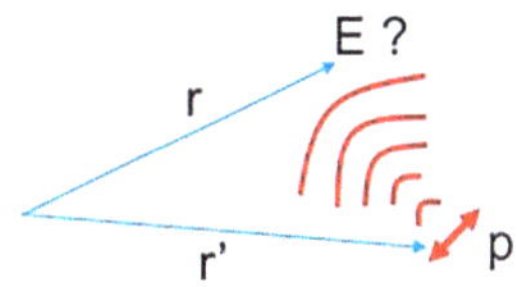

Fig. 4.2 Field $\boldsymbol{E}$ at $\boldsymbol{r}$ resulting from dipole excitation at $\mathbf{r}'$

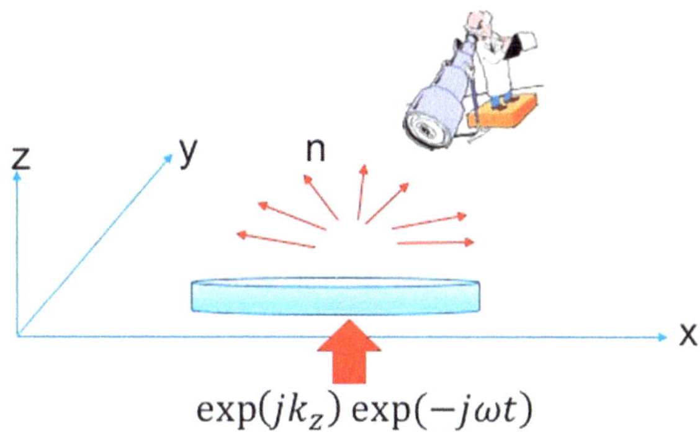

Fig. 4.3 Diffraction of light from a sample shined by a plane wave

where $\boldsymbol{j}$ is a source current density. Eq. 4.5 has actually the form of the Huygens-Fresnel principle, and G is a Huygens wavelet that is a spherical wave in case the Helmholtz operator is applied to G (first term of Eq. 4.3). We will get back to the above formalism for commenting the different approaches of near-field probing.

4.2.2 *Scalar Fourier Theory of Diffraction*

An efficient way of discussing the nature of the near-field is to use the scalar Fourier theory of diffraction [23, 125]. Let's consider a plane wave propagating along z and shining a semi-transparent sample of complex amplitude transmission $\tau(x,y)$ in the (x,y) plane (Fig. 4.3).

Light is diffracted into a medium of refractive index n and far-field detected. At $z = 0$, the electric field can be expressed as

$$E\left(x, y, z = 0^{+}\right) = \tau\left(x, y\right) \exp\left(-i\omega t\right) \tag{4.6}$$

Equation (4.6) is a boundary condition that will be used for discussing the nature of diffracted light. Let's express τ from its Fourier transform $\tilde{T}$:

$$\tau\left(x, y\right) = \iint_{-\infty}^{+\infty} \tilde{T}\left(u, v\right) \exp\left(2\pi i\left[xu + yv\right]\right) dudv \tag{4.7}$$

where u and v are the spatial frequencies of the sample, along x and y, respectively. The total diffracted field in a medium of refractive index n must obey the Helmholtz equation, a well-known canonical solution of which is:

$$A \exp\left(i\left[xk_x + yk_y + zk_z\right]\right)\exp\left(-i\omega t\right) \tag{4.8}$$

Equation 4.8 represents a plane wave of amplitude A and wave vector **k** whose components are k_x, k_y, k_z. The wave number, modulus of $\boldsymbol{k}$, is $k = 2\pi n/\lambda_0$, where λ_0 is the light wavelength in vacuum. Since the Helmholtz equation is linear, we can solve the diffraction problem for each spatial Fourier component and sum up the results to express the total diffracted field. Each Fourier component gives rise to a solution that is a plane wave. For a specific pair of in-plane spatial frequency (u,v), Eqs. (4.6, 4.7, and 4.8) and related identification lead to $A = \tilde{T}(u, v)$, $k_x = 2\pi u$, and $k_y = 2\pi v$. In other words, each spectral component of τ leads to a diffracted plane wave whose amplitude depends on the amplitude of the spectral components u, v and in-plane components of k are imposed by the in-plane spatial frequencies u, v. The z component of the diffracted wave number can then be expressed as:

$$k_z = \sqrt{k^2 - k_x^2 - k_y^2} = 2\pi\sqrt{\frac{n^2}{\lambda_0^2} - u^2 - v^2} \tag{4.9}$$

As a result, the total diffracted field is a spectrum of plane waves (Eq. 4.10), each of them being associated to a specific pair (u,v) of in-plane spatial frequency of the sample. This spectrum is also named “angular spectrum” because each plane wave propagates with an angle that is defined by the components of the wave vector:

$$E(x, y, z, t) = \exp(-i\omega t)\iint_{-\infty}^{+\infty}\tilde{T}(u, v)\exp(ik_z z)\exp(2\pi i\,[xu + yv])\,du\,dv, \tag{4.10}$$

where, k_z obeys Eq. 4.9. The exact nature of each plane wave depends on the nature of k_z.

There are actually two scenarios:

Scenario 1

“Big” details of the sample of typical size > λ_0/n are associated to low in-plane spatial frequencies u, v that are smaller than n/λ_0. In that case, $k_{//} < k$, where $k_{//}$ is the modulus of the wavevector projected on the sample plane ($k_{//}{}^2 = k_x^2 + k_y^2$) and $k_z{}^2 = k^2 - k_{//}{}^2 > 0$ in Eq. 4.9, resulting in k_z which is a purely real number. The associated plane wave is a propagating wave in medium n.

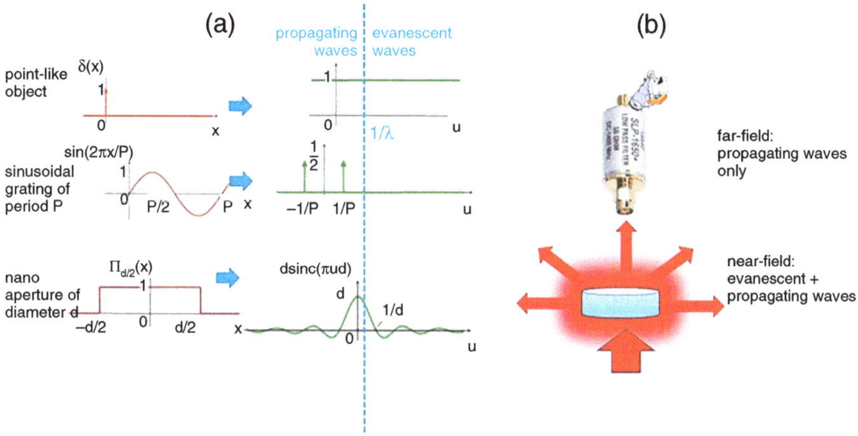

Fig. 4.4 Fourier theory of diffraction. (**a**) Different examples of diffracting objects and resulting waves. (**b**) Diffraction and propagation: different zones and low-pass filter effect

Scenario 2

"Small" details of the sample of typical size $< \lambda_0/n$ are associated to high in-plane spatial frequencies u, v that are larger than n/λ_0. In that case, $k_{//} > k$ and $k_z{}^2 = \mathrm{k}^2 - k_{//}{}^2 < 0$ in Eq. 4.9, resulting in k_z which is a purely imaginary number. The term $\exp(\mathrm{i}k_z z)$ in Eq. 4.10 becomes then $\exp(-\alpha z)$, and the result is an evanescent wave whose amplitude presents decay length $= 1/\alpha$ along z, where α is the modulus of k_z. This wave is a surface wave that cannot be detected in the far-field: it does not propagate in the n medium while it propagates on the sample surface with real x and y wavevector components that are imposed by the sample in-plane spatial frequencies. It should be stressed that each pair of in-plane spatial frequencies (u,$v > n/\lambda_0$) leads to a specific evanescent wave that is characterized by $\alpha = 2\pi\sqrt{u^2 + v^2 - \frac{n^2}{\lambda_0^2}}$, $k_x = 2\pi u$, and $k_y = 2\pi v$. In other words, there is a continuous superposition of evanescent waves at the sample surface, directly connected to the spatial spectrum of τ.

In order to illustrate the above considerations, let us consider Fig. 4.4a, which shows three examples of τ function. For simplicity, we consider 1-D objects (along x) within a 2-D space (x,z). In red and green are represented $\tau(\mathrm{x})$ and $\tilde{T}(u)$, respectively. The blue vertical dashed line represents the value $1/\lambda = n/\lambda_0$ in the u space. Obviously, $-1/\lambda$ could have been represented too, but the following discussion will be, for simplicity, restricted to the positive part of the spectrum.

The first example in Fig. 4.4a shows a Dirac function describing a point-like object (dipole, molecule). The corresponding spectrum contains all the possible spatial frequencies. Some of them ($< 1/\lambda$) lead to propagating waves detectable in the far-field, others ($> 1/\lambda$) produce evanescent waves that cannot be detected in the far-field. An observer in the far-field can observe the point-like object but with truncated information: the object's apparent size is λ. Information about the high spatial frequencies is lost.

The second example shows a sinusoidal grating of period P. The corresponding spectrum is very simple: it contains only three components: $1/P$ (and symmetrical $-1/P$) and 0. We considered the case $1/P < 1/\lambda$: the whole spectrum leads to one propagating wave (+its high order spatial harmonics) and specular transmitted light (0 frequency). The grating is visible in the far-field, and the resulting diffraction obeys the well-known grating behavior that is exploited for dispersion and spectroscopy. If the value of P is decreased until $1/P$ becomes $> 1/\lambda$, the diffracted wave suddenly becomes evanescent and the grating is no longer visible in the far-field: only the 0 frequency can be detected; it defines the contrast of an apparently homogenous sample.

The last example concerns a nano-aperture of diameter d. According to its spatial spectrum in green, only a part of the diffracted light is made of propagating waves that can be far-field detected. In the far-field, the nanoaperture looks like an aperture of diameter λ.

The above considerations allow one to define the concepts of "near-field" and "far-field" (Fig. 4.4b): the near-field corresponds to whole angular spectrum made of both evanescent and propagating waves. The far-field contains propagative waves only, and information about high spatial frequencies $> 1/\lambda$ is lost. Hence, diffraction and far-field detection act as a *low-pass filter* in the near-field.

4.3 Probing the Near-Field with a Physical Optical Nanoantenna

In order to access the whole angular spectrum, a physical nano-probe placed in the near-field can be used (Fig. 4.5).

Figure 4.5a illustrates the general principle. The yellow and blue nano-objects are the sample and the probe, respectively. Instead of letting the HF principle be manifested with the sample alone, a physical probe (size << λ) is approached at a

Fig. 4.5 Use of a physical nano-probe to access to the optical near-field

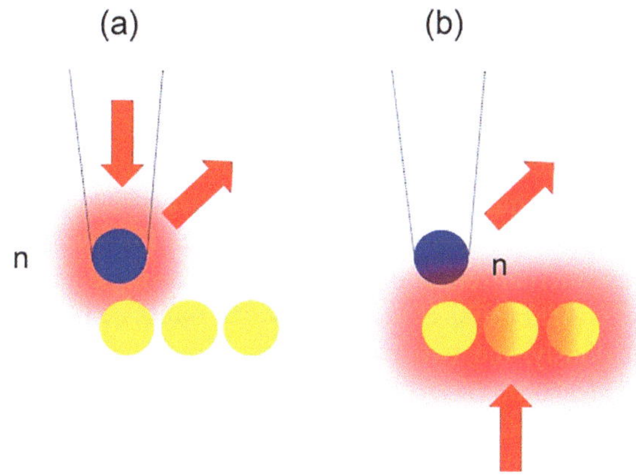

Fig. 4.6 Scanning near-field optical microscopy. (**a**) Illumination mode, (**b**) detection mode

very small distance (<< λ) from the sample. In that case, the HF principle is, of course, still valid, but the resulting far-field signal (Eq. 4.2) results from the local interaction between probe and sample, and it is expected to depend on the local properties of the sample (dielectric constant, polarizability, scattering/absorption cross section, etc.) interacting with the nanoscale probe. Obviously, for a given position of probe and sample, the far-field signal, albeit sensitive to the above-evoked coupling, is spatially diffraction limited. However, this nanoscale coupling sensitivity can be revealed if the probe (or sample) is scanned in the x,y plane with a nanometric precision. The signal $S(x,y)$ resulting from the HF principle at position x,y is very sensitive to the variation of the sample properties with a spatial resolution that depends on both probe size and probe-sample distance instead of the light wavelength. The diffraction limit can be broken by this approach, which constitutes the principle of *scanning near-field optical microscopy*, SNOM [158] / *near-field scanning optical microscopy* NSOM (Fig. 4.5b). Let us stress that the term "scanning" is important because the pixel-by-pixel reconstruction optical image at the nanoscale is made possible *via* the scanning.

Based on this general principle, different SNOM configurations have been developed. The optical probe is coupled to a scanning probe microscope (*atomic force microscopy* AFM, *scanning tunneling microscopy* STM, etc.), allowing for scanning and probe-sample distance control. Two general approaches have been considered: the illumination mode and the detection mode (Fig. 4.6).

In the illumination mode (Fig. 4.6a), the probe's evanescent near-field is exploited. Since the probe's size is nanometric, its spatial spectrum is rich in terms of high spatial frequencies > 1/λ. The resulting evanescent fields code for probe nanometer size and can be viewed, in a general point of view, as an optical nanosource that can be exploited in the probe's near-field. This nanosource is used for locally illuminating the sample. The local probe/sample interaction results in propagating waves that are far-field detected.

In the detection mode (Fig. 4.6b), the sample's evanescent near-field resulting from its high spatial frequencies > 1/λ is picked up by the local probe. Again, the local probe/sample interaction results in propagating waves that are far-field detected.

The generation of propagating waves from a local interaction can be understood from Eq. 4.9: approaching probe (sample) to sample (probe) results in an increase of the effective index surrounding the sample (probe), leading to an increase of the total wave number. In that case, negative k_z^2 can become positive and an evanescent wave can become a propagating wave. This principle can be used for frustrating total internal reflection in the case of a single evanescent wave generated beyond the critical angle of incidence at the interface between two dielectric media [194]. Obviously, the two above modes can be simultaneously active. In general, the physical probe is an optical antenna that converts the near-field into far-field and vice et versa [146]. It should be stressed that the probe is, in general, not only a probe. A priori, $S(x,y)$ in Fig. 4.5b does not describe directly the local optical properties of the sample without the presence of the probe. It rather results from a local interaction between probe and sample. Different situations are possible ranging from passive approach to active/perturbative approach. By "passive approach," one means that $S(x,y)$ is a good signature of the optical properties of the sample without the presence of the probe. By "active/perturbative approach," one means that the probe fully takes part in the physics of the probe/sample nanosystem. In other words, the sample is perturbed by the presence of the probe and its optical properties are locally modified.

An example of perturbative approach can be described. Let's suppose the sample to be a gold single nanoparticle to be characterized. The probe is another gold nanoparticle attached to the extremity of a tapered optical fiber [5]. Putting together both gold nanoparticles necessary leads to a plasmon hybridizing involving bonding and antibonding modes [144, 177]. As a result, $S(x,y)$ will give information about the gold coupled dimer rather than the initial sample. In [5], it should be noticed that the probe is used as a perturbative probe: its presence modifies the rate of deexcitation of a fluorescent single molecule. This effect will be further discussed in the following.

Based on the above general principles, many types of nano-antenna have been developed over the past 40 years, with many associated technological challenges. In particular, the following issues have been addressed (with examples of references among many):

- Probe fabrication (type, reliability, reproducibility) [37]
- Control of the probe-to-sample distance [192]
- Image formation and spatial resolution [20, 63]
- Spectral range [106, 164, 181]
- Condition of illumination and detection [113, 145]
- Study and exploitation of polarization [91, 100]
- Phase imaging [136]
- Time resolution [31]
- Artifacts [21, 79]

Obviously, these different numerous points will not be developed here. The first one (probe issue) has turned out to be crucial. Figure 4.7 illustrates the three

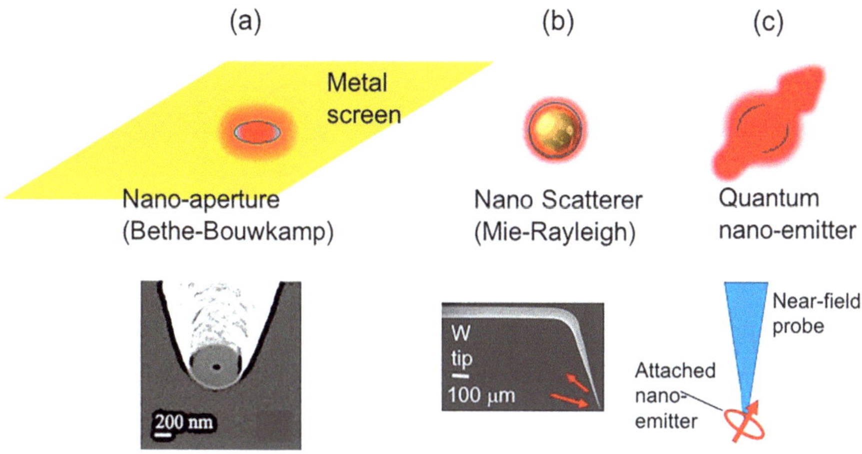

Fig. 4.7 Three main approaches of SNOM (top: principle, bottom: example of practical implementation). (**a**) Aperture SNOM. Bottom image reproduced from Ref. [190], with the permission of AIP. (**b**) Apertureless SNOM (alias scattering-type SNOM). The bottom image is similar to that from Ref. [203]. (**c**) Active SNOM probe based on a quantum nano-emitters

families of optical nano-antenna that have given rise to huge efforts of research, development, and demonstration.

Figure 4.7a illustrates a SNOM probe based on a nano-aperture in a metal screen, which is used to squeeze and confine light down to the nanoscale. This approach was initially suggested in 1928 by E. H. Synge [178]. The sub-wavelength confinement, which can be described by the Bethe-Bouwkamp solution [51], is associated to evanescent waves, which can be exploited in the near-field for either locally illuminating the sample or collecting the near-field through the aperture. In general, such a probe has been integrated at the extremity of metallized tapered optical fibers used for shining the sample or collecting the near-field from its surface. Figure 4.7b illustrates an apertureless optical nanoantenna. As a complementary approach, the probe is a nano scatterer whose physics can be described by the Mie-Rayleigh theory [90]. In that case, the nano-probe scatters light resulting from its local interaction with the sample surface and both illumination and collection are made through an objective lens. In general, such a probe can be a probe used for atomic force microscopy or scanning tunneling microscopy. Figure 4.7c shows an "active SNOM probe" made of a quantum emitter (molecule, nanocrystal). In that case, the probe itself acts as a nano-emitters interacting with the sample surface (e.g., [92]). In order to comment and compare these different approaches, let's get back to the Huygens-Fresnel principle (Fig. 4.8)

In Fig. 4.8, D represents the sample domain of integration of Eq. (4.2). It is diffraction-limited and its minimum size is $\lambda/2$. D' is the domain of interest that represents the zone of local optical probe/sample interaction. It is not diffraction-limited: it depends on probe size and probe-to-sample distance and is typically $<< \lambda/2$. The difference between different SNOM nanoantennas can be discussed

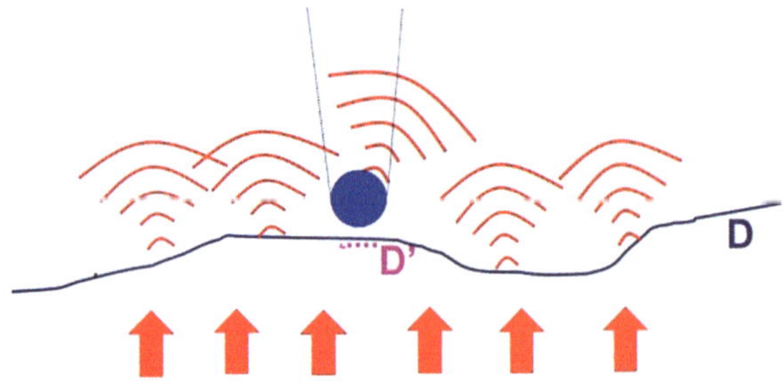

Fig. 4.8 SNOM description based on the Huygens-Fresnel principle

in terms of effective domain of integration in the HF principle. In aperture SNOM, since light at the sample plane is confined/trapped by the nanoaperture, we have:

$$\Psi(P) = \int_{D'} \cdots \tag{4.11}$$

and the far-field signal can describe the nanoscale properties of the sample. The above situation is also valid for SNOM using a nanoemitter because the light is detected at the wavelength of fluorescence/photoluminescence, which is different from the incident wavelength through the Stokes shift. Both this spectral filtering and the nanosize of the emitters limit the HF domain of integration to D'. Regarding the apertureless configuration using a Mie-Rayleigh scatterer as a near-field probe, the above spatial selection is not directly achieved because scattering is a linear process. Both illumination and detection are made through an external objective lens, and we have:

$$\Psi(P) = \int_{D} \cdots = \int_{D-D'} \cdots + \int_{D'} \cdots \tag{4.12}$$

The term integrated over D–D' is diffraction limited. It constitutes a far-field background that depends on the D' environment over $\lambda/2$. The term integrated over D' is the near-field term that is much weaker than the background. The challenge of apertureless SNOM is to become sensitive to the near-field term by extracting it from the background. For achieving this, three approaches have been successfully developed.

(i) In the case of probes having free charges at their surface (metal or semiconducting tips), it is possible to excite the lightning rod effect at the extremity of the tip to get an optical nanosource [12]. This effect is actually related to an electromagnetic singularity that happens at sharp edges and corners [16]. It can be described by the Gauss theorem, resulting from the discontinuity of the field component perpendicular to the tip/medium (1/2) interface: $\mathbf{n}_{\perp}.(\varepsilon_1 \mathbf{E_1} - \varepsilon_2 \mathbf{E_2}) = \rho_S/\varepsilon_0$, where $\mathbf{n}_{\perp}$ is the unit vector perpendicular to the metal surface

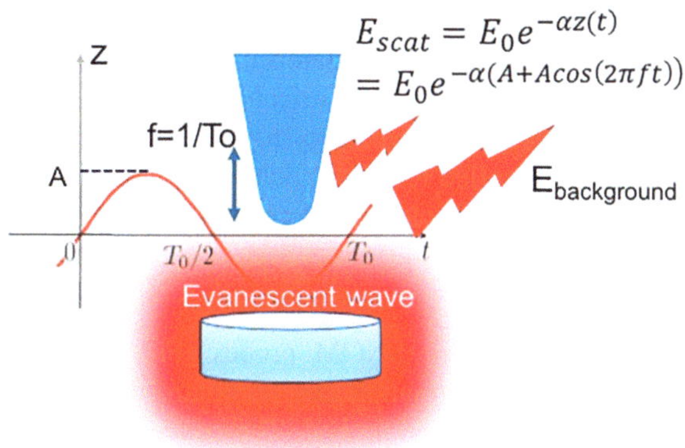

Fig. 4.9 Apertureless probe vibrating periodically above evanescent waves

and ρ_S is the tip surface charge density that can become very large for nanometer sized tips. Exciting this optical nanosource results in the decrease of the effective domain of integration in the HP principle: D becomes D' because light is mainly confined over D' in the sample/tip near-field. This effect has been successfully used for tip-enhanced near-field optical microscopy and spectroscopy (e.g., [76]).

(ii) The background can be considerably reduced by making vibrate the tip (frequency f, amplitude A) perpendicular to the sample, i.e., along the field gradient of the evanescent waves to be detected (e.g., [11]) (Fig. 4.9). This vibration is made possible with the noncontact or tapping modes of atomic force microscopy [189]. The amplitude of light scattered by the tip, E_{scat}, takes the form expressed in Fig. 4.9. The time Fourier series of this expression leads to many frequencies of modulation that can be selectively detected at a chosen frequency of demodulation f_{dem} by lock-in detection [13] to efficiently extract the near-field information from the background. Indeed, the $E_{\text{background}}$ is poorly modulated by tip vibration, while the evanescent waves are highly modulated due to their strong field gradient along z. In general, $f_{\text{dem}} = f$. However, higher harmonics of f can be used because the modulation of light of exponential field decay by sinus vertical motion is not linear: E_{scat} can present many high-harmonics of modulation depending on two key parameters: the amplitude A and the angular spectrum generated by the sample via diffraction. For example, $f_{\text{dem}} = 2f$ has been efficient for near-field characterization with both apertureless and aperture SNOM [11, 116].

(iii) The influence of the background can be further reduced by the use of a heterodyne approach. Let's first point out that in apertureless SNOM, the light from D' interferes with light from D: signal is always interferometric by nature. $E_{\text{background}}$ acts as a reference field for interferometry [6]. Potentially, this feature constitutes an asset because both amplitude and phase information can be obtained. However, since the background is a priori not controlled, the interferometry is not controlled. A solution is to use a third field that interferes with the other fields before detection. This field comes from the same initial laser source but has been frequency shifted by $\Delta\omega$. The interference between this new controlled reference field with E_{scat} and $E_{\text{background}}$ leads to various light modulation. In particular, $f_{\text{dem}} = nf \mp m\Delta\omega/2\pi$ (n and m are integers) permits a high sensitivity to the beating between the scattered near-field and the controlled reference field, resulting in high-resolution complex near-field image of complex fields (amplitude and phase, e.g. [62]). The uncontrolled background field is thus replaced by the frequency-shifted controlled reference field.

As far as active SNOM probes (Fig. 4.7c) are concerned, complementary valuable near-information can be obtained: the sample *photonic local density of states* (LDOS) can be probed. The concept of LDOS is well known in solid state physics, where the density of states ρ(E) is defined as:

$$\rho(E) \times \delta E = \delta N \tag{4.13}$$

where δN is the number of states whose energies lie in the range from E to $E + \delta E$. In general, one considers the states occupied by electrons or phonons. In nano-optics, an LDOS can be defined too [26]. Any photonic nano-system has eigenmodes e_n, solutions of homogeneous Helmholtz equation:

$$\nabla \times \nabla \times e_n(r) - \epsilon(r)\frac{\omega_n^2}{c^2}e_n(r) = 0 \tag{4.14}$$

e_n solutions, associated to eigenvalues of frequency ω_n, are intrinsic to the nanosystem and do not depend on the condition of excitation.

In the particular case of nanosystems for which a discrete set of eigenmodes can be introduced, the photonic LDOS at position $\boldsymbol{r}$ can be expressed as:

$$\rho\,(\boldsymbol{r}, \omega) = \sum\nolimits_n |e_n\,(\boldsymbol{r})|^2 \delta\,(\omega - \omega_n) \tag{4.15}$$

Initially described by P. A. M. Dirac [47], the Fermi golden rule (Eq. 4.16), established at the same time, describes the way an excited emitter de-excites in the presence of a surrounding LDOS [26].

$$\Gamma = \frac{2\omega}{3\varepsilon_0}\langle f\,|\hat{\mu}|\,i\rangle^2 \rho\,(r, \omega) \tag{4.16}$$

where Γ is the deexcitation rate. i and f stand for the initial (excited) state and the final lower state, respectively. $\hat{\mu}$ represents the emitter dipole moment related to the perturbation matrix element and the interaction Hamiltonian between both states. Hence, Γ (that includes both radiative and nonradiative relaxations) is proportional to the LDOS. In other words, the LDOS acts as channels of deexcitation of the quantum emitter. In general, in the presence of rich LDOS, deexcitation is boosted:

Γ increases and the life time $\tau = 1/\Gamma$ decreases. In the case of nanocavity presenting a defined photonic LDOS, Eq. 4.16 can be rewritten for each cavity eigenmode:

$$\Gamma/\Gamma_0 = \left(3/4\pi^2\right)\ \lambda_{\text{if}}{}^3 Q/V \tag{4.17}$$

where Γ_0 is the free-space emitter deexcitation rate (without the presence of the nanocavity); λ_{if} is the wavelength transition between states *i* and *f*; *Q* is the quality factor of the mode, related to the damping; and *V* is the mode volume. In the case of a plasmonic metal nanocavity, the *Q* factor is generally weak (rarely greater than 70) [195] but the mode volume can be very small, making the LDOS high, resulting in a strong decrease of τ. This effect is known as the "Purcell effect" because change in Γ/Γ_0 due to a structured environment was initially computed by E. M. Purcell in 1946 in case of a single mode cavity [161]. Ratio Γ/Γ_0 is named "Purcell factor." Consequently, a nanoemitter scanning above a nanostructure (SNOM configuration, Fig. 4.7c) can enable the imaging of the structure LDOS *via* lifetime measurement at each probe position [103, 172]. In Sect. 4.4, we will see that the LDOS can be also characterized with free electrons. Figure 4.10 shows selected examples of SNOM images illustrating the possibility to break the diffraction limit by probing the near-field with a physical optical nano-antenna. Images are briefly commented in the figure captions. Readers are invited to read the corresponding referenced articles for more details.

4.4 Probing the Near-Field with Electrons

General Imaging Principle

As formalized by the criterion mentioned above Eq. (4.1), the ultimate resolution of an optical microscope is related to the wavelength of the light. An electron with an energy of a few eV seen as a wave has a wavelength orders of magnitude smaller than that of a photon in the visible range. The de Broglie wavelength of a nonrelativistic particle is inversely proportional to its momentum, see Eq. (4.18).

$$\lambda = \frac{h}{mv} \tag{4.18}$$

where *m*, *v* are the mass and velocity of the particle, respectively. The wavelength of an electron of 1 eV of kinetic energy amounts to 1.2 nm, much smaller than that of the IR photon of the same energy, namely 1240 nm. Electron microscopies can therefore overcome the limitations of optical microscopies in terms of spatial resolution. Initially dedicated to high-resolution imaging, electron microscopies are now essential tools for the study of the near-field, especially in the context of plasmonic materials.

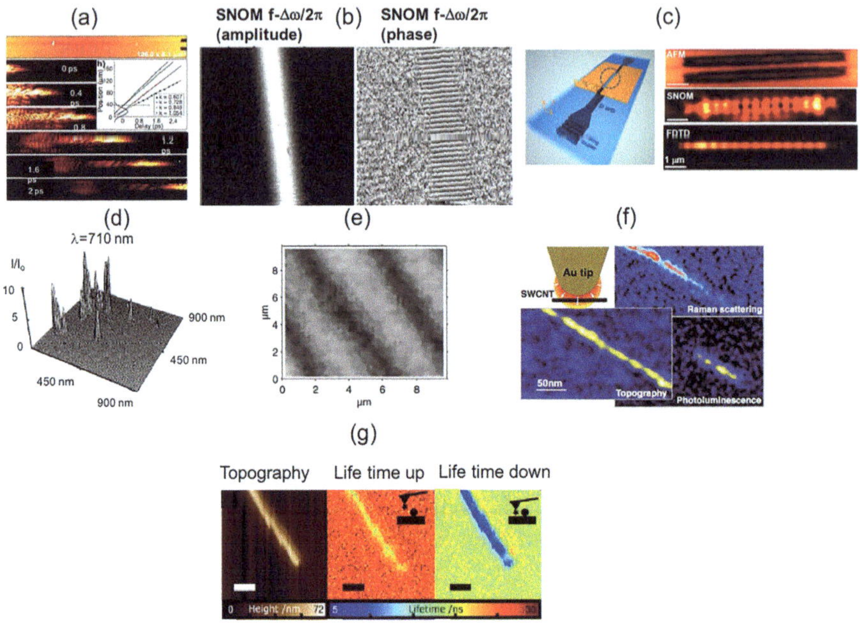

Fig. 4.10 Examples of SNOM images breaking the diffraction limit. (**a**) Time-resolved interferometric SNOM images on a Si photonic crystal waveguide (aperture probe, $\lambda = 1.25\ \mu$m). The different rectangular images have a size of $128 \times 8.1\ \mu m^2$. Top image: topography. The other images show the amplitude of a guided laser pulse at different times indicated on the images. Inset: measured position pulse spatial with different modes (represented by number k) as a function of the delay time. Local phase and group velocity were accessed. (Reprinted with permission from Ref. [61]. Copyright (2005) by the American Physical Society). (**b**) $10\ \mu m \times 10\ \mu m$ SNOM images of guided light in a buried ion-exchanged glass waveguide (heterodyne configuration, silicon probe, 1.55 μm). Left: amplitude, right: phase. Both mode confinement and propagation constant can be assessed. (Reprinted/Adapted with permission from Ref. [62] © The Optical Society). (**c**) SNOM study of Plasmonic/photonic hybrid modes on a Si-Cu waveguide (Si probe, $\lambda = 1.55\ \mu$m). Left: Scheme of the hybrid waveguide. Right: AFM image, SNOM image (near-field intensity), corresponding *finite difference time domain* FDTD simulation. The scale bar represents 1 μm. Spatial confinement and propagation of the hybrid modes were studied. (The figure is reprinted (adapted) with permission from Ref. [44], Copyright 2010, American Chemical Society). (**d**) SNOM observation of localized electromagnetic hot spots in random metal dielectric film (W tip, $\lambda = 700$–770 nm). Anderson localization of complex surface plasmon modes in a semicontinuous metal film was observed and studied. (Reprinted with permission from Ref. [64] Copyright (1999), American Physical Society). (**e**) SNOM imaging of bore implantation in silicon (W tip, $\lambda = 10.8\ \mu$m). Pure nanoscale dielectric contrast was revealed. (Reproduced from Ref. [107] with the permission of AIP Publishing). (**f**) SNOM investigation of photoluminescence and Raman scattering from carbon nanotubes (tip-enhanced configuration, gold tip, $\lambda_{exc} = 632$ nm, $\lambda_{emRam} = 700$ nm, $\lambda_{emPL} = 950$ nm. Several phenomena were studied at the nanoscale: e.g., phonon excitation, local band energy renormalization induced by charge carrier doping and inter-nanotube energy transfer. (From Ref. [76]). (**g**) Life near-field imaging, revealing the spatial distribution of the LDOS of a silver nanowire (use of an active tip). Scale bar is 200 nm. (The figure is reprinted (adapted) with permission from Ref. [172], Copyright 2014, American Chemical Society)

Like optical methods, the principle of imaging a plasmonic excitation by electron microscopy relies on the monitoring of either alterations of its excitation channels or changes associated with its damping physics. The excitation of a plasmon can be conducted by interaction with (i) photons and/or (ii) electrons. For a planar metal/dielectric interface under electromagnetic plane wave illumination, the plasmon dispersion relation lies right from the light line and the fulfillment of momentum conservation at a given photon frequency implies an increase of the light wave vector. Two methods are widely used, namely the grating coupler [198] and the *attenuated total internal reflection* ATR technique [104, 148]. For fast electron excitation, moment and energy are directly transferred to the solid by the electron beam. The component of the scattering vector parallel to the interface determines the *surface plasmon* (SP) wave vector and, together with the SP dispersion relation, the energy loss of the scattered particles [70]. Further down the excitation physics, damping processes of a surface plasmon constitute alternate access ways to the investigation of the plasmon near-field physics. Symmetrically to the light excitation process, a surface plasmon can (iii) radiate its energy back into free space (luminescence). Random roughness and/or specifically designed diffraction gratings constitute common momentum suppliers for momentum conservation. Plasmons can also decay nonradiatively (iv) by excitation of single (electron – hole) pairs, resulting in a population of nonthermal (hot) charged carriers. Within a few hundreds of fs (electron, electron) scattering events thermalize this hot electron population, which subsequently interacts with the phonon bath on a ps time scale and ultimately dissipates the plasmonic energy as heat. These ohmic losses constitute the main energy dissipation channel. Detection of the heat produced by plasmonic decay can be conducted by photoacoustic effect [127, 140, 168, 184]. Additional relaxation pathways are (v) the emission of electrons through either linear, nonlinear or plasmon field-assisted photoelectric effects and (vi) the coupling with secondary plasmonic excitations (*surface plasmons polaritons* SPPs and/or *localized surface plasmon* LSPs) [205].

In the context of electron microscopies, these different excitation and deexcitation schemes have paved the way for the development of efficient investigation methods of the optical near-field. The following paragraphs briefly address the main ones, namely the *scanning transmission microscopy in electron energy loss spectroscopy* STEM EELS [141], the parented *cathodoluminescence* CL [191], the *photon-induced near-field electron microscopy* PINEM [10, 155, 202] and the *photoemission electron microscopy* PEEM [81, 95, 130].

Additionally, panoramas of the near-field imaging techniques based on electron microscopies can be found in Vogelgesang and Dmitriev [188], Douillard and Charra [50], Rotenberg and Kuipers [165], Merlen and Lagugné-Labarthet [131], Colliex et al. [36], El-Khoury et al. [53], Losquin and Lummen [120], Coenen and Haegel [32], Wu et al. [200], Talebi [179], Polman et al. [160].

4.4.1 Electron Energy Loss Imaging Microscopy (Electron In, Electron Out)

Imaging Principle

The use of high-energy electrons in an electron microscope is the source of a wide range of (electron, matter) interactions, which in turn gives rise to a large number of imaging and spectrometric techniques. Generally, fast electron beams are mostly used for imaging at very high spatial resolution. The electron beam is scanned in a raster scan pattern, and the position of the beam is combined with the intensity of various detected signals to produce an image. Modern transmission instruments can easily focus currents of the order of 100 pA within less than 0.1 nm, so positions of atoms are accessible [36]. The structural information is provided by images or diffraction patterns mostly implying elastic electron scattering. Associated analytical methods are largely based on the inelastic interactions, sources of energy and moment transfers between the primary electrons and the target. In line with the high primary electron energy involved, a few tens or hundreds of keV, the different relaxation processes cover a broad spectral range typically from the X-ray above 150 eV down to the phononic energies in the tens of meV range (IR range). Tracking of the energy losses of the transmitted electron paves the way for the *electron energy loss spectrometries* EELS (*high resolution* HREELS, EELS, *extended energy loss fine structure* EXELFS). Tracking of the radiative relaxation signatures translates into additional analyses, such as the *energy dispersive X-ray spectroscopy* EDXS in the X-ray spectral domain or the *cathodoluminescence* CL in the visible and near infrared wavelength ranges. The impact of the EELS measurements in the low-loss domain, i.e., in the [0.05, 5 eV] energy loss interval, combined with an unmatched spatial resolution in the nanometer range opens solid capabilities for the investigation of the plasmon physics [36]. The following paragraphs focus on one of these (electron in, electron out) microscopies, namely the *scanning transmission electron microscopy* in the *electron energy spectrometric mode* EELS STEM. Alternatively, the *energy-filtered transmission electron microscopy* EFTEM can be used. Here the image is built from energy-selected electrons. In more details, the full field image is transformed into an EELS spectrum with the help of a magnetic prism. The latter is energy-filtered with energy-selecting windows centered on one particular energy loss and the selected part is backtransformed into an energy-filtered image.

From a mechanistic point of view, the long-range nature of the Coulomb interaction between primary and valence electrons in a metal particle is responsible for the excitations of coherent collective plasmon oscillations. Any primary electron carries an electromagnetic field that polarizes the target particle and generates an induced field. The EELS losses correspond to the work done by the electron in the induced field it has itself generated, i.e., an electron both excites and probes plasmons. All the energy absorbed or scattered by a plasmonic particle is provided by the incident electrons. Within a wave description, the fast electron beam generates a transient electric field equivalent to a wave pulse of white light strongly localized along its trajectory. In short, upon fast electron beam illumination, a plasmonic nanoparticle sees a pulse containing a large spectrum of frequencies. The EELS probability Γ_{EELS} depends on all resonance modes of the particles and reflects closely its extinction cross section [18, 57, 101]. It can be described as the sum over all the modes i of a product of two terms, the first $\Im\left(f_i\left(\omega\right)\right)$ reflects the spectral dependence of the mode, the second $\left|TF_z\left(\overrightarrow{E}_i^z\left(\overrightarrow{R}_\perp, \frac{\omega}{\upsilon_e}\right)\right)\right|^2$ its spatial distribution.

$$\Gamma_{\text{EELS}}\left(\omega, \overrightarrow{R}_\perp\right) = \frac{e^2}{\pi\hbar\omega^2}\sum\nolimits_i \Im\left(f_i\left(\omega\right)\right)\left|TF_z\left(\overrightarrow{E}_i^z\left(\overrightarrow{R}_\perp, \frac{\omega}{\upsilon_e}\right)\right)\right|^2 \qquad (4.19)$$

The $f_i(\omega)$ function can be understood as a generalized polarizability. It covers all the spectral aspects of the mode under consideration and, within a Drude description, converges to a Lorentzian function centered on the plasmon resonance energy. The second term corresponds to the Fourier transform along the e-beam direction z, at impact parameter $\overrightarrow{R}_\perp = (x, y)$ (e-beam to particle center distance) in real space and momentum $k_z = \omega/\upsilon_e$ in reciprocal space of the electron induced electric near-field $\overrightarrow{E}_i^z$. So, within a Fourier transform, the EELS losses map spectrally and spatially the *electromagnetic density of states* projected along the direction of the incident primary electrons, z-EMLDOS. Only the z-component of the EMLDOS is probed since the EELS losses scale as the dot product $\overrightarrow{\upsilon}_{e,z}.\overrightarrow{E}_{i,z}$ with $\overrightarrow{\upsilon}_{e,z}$ the velocity of the electrons along the normal specimen direction.

One additional aspect needs to be mentioned. Using a fast electron beam as excitation source, the EELS physics departs from a true optical excitation. Therefore, no photon selection rules apply and both dark and bright modes are available.

For the case of a sphere excited under the quasi-static regime, the EELS probability can be formally connected to optical characteristics measured in far-field, namely the extinction cross section.

$$\Gamma_{\text{EELS}}^{\text{sphere}}\left(\omega, \overrightarrow{R}_\perp\right) \propto \Im\left(\alpha_{\iota=1}\left(\omega\right)\right).\exp\left(-\frac{R_\perp}{\gamma\upsilon_e/\omega}\right) \qquad (4.20)$$

where $\alpha_{l=1}$ is the sphere polarizability for the mode $l = 1$ (the only mode to be considered in quasi-static approximation) and γ the Lorentz factor accounting for the relativistic nature of the fast electrons $\gamma = \sqrt{1 - v_e^2/c^2}$. The EELS probability of a nanosphere scales with its extinction cross section. Its value is position-dependent with a quasi-exponential decay with the impact parameter $R_\perp$. In EELS STEM imaging, the spatial resolution has its origin in the coupling between the transient field generated by the fast-moving electrons and the resonant plasmon modes. As noted in Eq. (4.20), the extension of the evanescent electron field is determined by the attenuation length $\gamma\upsilon_e/\omega$ ~ 10–100 nm and scales with the energy

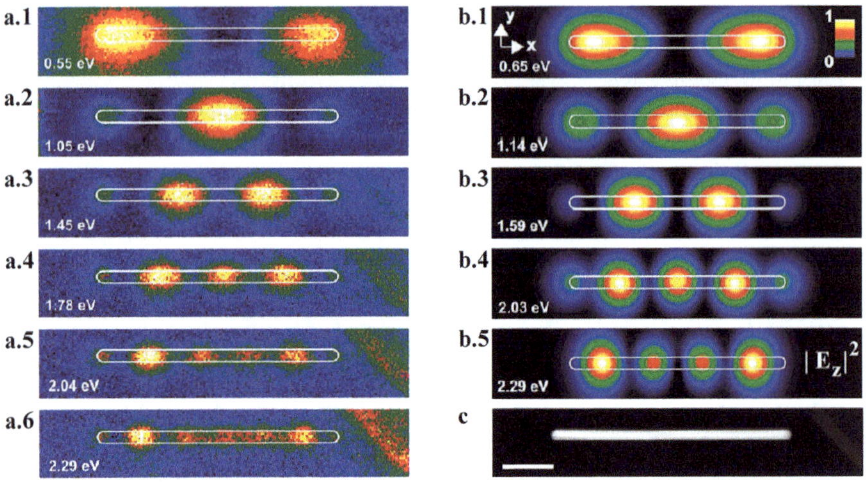

Fig. 4.11 Comparison of experimental EELS maps and z-projected electromagnetic density of states of Fabry Perot like resonances of a Ag nanorod. (**a**) (1–6) EELS maps of longitudinal plasmon modes; orders m = {1–6} from top to bottom. (**b**) (1–5) Simulated electromagnetic density of states projected onto the beam direction of the incident electrons z-EMLDOS. c. Nanorod imaging, annular dark field ADF image. Scale bar 100 nm. (Reprinted (adapted) with permission from Ref. [167]. Copyright 2011 American Chemical Society)

of the primary electron beam. Experimentally, for a 100 keV electron beam the spatial resolution of EELS imaging is close to 18 nm [141].

Near-Field Mapping and Engineering Numerous investigations have demonstrated the strong capabilities of the EELS *scanning transmission electron microscopy* EELS STEM and *energy-filtered transmission electron microscopy* EFTEM in high resolution mapping of the electric near-field associated to plasmonic resonances of metallic nanoparticles [74, 141, 167]. Figure 4.11 gives a representative illustration of such capabilities in the case of Ag nanorods of (length, diameter) = (470, 25 nm). Upon particle excitation, a metallic nanorod is the seat of Fabry-Perot type resonances related to stationary interferences of the short-range surface plasmons polaritons [49, 180]. Energy-selected EELS maps, extracted from windows centered on increasing energy loss peaks, reveal the different modes of a single nanorod. At variance with light excitation, both odd (m = 1, 3, 5) and even (m = 2, 4, 6) Fabry-Perot resonance modes are excited and detected (Fig. 4.11a). EELS maps show strong agreement with the corresponding electromagnetic density of states z-EMLDOS calculated along the direction of the incident electron velocity vector (Fig. 4.11b).

Instrumentation In its spectrum imaging configuration, an EELS STEM instrument consists of (i) an electron source of high brightness, generally a cold *field emission gun* FEG operated at primary voltages in the 60–200 kV range; (ii) electron optics, i.e., focusing lens, optionally coupled with an aberration corrector, scanning

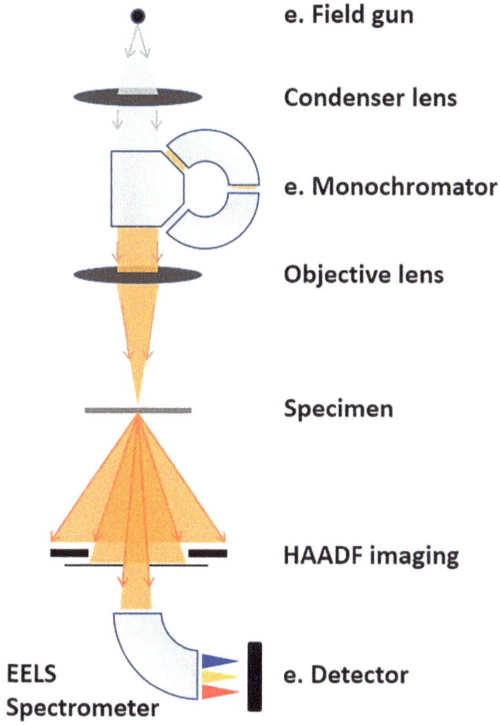

Fig. 4.12 Schematic representation of an EELS STEM imaging microscope. The specimen is excited by monochromatized fast electrons moving along its normal z axis. The e-beam is raster scanned in the specimen xy plane. Grey beam corresponds to nonmonochromatized electrons, orange beam to monochromatized electrons; simultaneous high-angle annular dark-field HAADF imaging is acquired from incoherently scattered electrons; the EELS signal is collected after the energy-dispersing magnetic prism. Beam energy is in the 60–200 kV range. Loss energy resolution is in the range 10–100 meV. Specimen angle relative to the beam direction can be adapted $\pm 60°$ to conduct EELS 3D tomography investigation. Use of a laser-driven cathode gives access to time-resolved EELS measurements

coils, to move the electron probe over the specimen area; (iii) a set of detectors which can collect in parallel several signals, such as the *annular dark field* ADF one for topography and structural information and an EELS spectrometer with its detection chain (magnetic prism), see Fig. 4.12. In *scanning electron microscopy* SEM imaging, image is built on backscattered, secondary and/or Auger electrons. In TEM imaging, image is built on transmitted electrons. The strength of STEM lies in its unmatched spatial resolution, imaging at subwavelength resolution of a few tenths of nm ($\delta R = 0.1$ nm) are readily possible.

Regarding EELS measurements, the major experimental issue is to get rid of the *zero-loss peak* ZLP. Indeed tracking electron losses in the [0.05, 5 eV] energy range close to the ZLP peak constitutes a major instrumental challenge. This pitfall has been circumvented by the developments of high brightness electron guns and advances in electron beam monochromation, so that modern instruments are able to image simultaneously at spatial- ($\delta R = 0.1$ nm) and energy- ($\delta E = 100$ meV) resolutions [36].

EELS Imaging – Variants and Perspectives

EELS 3D tomography. One of the paths of development of the EELS technique is the possibility of reconstructing in three dimensions the distribution of the electric near-field of an object [83, 84]. This tomographic reconstruction approach is obtained by acquisition of a loss signal for a series of incident angles and subsequent algorithm reconstruction. This type of approach has been successfully carried out on Ag nanocubes [143] and Ag bipyramids [34, 35].

Time-resolved EELS measurement. The provision of pulsed electron sources has opened the way to time-resolved electron microscopy measurements [25]. Most frequently, ultrafast pulsed electron sources are based on the concept of laser driven cathodes. In its simplest implementation a pulsed laser beam is directed onto an electron cathode and ultrafast bunches of electrons are generated by a photoelectric effect [54]. Additional use of a second light pulse can further shorten the electron time profile by optical gating [78]. The temporal profile of the electron pulses reflects its optical counterpart. Electron pulse durations in the fs–ps time range are accessible. Physical developments of interest in line with these advances are the search for ways to shape the electron pulse in both the temporal and spatial domains [17, 98, 169].

HREELS imaging. Thanks to the continuous development of technological key components such as bright cold field-emission electron sources, efficient electron monochromators, aberration-corrected beam optics, and electron-energy spectrometers, access to electron microscopes with an energy loss resolution of <10 meV is now possible. This energy window opens a new field of investigation for the electron loss microscopy, that of phononic vibrational excitations. Along this line, spatial-resolved investigations of (e, matter) excitations in the infrared windows range have been recently demonstrated. Examples include the mapping of surface phonons polaritons at the nanoscale in individual MgO nanocubes [108], ZnO nanorods [162], and the tracking of high multipolar resonance modes in Au nanowires [201].

4.4.2 *Cathodoluminescence (CL) Imaging Microscopy – (Electron In, Photon Out)*

Imaging Principle The cathodoluminescence imaging technique is closely parented to EELS in terms of physical processes. It exploits a (electron in, photon out) imaging principle. A fast electron beam is used as the primary excitation source and the specimen responses are studied by collecting in the far-field the subsequent radiative coherent emission resulting from its relaxation processes [68]. As for the case of EELS STEM, the interaction time of one fast moving electron with the sample is of the order of one femtosecond (100 keV, $v_e = 164$ nm/fs), so the transient light pulse experienced by the specimen corresponds to a broadband optical excitation spanning simultaneously the ultraviolet, visible, and infrared wavelength ranges. Beyond scattering interaction, a fast electron beam can also interact with matter through direct (electron – hole) pair creation and/or generation of energetic secondary electrons inside the material which in turn create secondary optical excitations. These additional (electron, matter) interaction channels give birth to an incoherent cathodoluminescence emission [31, 102]. Coherent and incoherent CL channels exhibit different signal emission distributions. Coherent CL signal corresponds to a transient dipole emission. It is strongly polarized and shows an anisotropic emission. Incoherent CL emission is spontaneous, unpolarized, and isotropic in nature. So, coherent and incoherent emissions can be disentangled through angle- and/or polarized-resolved measurements [134]. In addition, in metals, incoherent CL is damped by nonradiative processes, so that its coherent fraction makes a significant contribution to the overall signal.

The coherent fraction of the cathodoluminescence preserves the memory of the primary (electron, electron) interaction. As such, it constitutes a physical signature of any plasmon excitation and is exploited for near-field imaging. The concept of electromagnetic density of states introduced to rationalize the EELS technique is also relevant in the case of the coherent cathodoluminescence [117, 118]. However, radiative relaxation is only one of the many damping channels available to a plasmon to relax its energy. As a consequence, only part of it is radiated in the far-field. So, while the EELS signal can be closely connected to the full electromagnetic density of states projected along the electron beam trajectory, only the radiative part of it is presumed to describe the CL signature. Replaced within a far-field optical description, the EELS signal is proportional to the extinction cross section of the plasmon particle $\propto \Im(\alpha)$, while the CL signal scales with its scattering cross section $\propto |\alpha|^2$. In more details, CL can be computed by estimating the energy carried out by the induced field and propagating to the far-field. The CL probability Γ_{CL} depends on the radiative modes of the particles [Losquin [117]] according to Eq. (4.21) valid for an arbitrary shape object in the quasi-static approximation:

$$\Gamma_{\text{CL}}\left(\omega, \overrightarrow{R_\perp}\right) = \frac{\omega e^2}{4\pi^2 \hbar c^3} \sum_i \sum_j f_i(\omega) f_j^*(\omega) \int_{\Omega_D} d\Omega\ \overrightarrow{p_{i\perp}}(\Omega) . \overrightarrow{p_{j\perp}}^*(\Omega) . TF_z\left(\overrightarrow{E}_i^z\left(\overrightarrow{R_\perp}, \frac{\omega}{v_e}\right)\right) . TF_z\left(\overrightarrow{E}_j^{z*}\left(\overrightarrow{R_\perp}, \frac{\omega}{v_e}\right)\right) \tag{4.21}$$

$$\Gamma_{CL}\left(\omega, \overrightarrow{R_\perp}\right) = \frac{\omega e^2}{4\pi^2\hbar c^3}\sum\nolimits_i |f_i(\omega)|^2$$
$$\int_{\Omega_D} d\Omega \left|\overrightarrow{p_{i\perp}}(\Omega)\right|^2 . \left|TF_z\left(\overrightarrow{E}_i^z\left(\overrightarrow{R_\perp}, \frac{\omega}{\upsilon_e}\right)\right)\right|^2$$
$$+\sum_i \sum_{j<i} \text{coupled intermode terms}\left(\overrightarrow{p_{i\perp}}, \overrightarrow{p_{j\perp}}^*\right) \tag{4.22}$$

where $f_i(\omega)$ is the generalized polarizability function introduced for the EELS description, $\overrightarrow{p_{i\perp}}(\Omega)$ is the transverse dipole of mode i, and $TF_z\left(\overrightarrow{E}_i^z\left(\overrightarrow{R_\perp}, \frac{\omega}{\upsilon_e}\right)\right)$ the Fourier transform along the e-beam direction z at impact point $\overrightarrow{R_\perp} = (x, y)$ in real space and momentum $k_z = \omega/\upsilon_e$ in reciprocal space of the electron-induced electric near-field $\overrightarrow{E}_i^z$. As a first approximation, the formal descriptions of the EELS and CL signals show great similarities in both spatial and spectral distributions of the experimental signals, see Eqs. 4.19 and 4.22. The differences lie in two aspects. First, the CL response is weighted by the dipolar moment of the considered mode. Only modes with finite net dipole moments will radiate efficiently to far-field. Second, compared to EELS, CL resonant line shapes may be shifted and asymmetric [33, 118, 128]. The spectral shift is reminiscent of the shift between the absorption and scattering cross sections for objects of size in the quasi-static regime [96, 166]. Potential CL asymmetry is due to interferences in the far-field from spectrally and spatially overlapping modes (i, j), provided their respective dipole moments are nonorthogonal, see coupled intermode terms in Eq. (4.22). Note that as observed for EELS, the CL response is a function of the impact parameter $\overrightarrow{R_\perp}$, and so the light emission pattern depends not only on the geometry of the object but also on the electron impact geometry.

Connection with the scattering cross section can be formally established in the case of a sphere of diameter a in the quasi-static regime. In this case, the CL probability can be expressed as [101]:

$$\Gamma_{CL}^{sphere}\left(\omega, \overrightarrow{R_\perp}\right) \propto a^6 |\alpha_{l=1}(\omega)|^2 \exp\left(-\frac{R_\perp}{\gamma\upsilon_e/\omega}\right) \tag{4.23}$$

with again $\alpha_{l=1}$ the sphere polarizability for the mode $l = 1$ and γ the Lorentz factor. The CL signal scales with the scattering cross section of the target particle. Its probability is position-dependent, a characteristic shared by all electron-probe-based techniques. It follows a quasi-exponential decay with the impact parameter $R_\perp$ and a signal attenuation length $\gamma\upsilon_e/\omega$ ~ 10–100 nm. As for EELS, the latter accounts for the coupling between the transient field generated by the fast-moving electrons and the resonant plasmon modes and therefore mainly defines the spatial resolution.

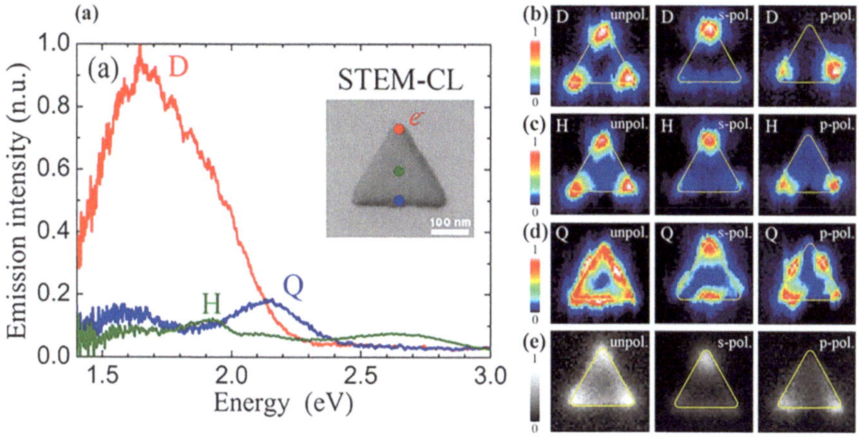

Fig. 4.13 Coherent CL-STEM near field imaging of a 266 nm edge Au prism at plasmon resonances. (**a**) Measured spectral dependencies of detected in-plane modes, unpolarized photon-emission intensity spectra collected for different electron-beam probe positions, as shown on the image inset. Three in-plane modes can be resolved, D dipolar, Q quadrupolar and H hexapolar modes. (**b–d**) Measured monochromatic CL photon maps of the modes D (**b**), H (**c**), and Q (**d**) at the peak energies for un-, s-, and p-polarized emission. Each pixel is associated with a different position of the electron beam. (**e**) Corresponding experimental energy-integrated photon maps. Polarization-resolved measurements allow direct imaging of energetically degenerate dipolar and quadrupolar plasmonic eigenstates. The color bar of the photon maps is linear and common to all emission polarizations. (Reprinted (adapted) with permission from Ref. [139]. Copyright 2018, American Chemical Society)

Near-Field Mapping and Engineering Figure 4.13 demonstrates the capabilities of the coherent CL imaging technique to spatially map degenerate states of plasmon modes localized on an individual metallic nanoparticle, namely a prism [139]. These results are obtained thanks to a combination of angle-, polarization-, and space-resolved cathodoluminescence imaging. For a regular prism of C_{3v} symmetry, the dipolar and quadrupolar in-plane plasmon modes exhibit two degenerate eigenmodes of symmetry E_1 and E_2 [7]. Experimentally, CL microscopy detects three different modes, corresponding to the dipolar, quadrupolar, and hexapolar modes (respectively, D, Q, and H in Fig. 4.13). These modes are efficiently excited for an electron beam focused at the corner (D, red spectrum), mid-edge (Q, blue spectrum), and center (H, green spectrum) of the nanoprism, respectively. For the two-fold degenerated dipolar and quadrupolar modes, the coherent CL photon maps collected under constant emission angle and (S, P) orthogonal polarizations exhibits signatures in close agreement with the expected nondegenerated near-field distributions $\left|TF_z\left(\vec{E}_i^z\left(\vec{R}_\perp, \frac{\omega}{v_e}\right)\right)\right|^2$ of the induced electric field projected on the z-direction of the electron beam [139]. These nondegenerated modes correspond to those that can be selected by proper polarization of the incident light [7] for a pure optical excitation scheme. The unpolarized mode H reflects a hexapolar dark out-of-plane mode of A_1 symmetry.

In addition to the latter example, numerous recent results demonstrate the great capabilities of the CL imaging technique as an efficient investigation tool of the near-field physics [99, 128, 170, 171, 185, 201, 210].

Instrumentation CL imaging can be implemented in both SEM (SEM – CL) and STEM (STEM – CL) microscopes. The electron microscope provides the excitation electron source. The main distinction between SEM and STEM setups lies in the primary electron energy range, i.e., typically a few 10 keV for a SEM and a few 100 keV for a STEM. The CL acquisition setup consists in a collecting off axis mirror (parabolic, elliptical half mirror) sending the emitted light to various detection devices: spectrometer and/or imaging *charge coupled device* CCD camera. Common available working modes are imaging, spectral and so called conoscopic (angle resolved) modes [31, 32]. In imaging mode, the electron beam is raster scanned over the specimen area and the CL signal is acquired at each beam position. In spectrometric mode, the collected photons are directed toward a spectrometer to generate a full CL spectrum. The angle-resolved mode gives access to the wave vector distribution of the light emitted from the specimen (momentum distribution). Geometrical optics applied to a parabolic off axis mirror demonstrates that light rays imaged on a flat detector reflects the wave vector distribution at the emission sites. Each point on the CCD flat detector corresponds to a unique emission angle described by a zenithal angle θ ([0°, 90°] range where $\theta = 0°$ refers to the normal to the surface) and an azimuthal angle φ running from 0° to 180°, see Fig. 4.14. The use of a moving pin-hole allows angle-resolved spectral mode. Additional use of half wave plate (polarizer) on the CL light beam allows extraction of the polarization of the light. As part of the many detection modules available on a SEM or STEM microscope, CL imaging can be acquired in combination with high spatial resolution image of the specimen. Beyond the electron beam diameter, the spatial CL imaging resolution has its source in the evanescent coupling between the primary electron field extension and the resonant plasmon modes. The extension of the electron field scales with the energy of the primary electron beam. For a 30 keV primary electron beam the spatial resolution of coherent CL imaging is close to 10 nm (7 at 10 keV) [170]. Spectral resolution is primary linked to the spectrometer entrance slit. Routine spectral resolutions in the nm are easily achieved.

CL Imaging – Variants and Perspectives

Spectral CL imaging. Hyperspectral CL imaging benefits from advances in light detection technology. Silicon based technology encountered in *charge coupled device* CCD and *complementary metal oxide semiconductor* CMOS imaging systems covers already a large wavelength interval from deep UV 200 nm to the NIR 1100 nm. In line with the developments observed in EELS, CL developments in light detection aim to extend the energy range from the visible to the infrared, opening CL imaging to the physics of phonons. Light detectors based on Ge (0.8–1.8 μm), InGaAs (0.7–2.6 μm), PbSe (1.5–5.2 μm), InAsSb (1.0–5.8 μm), and HgCdTe (2.0–10.6 μm) materials could cover the near-, short-, and mid-wavelength infrared ranges.

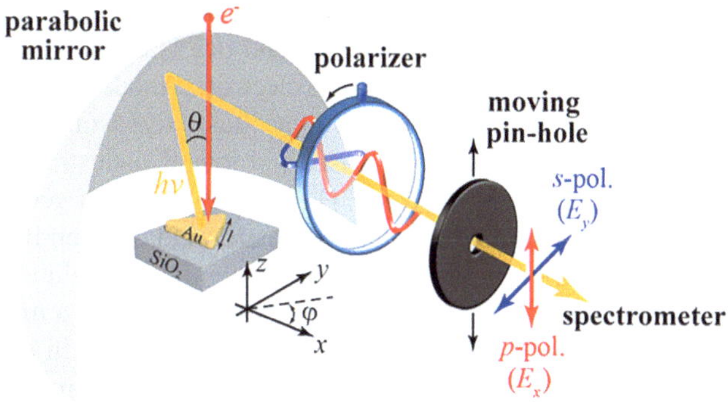

Fig. 4.14 Schematic representation of a CL- SEM, resp. STEM imaging setup allowing angle- and polarization-resolved acquisitions. Red arrow – The specimen is excited by fast electrons moving along its normal z axis. The e-beam is raster scanned in the xy plane. Orange arrow – A parabolic mirror collects and redirects the CL light to a spectrometer. Emission direction at angles (θ, φ) is selected by a moving pinhole mask. A polarizer allows to extract the polarization of the CL signal, p and s light polarization components. (Reprinted (adapted) with permission from Ref. [139]. Copyright 2018, American Chemical Society)

Chiral light-matter interactions. The spontaneous emission of a quantum emitter is the hallmark of its electromagnetic environment. The latter can be purposefully designed by using plasmonic nano-antennas, which allows a precise control of the light and matter interaction. Access to chiral light-matter interaction is a step in this direction. CL microscopy appears to be a tool of choice for the development of chiroptics [209]. As described above, polarization filtering of the CL light is one important degree of freedom in CL technique. By further resolving a CL signal into its left- and right-handed-circular polarization components, generalized chiral radiative EMLDOS that governs radiative decay can be further investigated. Several teams have distinguished themselves along this line, for example, Zu et al. [210] has studied the chiral photoluminescence of hybrid objects made of asymmetric Au nanoantennas. The choice of the impact position of the primary electrons on the antenna generates a set of chiral plasmonic environments that can be exploited at the level of quantum emitters.

CL 3D tomography. Following similar developments in EELS technique, 3D tomographic reconstruction of the near-field distribution of an object has been successfully completed in cathodoluminescence [9]. The 3D reconstruction is carried out by a numerical algorithm from a series of experimental data acquired under different ultrafast electron excitation angles. CL tomography imaging should benefit from the availability of tilt stage in electron microscopes.

Time-resolved CL measurement. The use of an ultrafast pulsed electron source gives access to time-resolved measurements and offers new opportunities for the study of the dynamics of the CL emission at small length scale. To date the developments of pulsed electron source follows two directions [133], (i) the ultrafast electrostatic beam blanker and (ii) the laser driven photocathode and its variants [78]. The first option exploits a set of electrostatic deflection plates that modulate rapidly the electron beam, the second one is based on photoemission events subsequent to the illumination of the electron cathode by laser light pulses operating in the UV wavelength range. One illustrative example is the study of carrier dynamics in single GaAs-based pyramidal nanostructures by *time-resolved cathodoluminecence* TR – CL [129]. The excitation geometry enables lifetime and pump–probe measurements with laser and electron beams as pump and probe, or vice versa. As of today (2020), the temporal resolution of the time-resolved CL technique is within the ps time range, a too long a time for the full investigation of plasmonic phenomena (1–10 fs time range). In complement to an ultrafast electron source, the temporal resolution is mostly limited by the light detection system, so it can be further improved by using a faster photon detector (*avalanche photodiode* APD) or a streak camera [129, 149].

4.4.3 Photon-Induced Near-Field Electron Imaging Microscopy (Photon In, Electron In, Electron Out)

Imaging Principle The use of an impinging fast electron beam to investigate the near-optical field guarantees access to an unmatched high spatial resolution. However, the provided excitation regime generates a transient electric field equivalent to a plane wave pulse of broadband frequencies and whose polarization cannot be controlled. It differs markedly from that generated by a monochromatic plane wave of well-defined energy, momentum, and polarization. This observation has led to the development of a combined approach based on the use of electron pulses of femtosecond (fs) duration synchronized with fs photon pulses, thus reaching the spatial, energy resolution and polarization characteristics of electron and optical microscopies, respectively. The achieved temporal resolution is on the time scale of the evanescent fields generated, namely the femtosecond domain. This combined microscopy is named *photon-induced near-field electron microscopy* PINEM [10]. The PINEM microscopy is based on a three-body interaction (photon, plasmon, electron). In free space, photons and electrons interact weakly because of a large momentum mismatch. Following an inelastic scattering event between a fast electron and n photons of energy $\hbar\omega_{\mathrm{p}}$, the change in the electron momentum can be expressed as $\Delta k \approx \pm\ n.\hbar\omega_{\mathrm{p}}/v_e$, where v_e is the electron velocity and the $\pm$ sign denotes the energy gain ("absorption") or loss ("emission") of photons by the electron. Whatever the electron beam energy, this quantity is larger than the momentum an ensemble of n photons can provide $\pm\ n.\hbar\omega_{\mathrm{p}}/\mathrm{c}$. In solids, plasmon near-fields constitute additional sources of momentum and so a (photon, electron) interaction is possible through plasmon exchange. In this three-body interaction scheme, the electron packets lose (*electron energy loss* EEL) and/or gain (*electron energy gain* EEG) energy in discrete quanta of the photon energy $\hbar\omega$. By energy-filtering only those electrons resulting from one specific photon absorption (resp.

electron energy gain), it is possible to (i) image in space the optically induced evanescent nearfield, i.e., map the (photon, electron) interaction, (ii) obtain its temporal profile on a femtosecond timescale and (iii) map its spatial polarization dependence (polarization control of bright modes).

In photon-induced near-field electron microscopy, an ultrashort optical pulse induces an evanescent oscillating field in the nanoparticle of interest. Simultaneously a fast electron pulse probes it inelastically along the electron beam trajectory [69, 86, 151–153].

$$\Gamma_{\text{PINEM}}\,(\mathrm{x},\mathrm{y},\Delta t) \propto \left| \int_{+\infty}^{-\infty} E_z^{\text{photon}}\,(x,y,z,\Delta t)\, e^{-i\left(\frac{\omega_p}{v_e}\right)z} dz \right|^2 = \left| TF_z\left(\overrightarrow{E}_z^{\text{photon}}\left(\overrightarrow{R}_\perp, \frac{\omega_p}{v_e}\right)\right)\right|^2 \tag{4.24}$$

The PINEM first order transition probability Γ_{PINEM} scales with the Fourier transform of E_z^{photon}, the photon induced electric field component along the e-beam direction at impact parameter $\overrightarrow{R_\perp} = (x, y)$ in real space, momentum value $k_z = \omega_{\text{p}}/v_e$ in reciprocal space and time delay Δt between photon and electron pulses. PINEM exhibits close similarities with EELS and CL signals. Main difference lies in the discrete energy exchange. In PINEM a single photon quantum $\hbar\omega_p$ is involved, whereas in EELS and coherent CL broadband excitation of the target particle takes place. Reformulated for a spherical object in the quasi-static regime, the PINEM transition probability is shown to scale with the scattering cross section of the object $\sigma_{\text{scattering}} \propto |\ \alpha\ |^2$. It is a position-dependent quantity with a quasi-exponential decay with e-beam to object distance $\overrightarrow{R_\perp}$ and an attenuation length $\gamma v_e/\omega$ ~ 10–100 nm which defines the spatial resolution [8].

$$\Gamma_{\text{PINEM}}^{\text{sphere}}\left(R_\perp, \omega_p, \Delta t = 0\right) \propto \left(\frac{\omega}{\gamma v_e^2}\right)^2 I_0 |\alpha_{l=1}(\omega)|^2\, \delta\left(\omega - \omega_p\right) \exp\left(-\frac{R_\perp}{\gamma v_e/\omega}\right) \tag{4.25}$$

with I_0 the laser irradiance, γ the Lorentz factor accounting for the relativistic nature of the fast electrons $\gamma = \sqrt{1 - v_e^2/c^2}$ and $\alpha_{l=1}$ the sphere polarizability for the mode $l = 1$.

The PINEM imaging principle constitutes a privileged place to finely probe the (photon, matter) interaction, consequently it is the object of continuous theoretical and experimental developments, see for instance strong coupling (photon, electron) regime in a photonic cavity [197], anomalous PINEM regime [150].

Near-Field Mapping and Engineering
The near-field imaging capabilities of the PINEM microscopy are demonstrated in Fig. 4.15 in the case of metallic nanoparticles [202]. The experiment aims to map the optical nearfield of a set of silver spheres under localized plasmon resonance. Two silver spheres, 37 nm in diameter and 70 nm apart, are excited by overlapping photon and electron pulses in both space and time. The light excitation is carried out at a resonance wavelength of 520 nm (green light, 2.4 eV) with two orthogonal linear polarizations, the light pulse duration is 200 fs. The average kinetic energy of the electron beam is 200 keV, its temporal profile is similar to the light pulse one. The picture is built by collecting electrons whose kinetic energy lies on the energy gain side of the reference zero loss peak. The energy gain windows is set at [1, 11 eV] to cover multiple photon absorption events $n.\hbar\omega_{\mathrm{p}}$, where n is an integer and $\hbar\omega_{\mathrm{p}}$ the plasmon energy quantum. The PINEM transition probability map exhibits clear features of a dipole shape field. The signal is distributed on either side of a nodal plane according to the polarization of the incident optical beam. The experimental PINEM signature rotates with the rotation of the incident in-plane field. While EELS and CL are mainly sensitive to the LDOS, this polarization control constitutes a clear evidence that the PINEM contrast is directly induced by the photon light pulse and the resulting optical near-field.

Plasmon Dynamics More recently, dynamical imaging of the propagation of surface plasmon polariton wavepackets has been demonstrated by PINEM [121]. The dynamics capture exploits the pump-probe character of the technique, see Fig. 4.16. A first optical pump pulse generates SPP waves at specific launch sites of a buried Ag/Si_3N_4 interface. The propagating plasmonic wave is then revealed by probing it at variable time delays by an unfocused (spatially dispersed) pulse of ultrafast electrons. The PINEM directly images the coherent interference pattern of the propagating SPPs and measures its group velocity. In addition, since PINEM imaging exploits scattered electron beams in transmission mode, it gives access to subsurface plasmonic signature entirely buried inside the material. This latter point open access to the characterization of active medium embedded in protective material, as encountered in 2D heterostructures and advanced multilayer devices.

Instrumentation Addition of the PINEM imaging mode in a TEM or STEM electron microscope implies access to an ultrafast pulsed electron source. The technology used is often that of a pulsed laser-driven cathode based on a pulsed laser light source. A femtosecond laser light beam is split into two coherent beams by passage in a beamsplitter. One of the beams is directed to the microscope electron source and generates ultrafast electron pulses via direct photoemission or field assisted photoemission of electrons [27, 54, 78, 137]. To overcome the work function of the electron cathode, the light pulse first passed through a nonlinear optical medium to increase the photon energy by harmonic generation. The electron pulse characteristics derive from those of the light pulse, in particular electron pulse durations in the tens of femtoseconds are accessible. In parallel, the second laser beam is used to pump the system, while being imaged by the electron pulse. In a time-resolved measurement, the relative delay between the electron and laser pulses

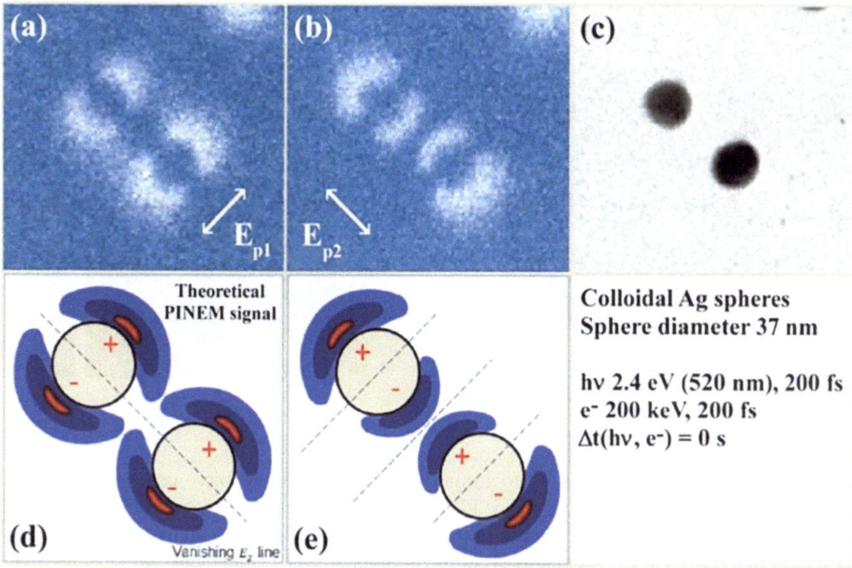

Fig. 4.15 PINEM imaging of two close-by Ag spheres, 37 nm in diameter. (**a**, **b**) Experimental PINEM signals for two orthogonal in-plane polarizations. The PINEM image contrast exhibits characteristics of a dipolar plasmon signature. Photon pulse 532 nm, 200 fs. Electron pulse 200 keV, 200 fs. Photon and electron pulses are overlapping in both space and time. PINEM image is built from scattered electrons that have gained energy in respect to the zero-loss peak (EEG mode); (**c**) bright field transmission electron microscope image; (**d**, **e**) theoretical PINEM field simulated according to the model developed in [151]. (Reprinted (adapted) with permission from Ref. [202]. Copyright 2012, American Chemical Society)

is controlled precisely by a linear delay stage. According to the physical need, the photon energy of the pump beam can be adapted by harmonic generation (Fig. 4.17).

PINEM Imaging – Variants and Perspectives

Time-resolved PINEM. The PINEM technique is intrinsically based on a pump-probe scheme approach where electrons and photons can play one or the other function. Its time resolution mainly depends on the duration of the electron pulse, which covers the range from several hundreds of femtoseconds to a few tens of picoseconds. Ongoing developments of electron sources aim to improve the temporal resolution and coherence length of the ultrafast electron beam in order to capture faster dynamics. Advanced developments based on optical gate technique have demonstrated PINEM imaging with electron pulses of time resolution < 30 fs [77]. In the near future, optical gating exploiting attosecond laser beam is expected to generated sub femtosecond electron pulses for electron imaging (“attomicroscopy”) [78, 137].

PINEM spectral range. Following EELS and CL development trends, imaging energy exchange in the infrared wavelength range to reveal low energy excitations such as phonons or bond vibrations constitutes a natural extension of the

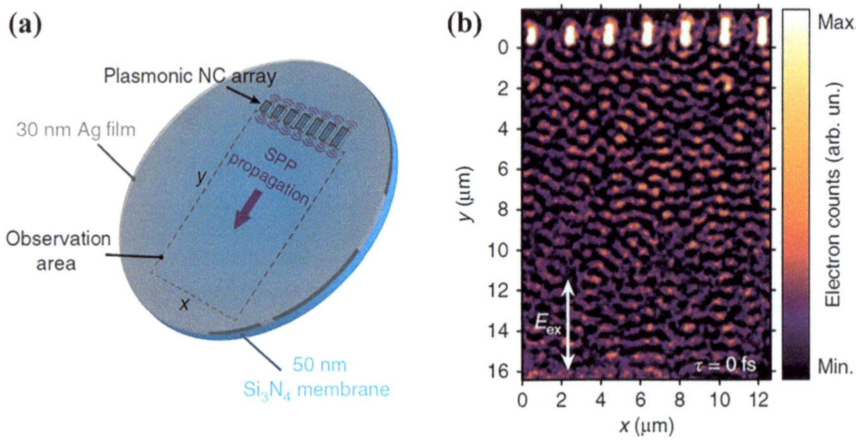

Fig. 4.16 Study of surface plasmon polariton dynamics by PINEM microscopy. (**a**) Schematic of the experiment. (**b**) PINEM imaging. Plasmonic interference pattern generated at a buried Ag/Si_3N_4 interface. The image is recorded at zero delay between light and electron pulses, using only electrons that have gained energy from the plasmonic near-field. (Reprinted (adapted) with permission from Ref. [121]. Copyright 2016, Springer Nature)

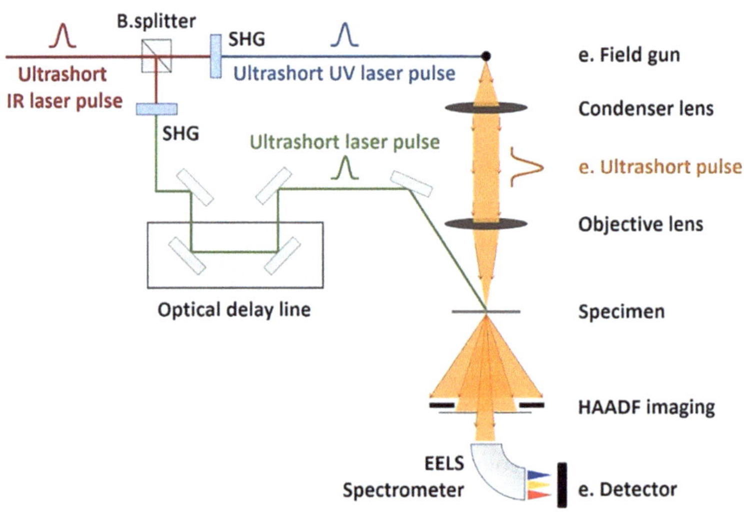

Fig. 4.17 Schematic representation of a PINEM STEM imaging setup. The specimen is simultaneously excited by a pulsed fast electron beam moving along its normal z axis and a pulsed light beam. The e-beam is raster scanned in the specimen xy plane. An energy-electron prism guarantees the collection of PINEM EEG and EEL spectra. Simultaneous high-angle annular dark-field HAADF imaging is acquired from incoherently scattered electrons. Laser driven cathode generates electron bunches in the tens of fs range duration. Beam energy is in the 60–200 kV range. Gain and loss of energy spectra in the range [+/−10 eV] in respect to the zero loss peak ZLP can be acquired

technique. A pioneering work along this line demonstrates clear feasibility in the near infrared range (1038 nm, 1.2 eV) [114].

Continuous wave resonant PINEM variant. In PINEM, the need for electron and photon pulsed sources imposes heavy instrumental developments, therefore only a few *ultrafast electron instruments* (*ultrafast transmission electron microscope* UTEM, *scanning ultrafast electron microscope* SUEM) are available worldwide. A simpler approach has recently been proposed [115], the continuous wave resonant photon stimulated electron imaging. It proceeds by the simultaneous irradiation of the specimen by a continuous wave laser beam and a continuous electron current. At specimen resonance, a plasmon can mediate energy transfer from the photon field to the electron probe and vice versa, allowing access to stimulated electron energy-gain and energy-loss responses. The experiment works in a low irradiance regime (<1 GW/m^2) and light excitation at various wavelengths can be implemented.

4.4.4 Photoemission Electron Imaging Microscopy PEEM (Photon In, Electron Out)

Imaging Principle

PEEM microscopy is a cathode lens electron microscopy, whose imaging principle consists in the collection of the electrons emitted by a surface regardless of the physical phenomenon causing it: thermal-, photo-, photoassisted field-, and/or field emissions. In short, emitted electrons are spatially imaged at the mesoscopic level [15]. It is a full field microscopy exploiting well-established electron optics, and no physical probe is involved. Most frequently, the primary photoelectric process is linear in nature and is produced by sample excitation with UV or X-ray light illumination. For low photon energies, the contrast corresponds to differences in surface work functions, electron density of states, and topographical relief. At higher excitation energies, element-selective imaging is possible.

By substituting the UV/X-ray light source with a femtosecond laser source operating in the visible and/or IR wavelength ranges, PEEM microscopy becomes a nonintrusive method of near-field mapping [28, 30, 48, 49, 81, 105, 138, 154, 199]. Optical near-field enhancement magnifies the photoelectric effect. For noble metals, the energies of plasmon resonances are in the visible and near infrared wavelength ranges, i.e., 520 nm (2.38 eV) – 1400 nm (0.885 eV), all energies lower than any metal work function (around 4.5 eV for Cu, Ag, Au [97, 175]) for which no direct photoelectric effect can take place. In contrast, the optical near-field associated to a plasmon resonance can initiate nonlinear photoemission processes. Therefore, PEEM images display maps of the near-field distribution at a subwavelength spatial resolution. The routine spatial resolution is 15 nm. As an instrumental corollary, the use of a coherent pulsed laser source provides PEEM microscopy with a wide range of optical spectrometric analyses (wavelength, polarization, irradiance, and pump probe analysis) so that a near-field investigation can be conducted at multiple scales: spatial (nm), spectral (meV), and temporal (fs).

For the emission of electrons from plasmonic nanoparticles, two mechanisms are considered: a surface and volume ones [132, 157, 187]. The surface photoeffect corresponds to a photon absorption event taking place at the particle surface boundary. The volume photoeffect is frequently described through a three-step model: (i) photon absorption inside the particle volume, (ii) transport of the excited electrons to the particle boundary, and (iii) electron emission at surface. The inelastic scattering between electrons during the transport step reduces the kinetic energy of the hot electrons. In fact, one simple collision of a hot electron with a cold one renders it unable to overcome the potential barrier at the metal boundary. For kinetic energies in the plasmon energy range, the electron mean free path amounts to tens of nm (= 41 nm for Au, 1 eV) [56]). So, in the general case, surface photoeffect is larger than the volume one, especially for large particles whose sizes are larger than the inelastic electron mean free path.

For a nanoparticle sustaining a localized surface plasmon resonance, the nonlinear photoemission yield associated to the surface photoeffect mechanism is proportional to the 2n-power of the normal-to-the-surface component of the electric field $E^{\perp}_{\text{int.}} = \overrightarrow{n_{\perp}}.\overrightarrow{E_{\text{Surf. int.}}}$. Correspondingly, the full photoelectron emission rate due to electron collisions with the nanoparticle surface is proportional to its surface integral where the integral extends over the entire nanoparticle surface Eq. (4.26):

$$R^{\text{Surface}}_{\text{Emission}} \approx \frac{2}{\pi^2}\frac{e^2E_F^2}{\hbar}\frac{1}{(\hbar\omega)^3}\int_{\text{Surface}}\left|E^{\perp}_{\text{Surf. int.}}\right|^{2n} dS \tag{4.26}$$

$R^{\text{Surface}}_{\text{Emission}}$ is the electron emission rate [e$^-$/s]; n the nonlinearity order, i.e., the number of photons absorbed to overcome the work function of the specimen; and $E^{\perp}_{\text{Surf. int.}}$ the normal to the surface component of the electric field inside the particle [55]. Note that normal surface components in the different dielectric media, i.e., inside and outside the metallic particle, are related by the conservation of the displacement vector $\varepsilon_{\text{int}}.\overrightarrow{E_{\text{Surf. int.}}} = \varepsilon_{\text{ext}}.\overrightarrow{E_{\text{Surf. ext.}}}$. In term of local density of states, the PEEM signal scales with the electromagnetic density of states projected on the object surface normal.

For spatially extended samples, propagative plasma waves, i.e., *surface plasmon polariton waves* SPPs, come into play. In a classical illumination geometry, the sample is under full field illumination. Both incident laser and plasmon electric fields are present at the sample surface and differences in group velocities between the wavepackets translate into beating patterns [24, 71, 72, 95, 105, 109, 112]. These interference patterns offer direct ways to extract physical parameters from the propagative plasma waves (wavelength, group velocity, dispersion curve) and their surroundings.

By natural extension, the use of a coherent ultra-short pulsed light source allows the implementation of spatially resolved pump probe experiments shining light on the spatio-temporal dynamics of plasmon polariton waves. This important variant is termed *time-resolved* (TR) nPPE – PEEM, sometimes referred as *interferometric time-resolved* ITR – PEEM. It demonstrates clear access to near-field dynamics within individual particles and solids [4, 38, 39, 73, 105, 112, 157].

Near-Field Mapping and Engineering Due to the multiple degrees of freedom offered by a laser light pulse (polarization, wavelength, irradiance, time profile, and pulse shape), PEEM microscopy is a powerful tool for mapping, nanoscale diagnosis and manipulation of the optical near-field. Figure 4.18a illustrates these possibilities in the case of localized plasmon resonances of sharp-edged metal nanoparticles of subwavelength dimensions [81]. The object considered is a multi-tipped Au sphere obtained by colloidal synthesis. The use of polarized light allows the selective excitation of any tip of the object by aligning the incident electric field on its axis of symmetry, see Fig. 4.18c–e. Due to its geometry, the plasmon excitation of a tip manifests itself by the appearance of a field hot spot at its end. The diameter of the hot spot quantifies the near-field resolution of the instrument, in this case about 25 nm. The enhancement factor of the amplitude of the electric field is of the order of a few tens. This method of selective excitation of a plasmon eigenstate by means of polarization is one of the simplest techniques for manipulating the optical near-field at small scales, other methods dealing with interference of near-field components or temporal pulse shaping have also been implemented [2, 40, 130].

Plasmon Dynamics In parallel to its near-field mapping capabilities, PEEM microscopy is also adapted to the study of the spatio-temporal dynamics of plasmon waves occurring at nanometric and femtosecond scales. The method is implemented in a pump probe approach. The primary laser beam is coherently divided into a pump and probe beams, whose time delay is adjusted by an optical delay line stage. The PEEM contrast then changes as a function of the time delay between the two pulses in a way that reflects the plasmon dynamics. An illustrative example conducted on a single crystal Au film is reported in Fig. 4.19 [95]. By illumination of a grating coupler, a first pulse ("pump") generates surface plasmon polariton waves. A second delayed pulse ("probe") interferes locally with these surface waves. The constructive interference between the surface wave packets and the probe pulse results in an exaltation of the electron photoemission yield at the position of the wave packet. Therefore, PEEM images obtained for different time delays between pump and probe pulses directly visualize the propagation of the surface plasmon polariton waves and provide access to its phase and group velocities. Addition of nonlinear media for harmonic generation opens ways to multicolor time-resolved measurements. Such an approach is at work in hot electron dynamics studies [58, 87, 122].

Instrumentation In its simplest configuration, an electron emission instrument is composed of an imaging column with appropriate electron optics [15]. The imaging column consists of an objective, a transfer lens, a field lens, an intermediate lens and several projective lenses. This lens arrangement allows imaging of either the image or the back focal (diffraction) plane of the objective lens over a large magnification

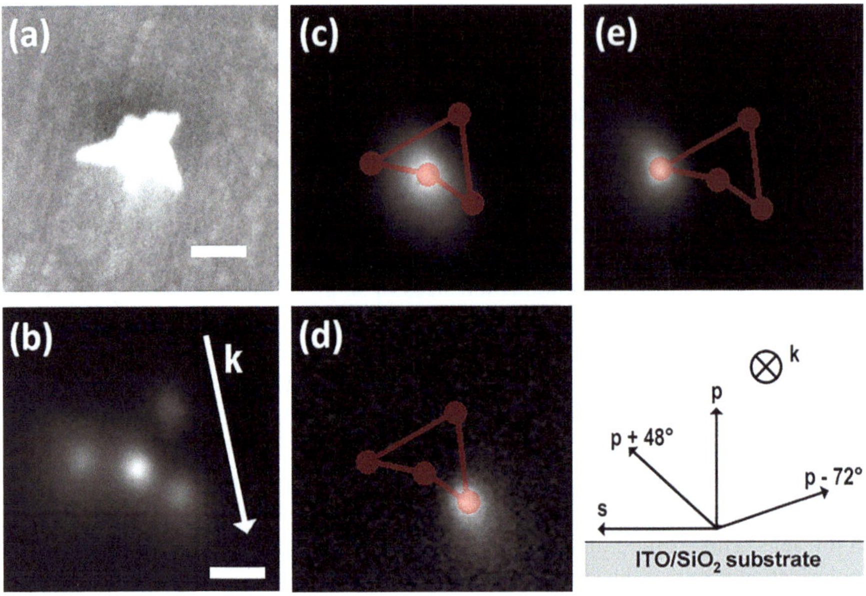

Fig. 4.18 PEEM. Near-field mapping and manipulation. Selective excitation of the tip of an Au nanostar by light polarization. Au nano star, 800 nm wavelength, variable polarizations. Scale bar 100 nm. SEM imaging (**a**), PEEM imaging (**b**–**e**). (Reprinted (adapted) with permission from Ref. [81]. Copyright 2011, American Chemical Society)

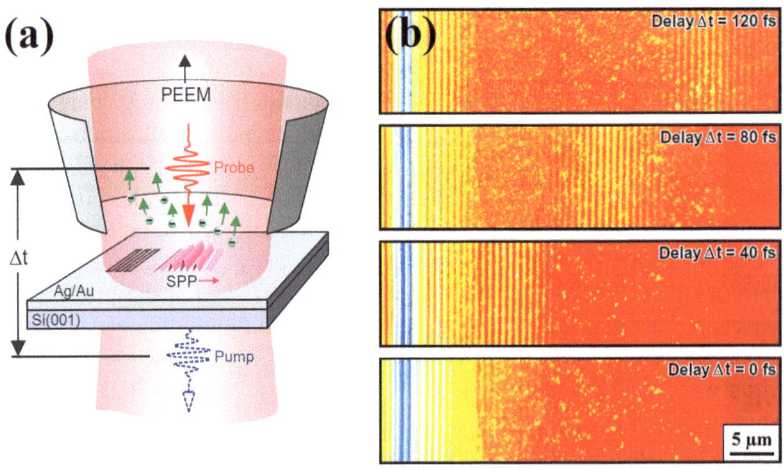

Fig. 4.19 Study of plasmon dynamics by time-resolved PEEM microscopy. Propagation of plasmon polariton waves at the surface of a single-crystal Au metal film. (Adapted from Ref. [95]. Copyright 2018, Springer)

range. In contrast to instruments based on a scanning probe (SEM, STEM), a PEEM instruments builds its images from a full field imaging principle equivalent to the one of a light microscope. The PEEM instrument is frequently equipped with an electron illuminating column which allows additional imaging modes, in particular the *low energy electron* imaging mode LEEM based on the collection of the backscattered electrons. The illuminating column consist of an electron gun (LaB_6 emitter, Shottky field gun, cold field gun) and a set of condenser lenses. Imaging and illuminating columns are mounted on a magnetic prism separator to separate incident and reflected electron beams according to their respective velocity vectors. Addition of an energy dispersive analyzer allows spectroscopic imaging at selected electron energies. The instruments operate in the 10–20 kV voltage range to reduce the aberrations of the electron optics. The imaging resolution is typically 5 nm and 15 nm in LEEM and PEEM imaging modes, respectively. The resolution of the instrument is limited by the spherical and chromatic aberrations of the objective lens. Aberration correction is possible by adding an electron mirror. The LEEM resolution on corrected instrument is < 2 nm.

At the crossroads between Solid State Physics and Optics, the nPPE PEEM microscopy inherits from considerable technical developments in ultrafast optics. The light source consists of an ultrashort pulsed laser. Common pulsed light sources are femtosecond Ti:Sapphire lasers operating in the near-infrared region in the wavelength range [700, 1000 nm] and pulse duration in the [10, 150 fs] time interval. Complementary wavelength ranges are accessible by frequency conversion processes based on nonlinear optical interactions (optical parametric oscillator, harmonic generation). The photon path is equipped with focusing and polarizing optics. Regarding illumination, the main degrees of freedom are the wavelength, the polarization of the light beam, its irradiance, and time profile. Illumination angles are discrete and determined by the positions of viewports on the UHV chamber. The energetic resolution in the excitation spectrometric mode amounts to 1 nm (2 meV) for a 150 fs pulse at 800 nm wavelength. Time-resolved investigations are possible for illumination in the pump probe configuration, see Fig. 4.20.

PEEM Imaging – Variants and Perspectives

On an instrumental level, the perspectives aim at further exploiting the multi-scale (nm, fs, meV) character of the method. Technical improvements concern both the instruments and the light sources.

Key elements regarding instrumental developments focus on (i) the correction of the aberrations of the electron optics (mirror aberration corrector); (ii) the implementation of new electron detector schemes, such as *time of flight* TOF or *complementary metal oxide semiconductor* CMOS-based detectors, the latter offering larger dynamic range, superior signal to noise ratio in comparison with the traditional *microchannel plate* MCP detector, and improved spatial resolution at lower magnifications; (iii) the integration of spin polarized electron sensitive devices as both electron source and detector; (iv) the development of bright and highly coherent field emission electron source of low energy spread for HREELS imaging, i.e., full field EELS imaging in LEEM imaging mode in the IR excitation range [124]. Lehr [111] illustrates such instrumental trends. The study monitors the distribution in moments and energy of the electrons emitted by a Au nanorod at

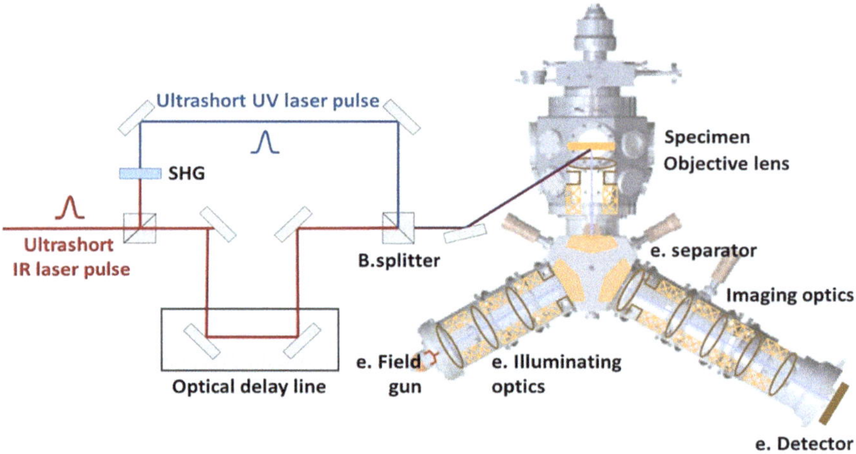

Fig. 4.20 nPPE PEEM/LEEM set up. In a time-resolved configuration, the specimen is successively excited by two femtosecond pulses delayed from each other through an optical delay line interferometer. The PEEM contrast (photoelectric yield contrast) changes as a function of the time delay between the two pulses in a way that reflects the plasmon dynamics. In a two-color "pump probe" approach, a first IR pulse generates a hot electron population followed by a second UV pulse that ejects it into vacuum. By analyzing the spatial distribution of the photoemitted electrons as a function of their energy and the delay between pulses, the transport of the electron population can be tracked. (Adapted from Elmitec GmbH)

plasmon resonance. It is conducted on a time-of-flight electron detection instrument. Examination of the electron distribution reveals the competition of two distinct emission processes. The first is a multiphoton process, the second one a field emission process. The balance from one to the other is driven by the incident irradiance. This study illustrates the interest of the technique for the investigation of high intensity electromagnetic fields and its associated physical phenomena, such as ponderomotive effects, or *above threshold photoemission* ATP, a process of electron emission by absorption of an unusually high number of photons (strong field regime).

On the optical excitation side, developments are mainly dedicated to the engineering of the temporal and spatial profiles of the light pulses.

Time-Resolved Imaging The use of *ultrashort pulse laser* USPL allows the capture of ultrafast spatio-temporal dynamics of the interaction between light and matter. In PEEM imaging, Ti:Al_2O_3 oscillator lasers are the most common light source for ultrafast excitation. These sources are technologically mature (commercially available) and with a pulse energy in the nJ range are well suited for imaging (high average count rate, no space charge effects). However, their use imposes limitations. First, the accessible wavelength range is intrinsically related to the nature of the active laser medium and covers at most the [650, 1100 nm] wavelength interval with an irradiance maximum centered on 800 nm.

Second, the pulse duration inherits from the 80 MHz oscillator characteristics and corresponds to a [50, 500 fs] duration range. As a result, important ongoing efforts are devoted to the development of broadband tunable ultrafast sources. These include specifically designed Ti:Al_2O_3 lasers, optical parametric oscillators, optical parametric amplifiers [87], and high order harmonic generators [29]. The use of few cycle laser source is illustrated in Lorek et al. [119], Mårsell et al. [123], Dai et al. [41].

Light Pulse Shaping and Wave Front Imaging Light pulse shaping is a natural extension of any nPPE PEEM setup. Pulse shaping opens avenues to coherent control of the near-field at the nanoscale. The method can be implemented in either time or spatial domain and has demonstrated solid results with nanoparticles [2, 3]. Light wave front imaging constitutes another domain of interest. For instance, light can carry an *orbital angular momentum* OAM. The physical properties of light with a helical wavefront can be probed by confining it onto a two-dimensional surface in the form of plasmonic vortices [176]. By extension, understanding and manipulation of the angular momentum degree of freedom of plasmonic fields open new perspectives in light matter interaction [40].

In parallel with the near-field optical aspects, the very nature of its signal opens up new research avenues for PEEM microscopy. It is particularly well suited to the study of the physics of hot electrons, a key element in research topics such as photochemistry and plasmonic photo- and electro-catalysis on particles and/or surfaces, solar energy conversion, and the production and transport of electrons on small scales studies [58, 87, 122].

Scanning photoionization microscopy SPIM constitutes a simpler variant of the PEEM microscopy [135, 173]. In SPIM, the specimen is illuminated by a diffraction limited laser spot generated by overfilling the back aperture of a microscope objective. The specimen is scanned across the fixed excitation spot, and an image is built by the collection of the emitted photoelectrons through either a Faraday cup or a channeltron electron multiplier. Single-electron counting level can be achieved. The microscope work under ultrahigh vacuum. Depending of the characteristics of the light source, the photoemission regime is either linear or nonlinear in nature. SPIM imaging can be easily combined with confocal fluorescence, dark field and Raman microscopies so as to obtain correlated signatures. This single particle microscopy has notably proved its worth in monitoring the plasmonics of individual metallic objects (nanocube [66], nanorods [67], and hole [14]).

4.5 Probing the Near-Field Through Nanochemistry

An alternative elegant way of probing the optical near-field was introduced in 2005 [88]. It consists of a nanoscale chemical imaging of the near-field. The general principle is illustrated in Fig. 4.21.

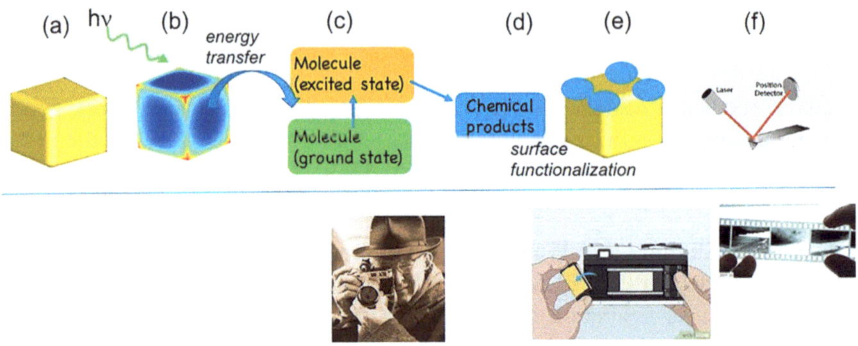

Fig. 4.21 Near-field photochemical imaging. (**a, b**) Nano-object and related optical near-field to be characterized. (**c, d**) The optical near-field induces a chemical reaction resulting in products. (**e**) These products lead to specific surface functionalization. (**f**) The modified nano-object is characterized by nanoscopy, e.g., atomic force microscopy. Bottom: illustration of the analogy with the silver photography

Figure 4.21a represents the nano-object to be characterized. Under illumination in the presence of a photochemical molecular material, the generated near-field (Fig. 4.21b) leads to excitation of molecules (Fig. 4.21c), resulting in chemical products (Fig. 4.21d) that initiate nanoscale surface functionalization on the nano-object surface (Fig. 4.21e). The resulting change of the nano-object morphology contains the signature of the optical near-field and can be visualized by well-controlled tools of nanoscopy, such as *scanning tunneling microscopy* (STM) or *atomic force microscopy* (AFM) (Fig. 4.21f). In other words, a near-field photo-chemical snapshot is taken, developed, and read. This analogy with silver photography is illustrated at the bottom of Fig. 4.21.

Different kinds of chemical reactions and associated nanoscale energy transfer have been used. The sections below describe some of them.

4.5.1 Use of Azoic Molecular Nanomotors Undergoing Photo-isomerization

This approach is illustrated in Fig. 4.22. The nano-object to be imaged is put in the presence of azobenzene molecules (Fig. 4.22a). Under illumination of the system, the nano-object's optical near-field energy is absorbed by the azo molecules that undergo trans ➔ cis isomerization [163]. For pseudo-stilbene type azo molecules, both isomer forms have the same absorption bands in the green: trans ➔ cis and cis ➔ trans isomerizations take place instantaneously through both absorption of the same near-field and thermal relaxation (the cis form is metastable). Multiple cycles of isomerization within the optical near-field lead to mechanical motion of the molecule (Fig.4.22b). In general, the azoic optical nanomotors are grafted to a

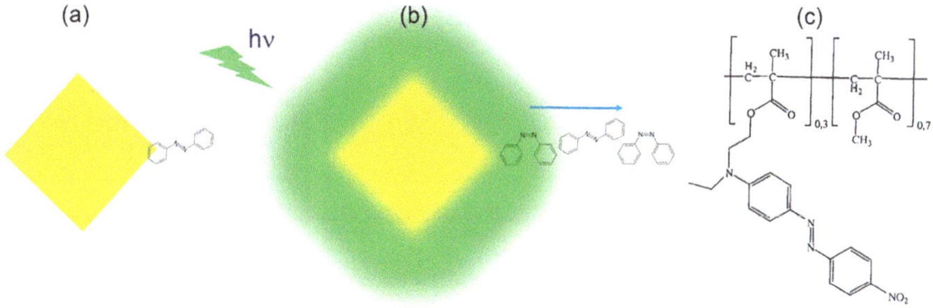

Fig. 4.22 (**a**, **b**) Principle of near-field imaging based on the use of azobenzene molecules that move under illumination through cycles of isomerization. (**c**) Azobenzene molecule grafted to polymethyl methacrylate (PMMA) matrix

polymer matrix (Fig. 4.22c). During their photo-induced motion, the azobenzene molecules push and pull the polymer matrix, which leads to a modification of the surface topography. This hybrid photopolymer is thus a self-developing material. Atomic force microscopy is used to characterize the new topography, which contains the signature of the optical field [156].

Figure 4.23 shows examples of photochemical near-field imaging in the visible wavelength range based on the approach described above. Figure 4.23a, b are AFM images of silver bowties that were coated with a 100-nm thick layer of the copolymer, illustrated in Fig. 4.22c. Figure 4.23d, e are AFM images of the same nanosystems after laser illumination at a wavelength of 514 nm and linear polarization aligned with the bowtie axis. A deep analysis of the photo-induced topography together with *Finite Difference Time Domain* (FDTD) calculations allowed for better understanding of the link between the optical near-field to be characterized and the associated displacement of matter [89]. In particular, it turned out that while the polymer moves laterally away from the field maxima for transverse (in-plane) polarizations, longitudinal field components tend to lift the matter vertically as a consequence of the free space requirement from the molecules. This strong polarization sensitivity offers the possibility to perform a vector analysis of the optical near-field. Figure 4.23c, f show another striking example of surface structuration from the near-field. An array of 100-nm Ag nanoparticles coated with copolymer (Fig. 4.23c) was illuminated by a circularly polarized laser beam. The resulting complex topography (Fig. 4.23f) is partly interpreted as the image of the negative near-field intensity of the in-plane field components; the azoic nanomotors tend to escape from light [88]. In order to further interpret the optically induced topography, a multiscale stochastic model was developed [93, 94].

This model relies of two main hypotheses: (i) In space, in the presence of a vector optical field, each azobenzene molecule has a probability of light absorption expressed as

$$P_{\text{abs}} \propto |\mathbf{E}|^2 \cos^2 \varphi \tag{4.27}$$

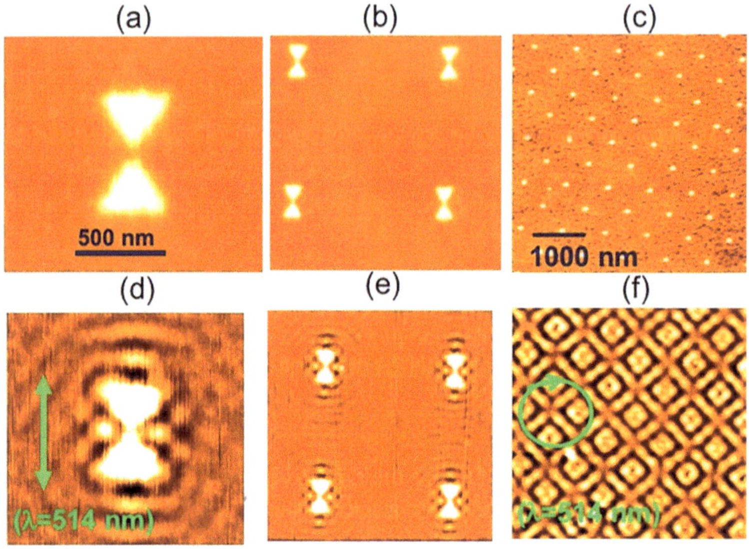

Fig. 4.23 Near-field photochemical imaging of silver nanostructures (AFM images). (**a**) Single Ag bow-tie covered with azo photopolymer. (**b**) Four Ag bow-ties covered with azo photopolymer. (**c**) Array of Ag circular nanoparticles covered with azo photopolymer. (**d**) Same as (a) after linearly polarized laser illumination (green arrow). (**e**) Same as (**b**) after laser illumination (same illumination condition as (**d**). (**f**) Same as (**c**) after circularly polarized laser illumination (green circle). (**a**), (**b**), (**d**), (**e**) are reprinted (adapted) with permission from Ref. [89], Copyright 2008, American Chemical Society. (**c**), (**f**) are reprinted (adapted) with permission from Ref. [88], Copyright 2005, American Chemical Society

where **E** is the local optical field and φ is the angle between the field direction and the azo molecule axis; (ii) if absorption eventually occurs, the molecule moves along its axis ("caterpillar" model [110]) and the perturbation can induce a subsequent rotation of the molecule. The molecule keeps moving until P_{abs} goes to zero. This model turned out to reproduce well the experimental data, including the time evolution of the topography during light exposure.

This photochemical approach of imaging was used for describing topical nanoscale optical effects in the visible (e.g., [75, 182]).

More recently, two-photon absorption was exploited to image/investigate near-field phenomena in the near-infrared. Figures 4.24 and 4.25 show two examples.

Figure 4.24 illustrates an important observation: a perfectly achiral nanostructure, a gold plasmonic trimer with trigonal D_{3h} symmetry (oligomer), exhibits a near-field response at $\lambda = 790$ nm that is sensitive to the handedness of circularly polarized light [85]. This effect stems from the near-field interference between the different plasmonic modes sustained by the oligomer under circularly polarized light excitation. Moreover, it was shown that the intensity of the local optical chirality can be tuned by the strength of the near-field coupling, as controlled by the gap distance between the individual nanodisks in the oligomer.

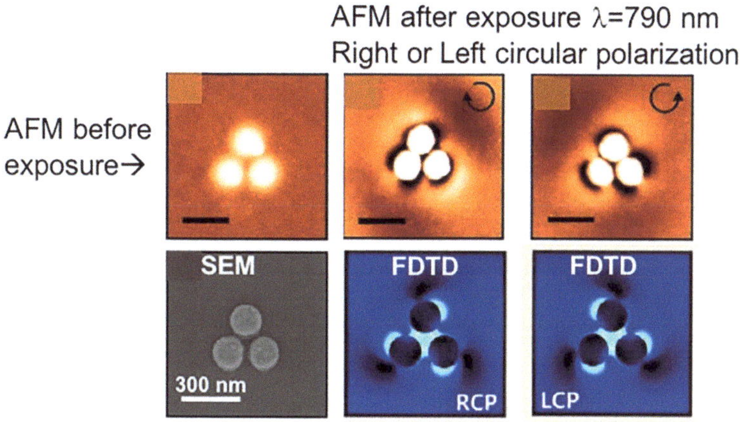

Fig. 4.24 Near-field optical NIR imaging of a triangle-shaped trimer of Au nanoparticles revealing near-field chirality on achiral nanostrutures (see text for comment) Top: AFM image of the achiral trimer covered with thin film and azopolymer before and after illumination with right handed and left handed circular polarizations. Bottom: SEM imaging of the achiral trimer and FDTD simulations for resp. right-handed (RCP) and left-handed (LCP) circular polarizations. (The figures are reprinted (adapted) with permission from [85], Copyright 2019, American Chemical Society)

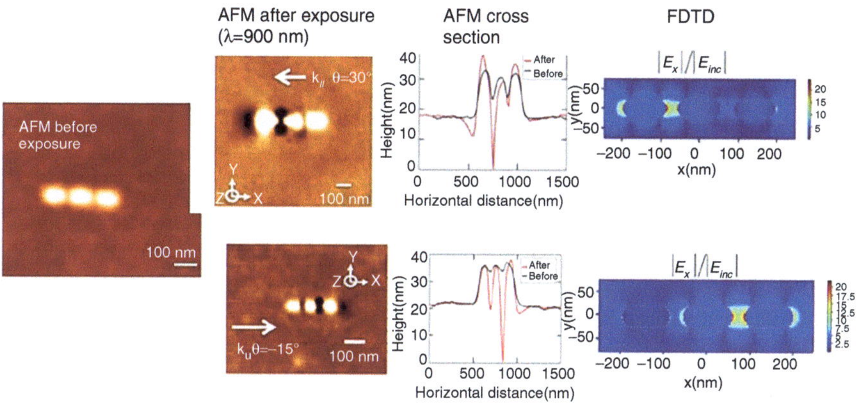

Fig. 4.25 Near-field optical NIR imaging of a linear trimer of Au nanoparticles revealing gap hot spot whose position can be control with the incident angle (see text for comment). AFM imaging before and after light exposure in azopolymer. Corresponding FDTD simulations. (Reprinted (adapted) with permission from Ref. [207], Copyright 2017, American Chemical Society)

Figure 4.25 shows an interesting coupling effect involving aligned gold nanorods [207]. Through near-field coupling between different plasmonic nanostructures it is possible to get hybridization of plasmonic modes that are hosted by individual particles. Both control and observation of the associated subwavelength-sized light spot are challenging. Using a two-photon absorption process in azopolymer, the near-field distribution around a linear trimer antenna was imprinted. The trimer antenna

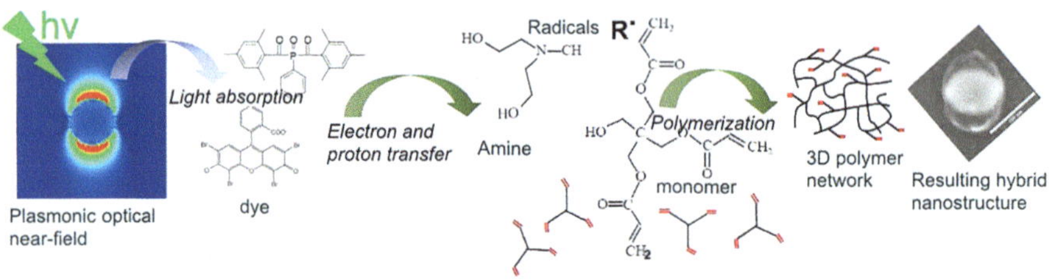

Fig. 4.26 Plasmonic near-field photopolymerization. The optical near-field energy triggers chemical reactions that result in a polymer molding of the near-field intensity. Characterizing this molding provides information about the spatial distribution of the near-field intensity

made of three aligned Au nanorods has two nanogaps. This geometry is designed to enable local switching of hot spots. Using oblique collimated and polarized illumination from the substrate at 900 nm wavelength, the near-field photochemical imaging reveals, with a 20-nm spatial resolution, the selective addressing of one single hot spot within the trimer structure. Switching between the two nanogap locations is achieved on demand by simply illuminating with an opposite oblique incidence. In particular, this work showed that corresponding bonding and anti-bonding gap modes can be selectively excited, resulting in controlled hot spot locations.

4.5.2 Nanoscale Photopolymerization

Figure 4.26 illustrates another approach of photochemical nano-imaging, namely the plasmonic photopolymerization. A metallic nanoparticle, whose near-field is to be characterized, is immersed in a photopolymerizable liquid formulation prior to its illumination with polarized light.

This monomer formulation is typically made of acrylate-type molecules. Through 1 or 2 photon processes, the local plasmonic optical near-field is locally absorbed by photosensitive dyes whose excited states interact with an amine to form radicals that interact with the monomer to trigger polymerization. As the result, polymer nanoparts are integrated to the nanoparticles, as a 3D replica of the plasmonic modes [1]. Characterization of the resulting polymer molds provides information about the spatial distribution of the nanoparticle near-field intensity. The approach relies on good control of the energy threshold of polymerization, which is the minimum energy required for triggering polymerization. The incident energy is always weaker than this threshold so that nothing happens in the surrounding monomer formulation, except in the nanoparticle near-field, where light is locally plasmon-enhanced. The right side of Fig. 4.26 shows a resulting hybrid gold/polymer nanoparticle that was fabricated with the plasmonic near-field dipole mode of a Au disk [59]. It is worth noticing that the process modifies the

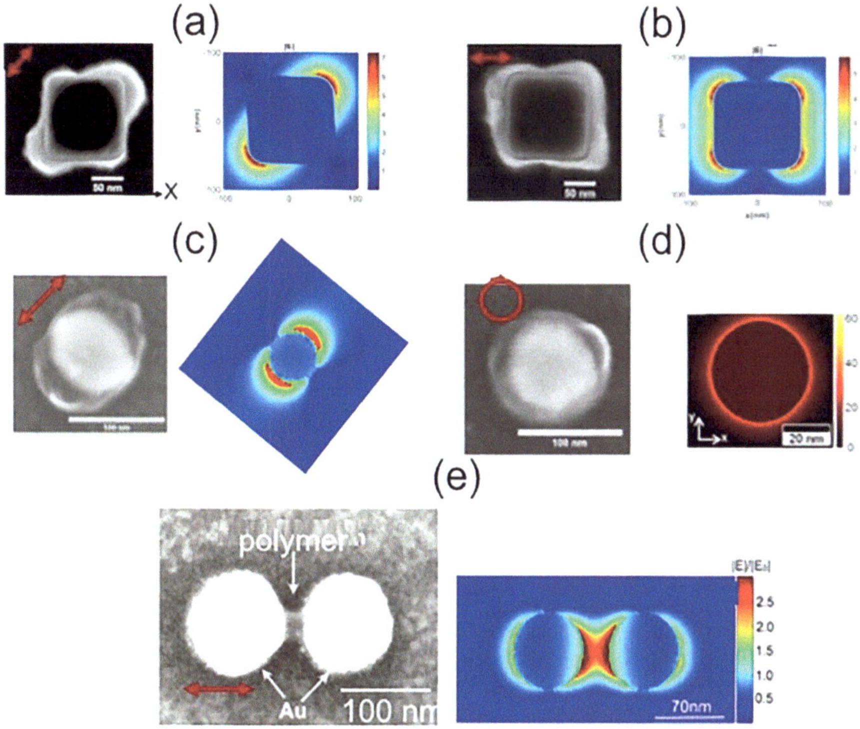

Fig. 4.27 Examples of photochemical imaging of metal nanostructures based on nano-scale photopolymerization. For each figure pair: the left-side image is a SEM image of the resulting hybrid nanostructure, the right-side figure is a FDTD calculation of the modulus of the electric near-field. The red arrow represents the state of polarization used for photopolymerization. The experiments were made at the 780 nm wavelength (two photon absorption by the polymer), except for (**e**), where 1-photon absorption at 747 nm was exploited (**a**, **b**): Au nanocube (from Ref. [59]), (**c**, **d**): Au nanodisks (from Ref. [59]), (**e**): Au dimer, reprinted (adapted) with permission from Ref. [208], Copyright 2014, American Chemical Society

local symmetry of the dielectric environment. In the case of a nanodisk under linear polarization illumination, the photopolymerization process associated with the particle dipolar resonance mode turns the $C_{\infty v}$ symmetry to C_{2v}. In a pioneering article, it was shown that such a symmetry modification can induce a spectral degeneracy breaking in the localized surface plasmons [1].

Figure 4.27 shows a set of examples of photochemical imaging of metal nanostructures based on nano-scale photopolymerization, other examples can be found in Shirai et al. [174], Wurtz et al. [204], Ueno et al. [186], Geldhauser et al. [60], Gruber et al. [65], Wang et al. [196], Ding et al. [46].

In Fig. 4.27a, the dipole eigenmode of a Au nanocube was molded through excitation of the cube with a linear polarization along its diagonal ($\lambda = 780$ nm). In Fig. 4.27b, excitation along the nanocube edge allowed for the excitation of a

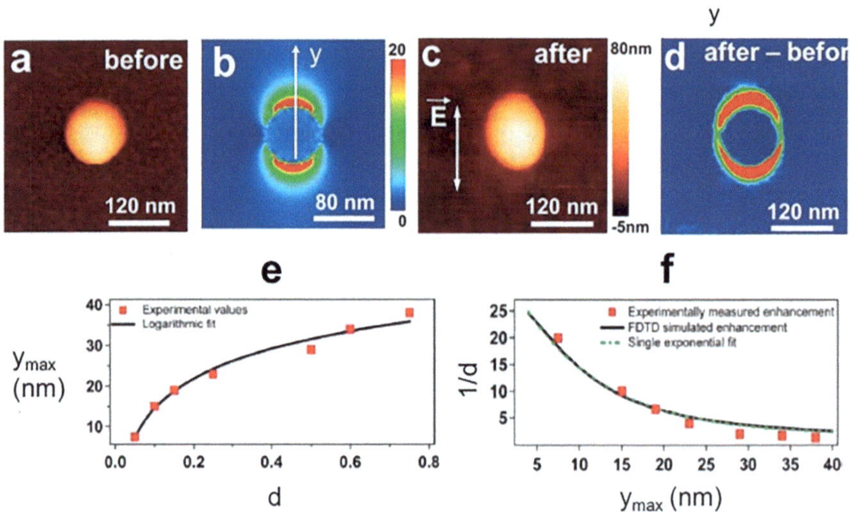

Fig. 4.28 Quantification of plasmonic fields by nanoscale photopolymerization. (**a**) AFM image of a 60-nm high Au nanoparticle. (**b**) calculated near-field intensity (y linear polarization $\lambda = 514$ nm). (**c**) AFM image of the hybrid nanoparticle resulting from photopolymerization induced by intensity. (**d**) differential images. (**e**): plasmon-induced polymer thickness as a function of the normalized dose. (**f**) Intensity enhancement factor as a function of the polymer thickness. (Reprinted (adapted) with permission from Ref. [43]. Copyright 2010, American Chemical Society)

combination of two eigenmodes, resulting in a four-lobe electromagnetic structure whose 3D-polymer replica was obtained. In Fig. 4.27c, d, a Au nanodisk was excited with linear or circular polarization, resulting in a near-field (and associated polymer molding) consisting, respectively, of two lobes (near-field dipole emission parallel to the incident field) or a ring. Figure 4.27e shows that the light hot spot at the gap of a dimer of coupled gold nanoparticles can be visualized by local selective polymerization.

Beyond imaging, this approach has shown to enable quantitative analysis of the optical near-field [43, 60]. An example is illustrated in Fig. 4.28.

Figure 4.28a is the AFM image of a 60-nm high Au disk. Figure 4.28b is the calculated near-field intensity resulting from an optical excitation, linearly polarized along y axis ($\lambda = 514$ nm wavelength). This is the near-field to be characterized experimentally. Figure 4.28c is the AFM image of the nanoparticle modified by photopolymerization resulting from the intensity distribution shown in Fig. 4.28b. In order to highlight the integrated polymeric nanostructure, a differential AFM image was calculated; it is shown in Fig. 4.28d. Such differential images are used for easy assessment of the polymer thickness along y direction incident (polarization direction). Quantification of the near-field is based on the following principle.

During plasmonic photopolymerization, the local dose D outside the particle can be expressed as:

$$D = D_o F_{\max} \exp(-\alpha y) \tag{4.28}$$

where D_o is the incident dose (in J/m^2), $F_{\max}$ is the maximum (surface) plasmon enhancement factor, and α is average plasmon spatial decay rate of the near-field generated by the Au nanoparticle (α^{-1} is the field attenuation length, i.e., the *size* of the nanosource to be characterized). Let's consider D_{th}, the energy threshold below which no polymerization can occur. It is an intrinsic parameter that can be experimentally assessed for any photopolymerizable formulation. Polymerization occurs if $D > D_{th}$, resulting in the following condition:

$$y < y_{\max} = \alpha^{-1} \ln(F_{\max}\, d) \tag{4.29}$$

where d is the normalized incident dose D_o/D_{th} always < 1.

In (Eq. 4.29), $y_{\max}$ is an observable: it is the plasmon-induced polymer thickness that can be assessed by AFM (see Fig. 4.28d). By performing plasmonic photopolymerization on many identical nanoparticles with different values for d, it is possible to experimentally reconstruct the logarithm behavior curve that is predicted by (Eq. 4.29). Figure 4.28e is an example of such a reconstruction. From fitting, it is then possible to extract numerical values for α^{-1} and $F_{\max}$. In Fig. 4.28e, one gets $\alpha^{-1} = 11$ nm and $F_{\max} = 39$. Hence, it is possible to quantify source size and intensity plasmon-enhancement factor at the metal surface. It is also worth representing $1/d$ as a function of $y_{\max}$ (Fig. 4.28f). Indeed, since polymerization occurs despite of an incident dose $D_o < D_{th}$, D_{th}/D_o (> 1) represents the minimum plasmon-induced intensity enhancement factor that allowed for polymerization. Hence, Fig. 4.28f is the signature of the spatial decay of the enhanced plasmon field along y.

A similar approach performed on a bowtie structure with the commercially available photopolymerizable resin SU-8 can be found in [60].

4.5.3 Other Chemical Reactions

Figure 4.29 shows other examples of plasmon-induced chemical reactions that permit near-field imaging. Figure 4.29a is a SEM image of a hybrid nanoparticle resulting from plasmon-induced reduction of diazonium salt on gold nanotriangles [142]. This specific plasmon-assisted nanochemistry involves charge transfers instead of photon transfers: the excitation of the localized surface plasmons can lead to local reduction of diazonium salts, resulting in polymerization and surface functionalization [183].

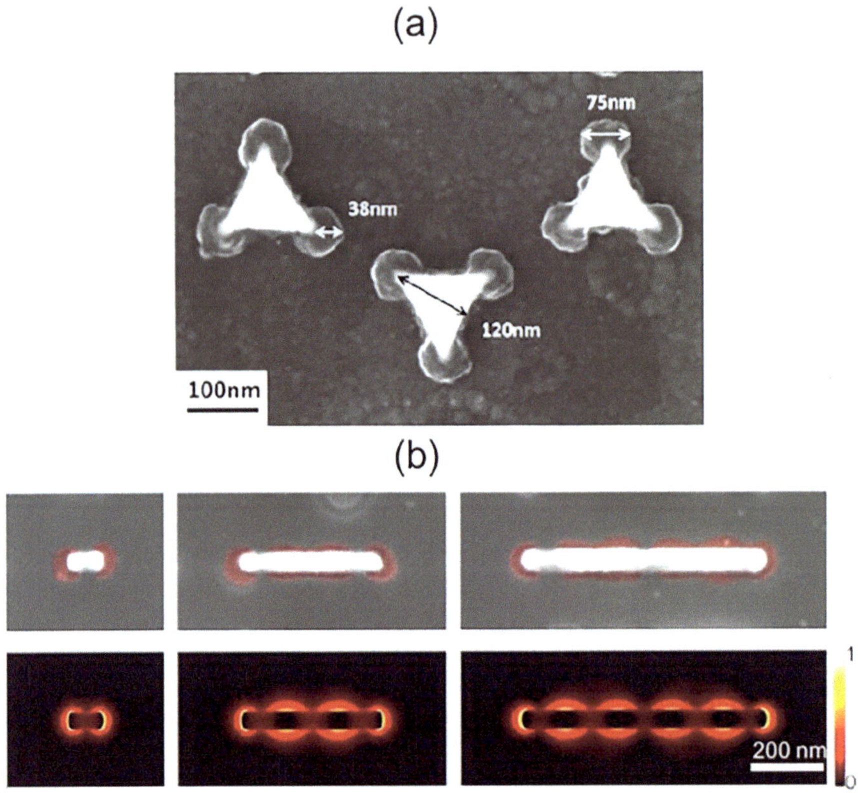

Fig. 4.29 Examples of other plasmon-induced chemical reactions permitting near-field imaging. (**a**) Reduction of diazonium salt on Au nanotriangles, resulting in surface polymerization (broad band IR illumination), from Ref. [142]. (**b**) Local long-chain scission in PMMA (photolysis) on Au nanorods of different sizes (860-nm wavelength illumination polarized along the rod long axe). (Reprinted (adapted) with permission from Ref. [193]. Copyright 2012, American Chemical Society)

The strength of this phenomenon depends on the density of excited hot charges. As a result, Fig. 4.29a can be considered as a visualization of the plasmonic hot spots at the sharp extremities of gold prisms. Figure 4.29b illustrates another photochemistry. It shows the near-field mapping of modes on Au nanorods. In that case, 4-photon photolysis of surrounded *poly(methyl methacrylate)* (PMMA) took place under plasmon excitation [193]. Such electron-beam lithography resists were also used for nanoscale mapping photoemission and hot-electron emission from plasmonic nanoantennas [82].

This photochemical reaction is associated to local long-chain scission making PMMA soluble in a developer. The removed polymer leaves an imprint that looks like the intensity of the near-field to be characterized. In that way, at $\lambda = 860$ nm, depending on the nanorod length, plasmon dipole mode and high order modes were observed by SEM. Equivalent standing plasmon modes of nanowire have also been imaged via two-photon plasmon polymerization of a thin film resist covering the wires and subsequent electron microscopy [65].

Beyond near-field analysis and imaging, it should be stressed that these photochemical approaches constitute powerful platforms for nanochemistry [206]. For example, nanoscale diffusion of species can be studied [42] and distinction between electron and photon nanosources for chemistry can be analyzed [82]. Furthermore, the above techniques pave the way to plasmon mediated surface functionalization of nanoscale structures [59, 183] to get new hybrid functional nano-objects presenting new optical properties.

4.6 Conclusion and Last Remarks

Since the description of diffraction from the seventeenth century, the first proposition of near-field optical microscopy in the 1920s, many approaches have been developed for breaking the diffraction limit and probing the optical near-field with a sub-wavelength resolution. The first experimental demonstrations have started in the 1980s through SNOM, boosted by the development of *scanning probe microscopy* SPM. The three main approaches developed in the chapter are very different from each other. It would not make any sense to compare them and pit them against each other. Each of them has assets and limits. Regarding SNOM, this approach quickly developed from the 1990s [158, 159] but had to face many technological challenges, especially concerning the probes. It has turned out to be very difficult to control morphology and time-stability of probes, while the SNOM signal and its nature are highly sensitive to this morphology. In general, it is not possible to know the actual state of the probe after its first landing on the sample. Apertureless probes are however more simple and can be controlled AFM or STM tips. This is probably why while most of the SNOM configurations were using nano-apertures in the 1990s, the "apertureless" SNOM configuration dominates the nano-optics landscape, as illustrated by the available powerful commercial apertureless SNOM systems that are currently available. On the other hand, SNOM can take advantage of the widely developed reliable SPM technologies and presents the advantage to be pretty easy to implement in air and room temperature and can be used, in principle, on any type of materials, including biological ones. In general, despite its limits, SNOM certainly opened the avenue to the domains of near-field optics and nanophotonics. As far as the use of electron microscopies is concerned, this approach, which took advantage of the development of TEM and SEM started in the 1930s, is powerful and reliable in terms of resolution and reproducibility. It requires, however, complex and expensive instrumentation and long-term expertise and is limited to samples presenting appropriate electrical properties (metal and semiconductors). This approach offers also a relevant platform for solid state physics involving photons. Finally, the photochemical approach is convenient and pretty easy to implement. Image interpretation requires the understanding of the chemical processes but, reciprocally, this approach offers a valuable platform for nanochemistry and opens new routes to surface functionalization and hybrid nanophotonics.

References

1. H.I.E. Ahrach, R. Bachelot, A. Vial, G. Lérondel, J. Plain, P. Royer, O. Soppera, Spectral degeneracy breaking in plasmon resonance of single metal nanoparticles by nanoscale near-field photopolymerization. Phys. Rev. Lett. **98**, 107402 (2007)
2. M. Aeschlimann, M. Bauer, D. Bayer, T. Brixner, F.J. de Abajo, W. Pfeiffer, M. Rohmer, C. Spindler, F. Steeb, Adaptive subwavelength control of nano-optical fields. Nature **446**, 301 (2007)
3. M. Aeschlimann, M. Bauer, D. Bayer, T. Brixner, S. Cunovic, F. Dimler, A. Fischer, W. Pfeiffer, M. Rohmer, C. Schneider, F. Steeb, C. Strüber, D.V. Voroninec, Spatiotemporal control of nanooptical excitations. PNAS **107**, 5329–5333 (2010). https://doi.org/10.1073/pnas.0913556107
4. M. Aeschlimann, M. Bauer, D. Bayer, T. Brixner, S. Cunovic, F. Dimler, A. Fischer, W. Pfeiffer, M. Rohmer, C. Schneider, F. Steeb, C. Strüber, D.V. Voroninec, Determination of local optical response functions of nanostructures with increasing complexity by using single and coupled Lorentzian oscillator models. Appl. Phys. B Lasers Opt. **122**, 199 (2016). https://doi.org/10.1007/s00340-016-6471-3
5. P. Anger, P. Bharadwaj, L. Novotny, Enhancement and quenching of single-molecule fluorescence. PRL **96**, 113002 (2006)
6. S. Aubert, A. Bruyant, S. Blaize, R. Bachelot, G. Lerondel, S. Hudlet, P. Royer, Analysis of the interferometric effect of the background light in apertureless scanning near-field optical microscopy. Opt. Soc. Am. B Optical Physics **20**(10), 2117–2124 (2003)
7. C. Awada, T. Popescu, L. Douillard, F. Charra, A. Perron, H. Yockell-Lelièvre, A.-L. Baudrion, P.-M. Adam, R. Bachelot, Selective excitation of plasmon resonances of single Au triangles by polarization dependent light excitation. J. Phys. Chem. C **16**, 14591–14598 (2012). https://doi.org/10.1021/jp303475c
8. A. Asenjo-Garcia, F. García de Abajo, Plasmon electron energy-gain spectroscopy. New J. Phys. **15**, 103021 (2013)
9. A.C. Atre, B.J.M. Brenny, T. Coenen, A. García-Etxarri, A. Polman, J.A. Dionne, Nanoscale optical tomography with cathodoluminescence spectroscopy. Nat. Nanotechnol. **10**, 429–436 (2015)
10. B. Barwick, D. Flannigan, A. Zewail, Photon-induced near-field electron microscopy. Nature **462**, 902 (2009)
11. R. Bachelot, G. Wurtz, P. Royer, An application of the apertureless scanning near-field optical microscopy: Imaging a GaAlAs laser diode in operation. Appl. Phys. Lett. **73**(23), 3333–3335 (1998)
12. R. Bachelot, F. H'Dhili, D. Barchiesi, G. Lerondel, R. Fikri, P. Royer, N. Landraud, J. Peretti, F. Chaput, G. Lampel, J.-P. Boilot, K. Lahlil, Apertureless near-field optical microscopy: A study of the local tip field enhancement using photosensitive azobenzene-containing films. J. Appl. Phys. **94**(3), 2060–2072 (2003)
13. D. Van Baak, G. Herold, Response of a lock-in amplifier to noise. Am. J. Phys. **82**(8), 785–797 (2014)
14. T.A. Baker, A. Grubisic, D.J. Nesbitt, Plasmon mediated multiphoton photoemission microscopy of Au nanoholes and nanohole dimers. J. Phys. Chem. C **118**, 6959–6971 (2014). https://doi.org/10.1021/jp411943f

15. E. Bauer, *Surface Microscopy with Low Energy Electrons* (Springer Science + Business Media. ISBN 978-1-4939-0934-6 ISBN 978-1-4939-0935-3 (eBook), New York/Heidelberg/Dordrecht/London, 2014). https://doi.org/10.1007/978-1-4939-0935-3
16. J.V. Bladel, *Singular Electromagnetic Fields and Sources* (IEEE Press, New York, 1991)
17. K.Y. Bliokh, I.P. Ivanov, G. Guzzinati, L. Clark, R. Van Boxem, A. Béché, R. Juchtmans, M.A. Alonso, P. Schattschneider, F. Nori, J. Verbeeck, Theory and applications of free-electron vortex states. Phys. Rep. **690**, 1 (2017)
18. G.D. Bernasconi, J. Butet, V. Flauraud, D. Alexander, J. Brugger, O.J.F. Martin, Where does energy go in electron energy loss spectroscopy of nanostructures? ACS Photonics **4**, 156–164 (2017). https://doi.org/10.1021/acsphotonics.6b00761
19. E. Betzig, J.K. Trautman, Near-field optics: microscopy, spectroscopy, and surface modification beyond the diffraction limit. Science **257**(5067), 189–195 (1992)
20. J.-L. Bijeon, P.-M. Adam, D. Barchiesi, P. Royer, A simple resolution criterion in an Apertureless Scanning Near-Field Optical Microscope (A-SNOM): Contribution of the tip vibration and lock-in detection. Def. Eur. Phys. J. Appl. Phys. **26**(1), 45–52 (2004)
21. L. Billot, M.L. De La Chapelle, D. Barchiesi, S.-H. Chang, S.K. Gray, J.A. Rogers, A. Bouhelier, P.-M. Adam, J.-L. Bijeon, G.P. Wiederrecht, R. Bachelot, P. Royer, Error signal artifact in apertureless scanning near-field optical microscopy. Appl. Phys. Lett. **89**(2), art.no.023105 (2006)
22. M. Born, E. Wolf, *Principles of Optics*, 6th edn. (Pergamon Press, Oxford, 1993)
23. O. Bryngdhal, Evanescent waves in optical imaging, in *Progress in Optics XI, Chap IV*, ed. by E. Wolf, (North Holland, Amsterdam, 1973)
24. N.M. Buckanie, P. Kirschbaum, S. Sindermann, F.-J. Meyer zu Heringdorf, Interaction of light and surface plasmon polaritons in Ag islands studied by nonlinear photoemission microscopy. Ultramicroscopy **130**, 49–53 (2013). https://doi.org/10.1016/j.ultramic.2013.03.007
25. F. Carbone, O.-H. Kwon, A.H. Zewail, Dynamics of chemical bonding mapped by energy-resolved 4D electron microscopy. Science **325**, 181 (2009)
26. R. Carminati, A. Cazé, D. Cao, F. Peragut, V. Krachmalnicoff, R. Pierrata, Y. De Wilde, Electromagnetic density of states in complex plasmonic systems. Surf. Sci. Rep. **70**, 1–41 (2015)
27. G.M. Caruso, F. Houdellier, S. Weber, M. Kociak, A. Arbouet, High brightness ultrafast transmission electron microscope based on a laser-driven cold field emission source: Principle and applications. Adv. Phys. X **4**, 1660214 (2019). https://doi.org/10.1080/23746149.2019.1660214
28. L. Chelaru, F. Meyer zu Heringdorf, In situ monitoring of surface plasmons in single-crystalline Ag-nanowires. Surf. Sci. **601**, 4541 (2007). https://doi.org/10.1016/j.susc.2007.04.146
29. S.H. Chew, F. Süßmann, C. Späth, A. Wirth, J. Schmidt, S. Zherebtsov, A. Guggenmos, A. Oelsner, N. Weber, J. Kapaldo, A. Gliserin, M.I. Stockman, M.F. Kling, U. Kleineberg, Time-of-flight-photoelectron emission microscopy on plasmonic structures using attosecond extreme ultraviolet pulses. Appl. Phys. Lett. **100**, 051904 (2012)
30. M. Cinchetti, A. Gloskovskii, S.A. Nepjiko, G. Schönhense, H. Rochholz, M. Kreiter, Photoemission electron microscopy as a tool for the investigation of optical near fields. Phys. Rev. Lett. **95**(47601), 10.1103/PhysRevLett.95.047601 (2005)
31. T. Coenen, B.J.M. Brenny, E. Jan Vesseur, A. Polman, Cathodoluminescence microscopy: Optical imaging and spectroscopy with deep-subwavelength resolution. MRS Bull. **40**, 359 (2015). https://doi.org/10.1557/mrs.2015.64
32. T. Coenen, N.M. Haegel, Cathodoluminescence for the 21st century: Learning more from light featured. Appl. Phys. Rev. **4**, 031103 (2017). https://doi.org/10.1063/1.4985767
33. S.M. Collins, O. Nicoletti, D. Rossouw, T. Ostasevicius, P.A. Midgley, Excitation dependent Fano-like interference effects in plasmonic silver nanorods. Phys. Rev. B Condens. Matter Mater. Phys. **90**, 155419 (2014)

34. S.M. Collins, E. Ringe, M. Duchamp, Z. Saghi, R.E. Dunin-Borkowski, P.A. Midgley, Eigenmode tomography of surface charge oscillations of plasmonic nanoparticles by electron energy loss spectroscopy. ACS Photonics **2**, 1628–1635 (2015). https://doi.org/10.1021/acsphotonics.5b00421
35. S.M. Collins, P.A. Midgley, Progress and opportunities in EELS and EDS tomography. Ultramicroscopy **180**, 133–141 (2017). https://doi.org/10.1016/j.ultramic.2017.01.003
36. C. Colliex, M. Kociak, O. Stéphan, Electron Energy Loss Spectroscopy imaging of surface plasmons at the nanometer scale. Ultramicroscopy **162**, A1–A24 (2016)
37. D. Courjon, C. Bainier (eds.), *Champ Proche Optique : Théorie et Applications*, Collection Technique et Scientifique des Télécommunications, Chapter 8 (Springer, Cham, 2001), pp. 147–176
38. Y. Dai, M. Dąbrowski, V.A. Apkarian, H. Petek, Ultrafast microscopy of spin-momentum locked surface plasmon polaritons. ACS Nano **12**, 6588–6596 (2018)
39. Y. Dai, H. Petek, Plasmonic spin-hall effect in surface plasmon polariton focusing. ACS Photonics **6**, 2005–2013 (2019)
40. Y. Dai, M. Dabrowski, H. Petek, Optical field tuning of localized plasmon modes in Ag microcrystals at the nanofemto scale. J. Chem. Phys. **152**, 054201 (2020). https://doi.org/10.1063/1.5139543
41. Y. Dai, Z. Zhou, A. Ghosh, S. Yang, H. Chen-Bin Huang, Petek., Ultrafast nanofemto photoemission electron microscopy of vectorial plasmonic fields. MRS Bull. **46** (2021)
42. C. Deeb, C. Ecoffet, R. Bachelot, J. Plain, A. Bouhelier, O. Soppera, Plasmon-based free-radical photopolymerization: Effect of diffusion on nanolithography processes. Am. Chem. Soc. **133**(27), 10535–10542 (2011)
43. C. Deeb, R. Bachelot, J. Plain, A.-L. Baudrion, S. Jradi, A. Bouhelier, O. Soppera, P.-K. Jain, L. Huang, C. Ecoffet, L. Balan, P. Royer, Quantitative analysis of localized surface plasmons based on molecular probing. ACS Nano **4**, 4579–4586 (2010)
44. C. Delacour, S. Blaize, P. Grosse, J.M. Fedeli, A. Bruyant, R. Salas-Montiel, G. Lerondel, A. Chelnokov, Efficient directional coupling between silicon and copper plasmonic nanoslot waveguides: Toward metal−oxide−silicon nanophotonics. Nano Lett. **10**, 2922 (2009)
45. B. Di Bartolo, J. Collins, L. Silvestri, *Nano-Optics: Principles Enabling Basic Research and Applications*. Part of the book series: NATO Science for Peace and Security Series B: Physics and Biophysics (NAPSB), conference proceeding (Springer, 2016)
46. T. Ding, J. Mertens, A. Lombardi, O.A. Scherman, J.J. Baumberg, Light-directed tuning of plasmon resonances via plasmon-induced polymerization using hot electrons. ACS Photonics **4**(6), 1453–1458 (2017)
47. P.A.M. Dirac, The quantum theory of emission and absorption of radiation. Roy. Soc. Lond. A **114**(767), 243–265 (1927). https://doi.org/10.1098/rspa.1927.0039
48. L. Douillard, F. Charra, C. Fiorini, P.-M. Adam, R. Bachelot, S. Kostcheev, G. Lerondel, M. Lamy de la Chapelle, P. Royer, Optical properties of metal nanoparticles as probed by photoemission electron microscopy. J. Appl. Phys. **101**, 83518–83522 (2007). https://doi.org/10.1063/1.2719282
49. L. Douillard, F. Charra, Z. Korczak, P.M. Adam, R. Bachelot, S. Kostcheev, G. Lerondel, P. Royer, Short range plasmon resonators probed by photoemission electron microscopy. Nano Lett. **8**, 935–940 (2008). https://doi.org/10.1021/nl080053v
50. L. Douillard, F. Charra, High-resolution mapping of plasmonic modes: Photoemission and scanning tunnelling luminescence microscopies. J. Phys. D. Appl. Phys. **44**, 464002 (2011). https://doi.org/10.1088/0022-3727/44/46/464002
51. A. Drezet, M.J. Nasse, S. Huant, J.C. Woehl, The optical near-field of an aperture tip. Europhys. Lett. **66**, 41 (2004)
52. B. Egger, S.G. Sprecher, *Super-Resolution STED and STORM/PALM Microscopy for Brain Imaging*. Part of the Progress in Optical Science and Photonics book series POSP, vol. 5 (2019). https://doi.org/10.1007/978-981-10-9020-2_12
53. P.Z. El-Khoury, P. Abellan, B.Y. Gong, A.F.S. Hage, J. Cottom, A.G. Joly, R. Brydson, C.Q.M. Ramasseb, W.P. Hessa, Visualizing surface plasmons with photons, photoelectrons, and electrons. Analyst **141**, 3562 (2016). https://doi.org/10.1039/c6an00308g

54. A. Feist, N. Bach, N.R. da Silva, T. Danz, M. Möller, K.E. Priebe, T. Domröse, J.G. Gatzmann, S. Rost, J. Schauss, S. Strauch, R. Bormann, M. Sivis, S. Schäfer, C. Ropers, Ultrafast transmission electron microscopy using a laser-driven field emitter: Femtosecond resolution with a high coherence electron beam. Ultramicroscopy **176**, 63–73 (2017). https://doi.org/10.1016/j.ultramic.2016.12.005
55. G. Ferrini, F. Banfi, C. Giannetti, F. Parmigiani, Non-linear electron photoemission from metals with ultrashort pulses. Nuclear Inst. Methods Phys. Res. A **601**, 123–131 (2009). https://doi.org/10.1016/j.nima.2008.12.107
56. K.W. Frese, C. Chen, Theoretical models of hot carrier effects at metal-semiconductor electrodes. J. Electrochem. Soc. **139**, 3234 (1992)
57. F.J. Garcia de Abajo, Relativistic energy loss and induced photon emission in the interaction of a dielectric sphere with an external electron beam. Phys. Rev. B **59**, 3095 (1999)
58. K. Fukumoto, K. Onda, Y. Yamada, T. Matsuki, T. Mukuta, S.-i. Tanaka, S.-y. Koshihara, Femtosecond time-resolved photoemission electron microscopy for spatiotemporal imaging of photogenerated carrier dynamics in semiconductors. Rev. Sci. Instrum. **85**, 083705 (2014)
59. D. Ge, S. Marguet, A. Issa, S. Jradi, T. Hoa Nguyen, M. Nahra, J. Béal, R. Deturche, H. Chen, S. Blaize, J. Plain, C. Fiorini, L. Douillard, O. Soppera, X.Q. Dinh, C. Dang, X. Yang, T. Xu, B. Wei, X.W. Sun, C. Couteau, R. Bachelot, Hybrid plasmonic nano-emitters with controlled single quantum emitter positioning on the local excitation field. Nat. Commun. **11**, 3414 (2020)
60. T. Geldhauser, A. Kolloch, N. Murazawa, K. Ueno, J. Boneberg, P. Leiderer, E. Scheer, H. Misawa, Quantitative measurement of the near-field enhancement of nanostructures by two-photon polymerization. Langmuir **28**, 9041–9046 (2012)
61. H. Gersen, T.J. Karle, R.J.P. Engelen, W. Bogaerts, J.P. Korterik, N.F. van Hulst, T.F. Krauss, L. Kuipers, Real-space observation of ultraslow light in photonic crystal waveguides. Phys. Rev. Lett. **94**, 073903 (2005)
62. L. Gomez, R. Bachelot, A. Bouhelier, G.P. Wiederrecht, S.-H. Chang, S.K. Gray, F. Hua, S. Jeon, J.A. Rogers, M.E. Castro, S. Blaize, I. Stefanon, G. Lerondel, P.J. Royer, Apertureless scanning near-field optical microscopy: A comparison between homodyne and heterodyne approaches. Opt. Soc. Am. B Optical Physics **23**(5), 823–833 (2006)
63. J.-J. Greffet, R. Carminati, Image formation in near-field optics. Prog. Surf. Sci. **56**(3), 133–237 (1997)
64. S. Grésillon, L. Aigouy, A.C. Boccara, J.C. Rivoal, X. Quelin, C. Desmarest, P. Gadenne, V.A. Shubin, A.K. Sarychev, V.M. Shalaev, Experimental observation of localized optical excitations in random metal-dielectric films. Phys. Rev. Lett. **82**, 4529 (1999)
65. C. Gruber, A. Hirzer, V. Schmidt, A. Trügler, U. Hohenester, H. Ditlbacher, A. Hohenau, J.R. Krenn, Imaging nanowire plasmon modes with two-photon polymerization. Appl. Phys. Lett. **106**, 081101 (2015). https://doi.org/10.1063/1.4913470
66. A. Grubisic, E. Ringe, C.M. Cobley, Y. Xia, L.D. Marks, R.P. Van Duyne, D.J. Nesbit, Plasmonic near-electric field enhancement effects in ultrafast photoelectron emission: Correlated spatial and laser polarization microscopy studies of individual Ag nanocubes. Nano Lett. **12**, 4823–4829 (2012). https://doi.org/10.1021/nl302271u
67. A. Grubisic, V. Schweikhard, T.A. Baker, D.J. Nesbitt, Multiphoton photoelectron emission microscopy of single Au nanorods: Combined experimental and theoretical study of rod morphology and dielectric environment on localized surface plasmon resonances. Phys. Chem. Chem. Phys. **15**, 10616 (2013)
68. F.J. García de Abajo, Optical excitations in electron microscopy. Rev. Mod. Phys. **82**, 209 (2010)
69. F.J. Garcia de Abajo, A. Asenjo-Garcia, M. Kociak, Multiphoton absorption and emission by interaction of swift electrons with evanescent light fields. Nano Lett. **10**, 1859 (2010)
70. A. Garcia-Etxarri, I. Romero, F.J.G. de Abajo, R. Hillenbrand, J. Aizpurua, Influence of the tip in near-field imaging of nanoparticle plasmonic modes: Weak and strong coupling regimes. Phys. Rev. B **79**(2009), 045111 (2009)

71. Y. Gong, A.G. Joly, P.Z. El-Khoury, W.P. Hess, Nonlinear photoemission electron micrographs of plasmonic nanoholes in gold thin films. J. Phys. Chem. C **118**, 25671–25676 (2014). https://doi.org/10.1021/jp509900h
72. Y. Gong, A. Joly, D. Hu, P. El-Khoury, W. Hess, Ultrafast imaging of surface plasmons propagating on a gold surface. NanoLetters **15**, 3472 (2015)
73. A. Gliserin, S.H. Chew, S. Choi, K. Kim, D.T. Hallinan, O. Jin-Woo, S. Kim, D.E. Kim, Interferometric time- and energy-resolved photoemission electron microscopy for few-femtosecond nanoplasmonic dynamics. Rev. Sci. Instrum. **90**, 093904 (2019). https://doi.org/10.1063/1.5110705
74. B. Goris, G. Guzzinati, C. Fernández-López, J. Pérez-Juste, L.M. Liz-Marzán, A. Trügler, U. Hohenester, J. Verbeeck, S. Bals, G. Van Tendeloo, Plasmon mapping in Au@Ag nanocube assemblies. J. Phys. Chem. C. Nanomater. Interf. **118**, 15356 (2014)
75. M. Haggui, M. Dridi, J. Plain, S. Marguet, P. Henri, G. Schatz, G. Wiederrecht, S. Gray, R. Bachelot, Spatial confinement of electromagnetic hot and cold spots in gold nanocubes. ACS Nano **6**, 1299–1307 (2012)
76. A. Hartschuh, H. Qian, C. Georgi, M. Böhmler, L. Novotny, Tip-enhanced near-field optical microscopy of carbon nanotubes. Anal. Bioanal. Chem. **394**, 1787–1795 (2009). https://doi.org/10.1007/s00216-009-2827-4A
77. M.T. Hassan, J.S. Baskin, B. Liao, A.H. Zewail, High-temporal-resolution electron microscopy for imaging ultrafast electron dynamics. Nat. Photonics **11**, 425 (2017). https://doi.org/10.1038/NPHOTON.2017.79
78. M. Hassan, Attomicroscopy: From femtosecond to attosecond electron microscopy. J. Phys. B Atomic Mol. Phys. **51**, 032005 (2018)
79. B. Hecht, H. Bielefeldt, Y. Inouye, D.W. Pohl, L. Novotny, Facts and artifacts in near-field optical microscopy. J. Appl. Phys. **81**, 2492 (1997)
80. P.N. Hedde, G.U. Nienhaus, Sub-wavelength optical fluorescence microscopy for biological applications, in *Nano-Optics for Enhancing Light-Matter Interactions on a Molecular Scale: Plasmonics, Photonic Crystals, Metamaterials and Sub-Wavelength Resolution*, Springer Series: NATO Science for Peace and Security Series B: Physics and Biophysics, ed. by B. di Bartolo, J. Collins, (Springer Science + Business Media B.V, Dordrecht, 2013), pp. 47–71
81. C. Hrelescu, T.K. Sau, A.L. Rogach, F. Jäckel, G. Laurent, L. Douillard, F. Charra, Selective excitation of individual plasmonic hotspots at the tips of single gold nanostars. Nano Lett. **11**, 402–407 (2011). https://doi.org/10.1021/nl103007m
82. R.G. Hobbs, W.P. Putnam, A. Fallahi, Y. Yang, F.X. Kärtner, K.K. Berggren, Mapping photoemission and hot-electron emission from plasmonic nanoantennas. Nano Lett. **17**, 6069–6060 (2017)
83. A. Hörl, A. Trügler, U. Hohenester, Tomography of particle plasmon fields from electron energy loss spectroscopy. Phys. Rev. Lett. **111**, 076801 (2013)
84. A. Hörl, G. Haberfehlner, A. Trügler, F.-P. Schmidt, U. Hohenester, G. Kothleitner, Tomographic imaging of the photonic environment of plasmonic nanoparticles. Nat. Commun. **8**, 37 (2017). https://doi.org/10.1038/s41467-017-00051-3
85. A. Horrer, Y. Zhang, D. Gerard, J. Béal, M. Kociak, J. Plain, R. Bachelot, Local optical chirality induced by near-field mode interference in achiral plasmonic metamolecules. Nano Lett. **20**, 509–516 (2019)
86. A. Howie, Photon interactions for electron microscopy applications. Eur. Phys. J.-Appl. Phys. **54**, 33502 (2011)
87. B. Huber, S. Pres, E. Wittmann, L. Dietrich, J. Lüttig, D. Fersch, E. Krauss, D. Friedrich, J. Kern, V. Lisinetskii, M. Hensen, B. Hecht, R. Bratschitsch, E. Riedle, T. Brixner, Space- and time-resolved UV-to-NIR surface spectroscopy and 2D nanoscopy at 1 MHz repetition rate. Rev. Sci. Instrum. **90**, 113103 (2019). https://doi.org/10.1063/1.5115322
88. C. Hubert, A. Rumyantseva, G. Lerondel, J. Grand, S. Kostcheev, L. Billot, A. Vial, R. Bachelot, P. Royer, S.-H. Chang, S.K. Gray, G.P. Wiederrecht, G.C. Schatz, Near-field photochemical imaging of noble metal nanostructures. Nano Lett. **5**(4), 615–619 (2005)

89. C. Hubert, R. Bachelot, J. Plain, S. Kostcheev, G. Lerondel, M. Juan, P. Royer, S. Zou, G.C. Schatz, G.P. Wiederrecht, S.K. Gray, Near-field polarization effects in molecular-motion-induced photochemical imaging. J. Phys. Chem. C **112**, 4111–4116 (2008)
90. H.C.V.d. Hulst Dover, *Light Scattering by Small Particles* (Wiley, 1981)
91. A. Jalocha, N.F. van Hulst, Polarization contrast in fluorescence scanning near-field optical reflection microscopy. J. Opt. Soc. Am. B **12**, 1577 (1995)
92. R. Jazi, T.P.L. Ung, P. Maso, G.C.D. Francs, M. Nasilowski, B. Dubertret, J.-P. Hermier, X. Quélina, S. Bui, Measuring the orientation of a single CdSe/CdS nanocrystal at the end of a near-field tip for the realization of a versatile active SNOM probe. Phys. Chem. Chem. Phys. **20**, 16444–16448 (2018)
93. M. Juan, J. Plain, R. Bachelot, P. Royer, S.K. Gray, G.P. Wiederrecht, Stochastic model for photoinduced surface relief grating formation through molecular transport in polymer films. Appl. Phys. Lett. **93**, 153304–153306 (2008)
94. M.L. Juan, J. Plain, R. Bachelot, P. Royer, S.K. Gray, G.P. Wiederrecht, Multiscale model for photoinduced molecular motion in azo polymers. ACS Nano **3**(6), 1573–1579 (2009)
95. P. Kahl, D. Podbiel, C. Schneider, A. Makris, S. Sindermann, C. Witt, D. Kilbane, M.H.-v. Hoegen, M. Aeschlimann, F. Meyer zu Heringdorf, Direct observation of surface plasmon polariton propagation and interference by time-resolved imaging in normal-incidence two photon photoemission microscopy plasmonics. Plasmonics **13**, 239–246 (2018)
96. M.A. Kats, N. Yu, P. Genevet, Z. Gaburro, F. Capasso, Effect of radiation damping on the spectral response of plasmonic components. Opt. Express **19**, 21748 (2011)
97. H. Kawano, Effective work functions of the elements. Prog. Surf. Sci. **97**, 100583 (2022)
98. C. Kealhofer, W. Schneider, D. Ehberger, A. Ryabov, F. Krausz, P. Baum, All-optical control and metrology of electron pulses. Science **352**, 429 (2016)
99. M.W. Knight, L. Liu, Y. Wang, L. Brown, S. Mukherjee, N.S. King, H.O. Everitt, P. Nordlander, N.J. Halas, Aluminum plasmonic nanoantennas. Nano Lett. **12**, 6000–6004 (2012). https://doi.org/10.1021/nl303517v
100. A.E. Klein, N. Janunts, M. Steinert, A. Tünnermann, T. Pertsch, Polarization-resolved near-field mapping of plasmonic aperture emission by a dual-SNOM system. Nano Lett. **14**, 5010 (2014)
101. M. Kociak, O. Stéphan, Mapping plasmons at the nanometer scale in an electron microscope. Chem. Soc. Rev. **43**, 3865 (2014)
102. M. Kociak, L.F. Zagonel, Cathodoluminescence in the scanning transmission electron microscope. Ultramicroscopy **176**, 112–131 (2017)
103. V. Krachmalnicoff et al., Towards a full characterization of a plasmonic nanostructure with a fluorescent near-field probe. Opt. Express **21**, 11536 (2013)
104. E. Kretschmann, H. Raether, Zeitschrift Fur Naturforschung Part A-Astrophysik Physik Und Physikalische Chemie A. **23**, 2135–2136 (1968)
105. A. Kubo, N. Pontius, H. Petek, Femtosecond microscopy of surface plasmon polariton wave packet evolution at the silver/vacuum interface. Nano Lett. **7**, –470 (2007). https://doi.org/10.1021/nl0627846
106. A. Lahrech, R. Bachelot, P. Gleyzes, A.C. Boccara, Infrared-reflection-mode near-field microscopy using an apertureless probe with a resolution of λ/600. Opt. Lett. **21**(17), 1315–1317 (1996)
107. A. Lahrech, R. Bachelot, P. Gleyzes, A.C. Boccara, Infrared near-field imaging of implanted semiconductors: Evidence of a pure dielectric contrast. Appl. Phys. Lett. **71**(5), 575–577 (1997)
108. M.J. Lagos, A. Trügler, U. Hohenester, P.E. Batson, Mapping vibrational surface and bulk modes in a single nanocube. Nature **543**, 529–532 (2017)
109. H.S. Lee, C. Awada, S. Boutami, F. Charra, L. Douillard, R. de Lamaestre, Espiau., Loss mechanisms of surface plasmon polaritons propagating on a smooth polycrystalline Cu surface. Opt. Express **20**, 8974–8981 (2012). https://doi.org/10.1364/OE.20.008974
110. P. Lefin, C. Fiorini, J.-M. Nunzi, Anisotropy of the photo-induced translation diffusion of azobenzene dyes in polymer matrices. J. Optics A Pure Appl. Optics **7**(1), 71–82 (1998)

111. M. Lehr, B. Foerster, M. Schmitt, K. Krüger, C. Sönnichsen, G. Schönhense, H.-J. Elmers, Momentum distribution of electrons emitted from resonantly excited individual gold nanorods. Nano Lett. **17**, 6606–6612 (2017). https://doi.org/10.1021/acs.nanolett.7b02434
112. C. Lemke, T. Leißner, S. Jauernik, A. Klick, J. Fiutowski, J. Kjelstrup-Hansen, H.-G. Rubahn, M. Bauer, Mapping surface plasmon polariton propagation via counter-propagating light pulses. Opt. Express **20**, 12877–12884 (2012). https://doi.org/10.1364/OE.20.012877
113. M.A. Lieb, A.J. Meixner, A high numerical aperture parabolic mirror as imaging device for confocal microscopy. Opt. Express **8**, 458–474 (2001)
114. H. Liu, J.S. Baskin, A.H. Zewail, Infrared PINEM developed by diffraction in 4D UEM. PNAS **113**, 2041–2046 (2016). https://doi.org/10.1073/pnas.1600317113
115. C. Liu, W. Yueying, H. Zhongwei, J.A. Busche, E.K. Beutler, N.P. Montoni, T.M. Moore, G.A. Magel, J.P. Camden, D.J. Masiello, G. Duscher, P.D. Rack, Photon stimulated electron energy-gain and electron energy-loss spectroscopy of individual plasmonic nanoparticles. ACS Photonics **6**, 2499–2508 (2019)
116. G. Longo, M. Girasole, G. Pompeo, A. Cricenti, Optical super resolution using higher harmonics and difference acquisition modes in an aperture tapping SNOM. Phys. Status Solidi **247**, 2056 (2010)
117. A. Losquin, M. Kociak, Link between cathodoluminescence and electron energy loss spectroscopy and the radiative and full electromagnetic local density of states. ACS Photonics **2**, 1619–1627 (2015). https://doi.org/10.1021/acsphotonics.5b00416
118. A. Losquin, L.F. Zagonel, V. Myroshnychenko, B. Rodríguez-González, M. Tencé, L. Scarabelli, J. Förstner, L.M. Liz-Marzán, F.J. García de Abajo, O. Stéphan, M. Kociak, Unveiling nanometer scale extinction and scattering phenomena through combined electron energy loss spectroscopy and cathodoluminescence measurements. Nano Lett. **15**, 1229–1237 (2015)
119. E. Lorek, E. Mårsell, A. Losquin, M. Miranda, A. Harth, C. Guo, R. Svärd, C.L. Arnold, A. L'Huiller, A. Mikkelsen, J. Mauritsson, Size and shape dependent few-cycle near-field dynamics of bowtie nanoantennas. Opt. Express **23**, 31460 (2015)
120. A. Losquin, T.T.A. Lummen, Electron microscopy methods for space-, energy-, time-resolved plasmonics. Front. Phys. **12**, 127301 (2017). https://doi.org/10.1007/s11467-016-0605-2
121. T.T.A. Lummen, R.J. Lamb, G. Berruto, T. LaGrange, F. Luca Dal Negro, J.G. de Abajo, D. McGrouther, B. Barwick, F. Carbone, Imaging and controlling plasmonic interference fields at buried interfaces. Nat. Commun. **7**, 13156 (2016). https://doi.org/10.1038/ncomms13156
122. M.K.L. Man, A. Margiolakis, S. Deckoff-Jones, T. Harada, E. Laine Wong, M.B.M. Krishna, J. Madéo, A. Winchester, S. Lei, R. Vajtai, P.M. Ajayan, K.M. Dani, Imaging the motion of electrons across semiconductor heterojunctions. Nature Nanotechnol. **12**, 36 (2017)
123. E. Mårsell, A. Losquin, R. Svärd, M. Miranda, C. Guo, A. Harth, E. Lorek, J. Mauritsson, C.L. Arnold, X. Hongxing, A. L'Huillier, A. Mikkelsen, Nanoscale imaging of local few-femtosecond near-field dynamics within a single plasmonic nanoantenna. Nano Lett. **15**, 6601–6608 (2015). https://doi.org/10.1021/acs.nanolett.5b02363
124. M. Mankos, K. Shadman, R. Hahn, Y.J. Picard, D. Comparat, O. Fedchenko, G. Schönhense, L. Amiaud, A. Lafosse, N. Barrett, Design for a high resolution electron energy loss microscope. Ultramicroscopy **207**, 112848 (2019). https://doi.org/10.1016/j.ultramic.2019.112848
125. G.A. Massey, Microscopy and pattern generation with scanned evanescent waves. Appl. Opt. **23**, 658 (1984)
126. B.R. Masters, History of the optical microscope in cell biology and medicine, in *Encyclopedia of Life Sciences*, (Wiley, New York, 2008)
127. R.J. Matelon, D.M. Newman, M.L. Wears, Photoacoustic determination of the plasmon enhanced electric field at a corrugated metal interface. Rev. Sci. Instrum. **75**, 2560–2563 (2004)
128. T. Matsukata, C. Wadell, N. Matthaiakakis, N. Yamamoto, T. Sannomiya, Selected mode mixing and interference visualized within a single optical nanoantenna. ACS Photonics **5**, 4986–4992 (2018). https://doi.org/10.1021/acsphotonics.8b01231

129. M. Merano, S. Sonderegger, A. Crottini, S. Collin, P. Renucci, E. Pelucchi, A. Malko, M.H. Baier, E. Kapon, B. Deveaud, J.-D. Ganière, Probing carrier dynamics in nanostructures by picosecond cathodoluminescence. Nature **438**, 479 (2005). https://doi.org/10.1038/nature04298
130. P. Melchior, D. Bayer, C. Schneider, A. Fischer, M. Rohmer, W. Pfeiffer, M. Aeschlimann, Optical near-field interference in the excitation of a bowtie nanoantenna. Phys. Rev. B **83**, 235407 (2011)
131. A. Merlen, F. Lagugné-Labarthet, Imaging the optical near field in plasmonic nanostructures. Appl. Spec. OA **68**, 1307 (2014). https://doi.org/10.1366/14-07699
132. M. Merschdorf, C. Kennerknecht, W. Pfeiffer, Collective and single-particle dynamics in time-resolved two-photon photoemission. Phys. Rev. B **70**, 193401 (2004)
133. S. Meuret, M. Solà Garcia, T. Coenen, E. Kieft, H. Zeijlemaker, M. Lätzel, S. Christiansen, S.Y. Woo, Y.H. Ra, Z. Mi, A. Polman, Complementary cathodoluminescence lifetime imaging configurations in a scanning electron microscope. Ultramicroscopy **197**, 28–38 (2019)
134. S. Mignuzzi, M. Mota, T. Coenen, Y. Li, A.P. Mihai, P.K. Petrov, R.F.M. Oulton, S.A. Maier, R. Sapienza, Energy−Momentum cathodoluminescence spectroscopy of dielectric nanostructures. ACS Photonics **5**, 1381–1387 (2018). https://doi.org/10.1021/acsphotonics.7b01404
135. O.L.A. Monti, T.A. Baker, D.J. Nesbitt, Imaging nanostructures with scanning photoionization microscopy. J. Chem. Phys. **125**, 154709 (2006). https://doi.org/10.1063/1.2354478
136. C. Moreno, J. Alda, E. Kinzel, G. Boreman, Phase imaging and detection in pseudo-heterodyne scattering scanning near-field optical microscopy measurements. Appl. Opt. **56**, 1037 (2017)
137. Y. Morimoto, P. Baum, Diffraction and microscopy with attosecond electron pulse trains. Nat. Phys. **14**, 252–256 (2018). https://doi.org/10.1038/s41567-017-0007-6
138. M. Munzinger, C. Wiemann, M. Rohmer, L. Guo, M. Aeschlimann, M. Bauer, The lateral photoemission distribution from a defined cluster/substrate system as probed by photoemission electron microscopy. New J. Phys. **7**, 68–83 (2005)
139. V. Myroshnychenko, F. Natsuki Nishio, J. García, J. de Abajo, N.Y. Förstner, Unveiling and imaging degenerate states in plasmonic nanoparticles with nanometer resolution. ACS Nano **12**, 8436–8446 (2018). https://doi.org/10.1021/acsnano.8b03926
140. S. Negm, H. Talaat, Surface plasmon resonance halfwidths as measured using attenuated total reflection, forward scattering and photoacoustics. J. Phys. Condens. Matter **1**, 10201–10205 (1989)
141. J. Nelayah, M. Kociak, O. Stephan, F.J.G. de Abajo, M. Tence, L. Henrard, D. Taverna, I. Pastoriza-Santos, L.M. Liz-Marzan, C. Colliex, Mapping surface plasmons on a single metallic nanoparticle. Nat. Phys. **3**, 348–353 (2007)
142. V.-Q. Nguyen, Y. Ai, P. Martin, J.-C. Lacroix, Plasmon-induced nanolocalized reduction of diazonium salts. ACS Omega **2**(5), 1947–1955 (2017)
143. O. Nicoletti, F. de la Peña, R.K. Leary, D.J. Holland, C. Ducati, P.A. Midgley, Three-dimensional imaging of localized surface plasmon resonances of metal nanoparticles. Nature **502**, 80 (2013). https://doi.org/10.1038/nature12469
144. P. Nordlander, C. Oubre, E. Prodan, K. Li, M.I. Stockman, Plasmon hybridization in nanoparticle dimers. Nano Lett. **4**, 899–903 (2004)
145. L. Novotny, E.J. Sánchez, X. Sunney, J. Xie, Near-field optical imaging using metal tips illuminated by higher-order Hermite–Gaussian beams. Ultramicroscopy **71**, 21–29 (1998)
146. L. Novotny, N.F. Van Hulst, Antennas for light. Nat. Photonics **5**, 83 (2011)
147. L. Novotny, B. Hecht, *Principle in nano-optics*, 2nd edn. (Cambridge University Press, Cambridge, 2013)
148. A. Otto, Eine neue Methode der Anregung nichtstrahlender Oberflächenplasmaschwingungen. Phys. Status Solidi **26**, K99–K102 (1968)
149. T. Onuma, Y. Kagamitani, K. Hazu, T. Ishiguro, T. Fukuda, S.F. Chichibu, Femtosecond-laser-driven photoelectron gun for time-resolved cathodoluminescence measurement of GaN. Rev. Sci. Instrum. **83**, 043905 (2012). https://doi.org/10.1063/1.3701368

150. Y. Pan, B. Zhang, A. Gover, Anomalous photon-induced near-field electron microscopy. Phys. Rev. Lett. **122**, 183204 (2019)
151. S.T. Park, M. Lin, A.H. Zewail, Photon-induced near-field electron microscopy (PINEM): Theoretical and experimental. New J. Phys. **12**, 123028 (2010). https://doi.org/10.1088/1367-2630/12/12/123028
152. S.T. Park, A.H. Zewail, Relativistic effects in photon-induced near-field electron microscopy. J. Phys. Chem. A **116**, 11128 (2012)
153. S.T. Park, A.H. Zewail, Photon-induced near field electron microscopy. Proc. SPIE Ultrafast Imag. Spectrosc. **8845**, 884506 (2013). https://doi.org/10.1117/12.2023082
154. S.J. Peppernick, A.G. Joly, K.M. Beck, W.P. Hess, Plasmonic field enhancement of individual nanoparticles by correlated scanning and photoemission electron microscopy. J. Chem. Phys. **134**, 034507 (2011). https://doi.org/10.1063/1.3543714
155. L. Piazza et al., Simultaneous observation of the quantization and the interference pattern of a plasmonic near-field. Nat. Commun. **6**, 6407 (2015)
156. J. Plain, G.P. Wiederrecht, S.K. Gray, P. Royer, R. Bachelot, Multiscale optical imaging of complex fields based on the use of azobenzene nanomotors. J. Phys. Chem. Lett. **4**(13), 2124–2132 (2013)
157. D. Podbiel, P. Kahl, A. Makris, B. Frank, S. Sindermann, T.J. Davis, H. Giessen, M.H.-v. Hoegen, F.J. Meyer zu Heringdorf., Imaging the nonlinear plasmoemission dynamics of electrons from strong plasmonic fields. Nano Lett. **17**(11), 6569–6574 (2017)
158. D.W. Pohl. Scanning near-field optical microscopy (SNOM). Adv. Optic. Electr. Microsc. **12** (1991). ISBN 0-12-029912 Copyright © 1991 Academic
159. D. Pohl, D. Courjon, in *Near Field Optics*, NATO ASI Series E: Applied Sciences, vol. 242. Proceedings of NFO I: Besançon/Arc & Senans, France, 26–28 October 1992 (Kluwer Academic, Dordrecht, 1993)
160. A. Polman, F. Mathieu Kociak, J. García, de Abajo., Electron-beam spectroscopy for nanophotonics. Nat. Mater. **18**, 1158–1171 (2019). https://doi.org/10.1038/s41563-019-0409-1
161. M. Purcell, Spontanneaous emission probability at radio frequencies. Phys. Rev. **69**, 681 (1946)
162. R. Qi, R. Wang, Y. Li, Y. Sun, S. Chen, B. Han, N. Li, Q. Zhang, X. Liu, Y. Dapeng, P. Gao, Probing far-infrared surface phonon polaritons in semiconductor nanostructures at nanoscale. Nano Lett. **19**, 5070–5076 (2019). https://doi.org/10.1021/acs.nanolett.9b01350
163. H. Rau, Photo isomerization of azobenzenes. Photoreact. Organic Thin Films, 3–47 (2002)
164. H.-G. von Ribbeck, M. Brehm, D.W. van der Weide, S. Winnerl, O. Drachenko, M. Helm, F. Keilmann, Spectroscopic THz near-field microscope. Opt. Express **16**, 3430 (2008)
165. N. Rotenberg, L. Kuipers, Mapping nanoscale light fields. Nature Photos **8**, 919 (2014)
166. B.M. Ross, L.P. Lee, Comparison of near- and far-field measures for plasmon resonance of metallic nanoparticles. Opt. Lett. **34**, 896–898 (2009)
167. D. Rossouw et al., Multipolar plasmonic resonances in silver nanowire antennas imaged with a subnanometer electron probe. Nano Lett. **11**, 1499 (2011)
168. B. Rothenhäusler, J. Rabe, P. Korpiun, W. Knoll, On the decay of plasmon surface polaritons at smooth and rough Ag-air interfaces: A reflectance and photo-acoustic study. Surf. Sci. **137**, 373–383 (1984). https://doi.org/10.1016/0039-6028(84)90696-4
169. A. Ryabov, P. Baum, Electron microscopy of electromagnetic waveforms. Science **353**, 374 (2016)
170. J. Schefold, S. Meuret, N. Schilder, T. Coenen, H. Agrawal, E.C. Garnett, A. Polman, Spatial resolution of coherent cathodoluminescence super-resolution microscopy. ACS Photonics **6**, 1067–1072 (2019). https://doi.org/10.1021/acsphotonics.9b00164
171. F.-P. Schmidt, A. Losquin, F. Hofer, A. Hohenau, J.R. Krenn, M. Kociak, How dark are radial breathing modes in plasmonic nanodisks? ACS Photonics **5**, 861–866 (2018). https://doi.org/10.1021/acsphotonics.7b01060
172. A.W. Schell, P. Enge, J.F.M. Werra, C. Wolff, K. Busch, O. Benson, Scanning single quantum emitter fluorescence lifetime imaging: Quantitative analysis of the local density of photonic states. Nano Lett. **14**, 2623 (2014)

173. V. Schweikhard, A. Grubisic, T.A. Baker, D.J. Nesbitt, Multiphoton scanning photoionization imaging microscopy for single-particle studies of plasmonic metal nanostructures. J. Phys. Chem. C **115**, 83–91 (2011). https://doi.org/10.1021/jp1075143
174. E. Shirai, Y. Urai, K. Itoh, Surface-enhanced photopolymerization of a diacetylene derivative in langmuir−blodgett films on a silver island film. J. Phys. Chem. B **102**(19), 3765–3772 (1998)
175. H.L. Skriver, N.M. Rosengaard, Surface energy and work function of elemental metals. Phys. Rev. B **46**, 7157–7168 (1992)
176. G. Spektor, D. Kilbane, A.K. Mahro, B. Frank, S. Ristok, L. Gal, P. Kahl, D. Podbiel, S. Mathias, H. Giessen, F.-J. Meyer zu Heringdorf, M. Orenstein, M. Aeschlimann, Revealing the subfemtosecond dynamics of orbital angular momentum in nanoplasmonic vortices. Science **355**, 1187–1191 (2017)
177. K.-H. Su, Q.-H. Wei, X. Zhang, J.J. Mock, D.R. Smith, S. Schultz, Interparticle coupling effects on plasmon resonances of nanogold particles. Nano Lett. **3**, 1087–1090 (2003)
178. E.H. Synge, A suggested method for extending microscopic resolution into the ultramicroscopic region. Lond. Edinb. Dublin Philos. Mag. J. Sci **6**(35), 356–362 (1928)
179. N. Talebi, Electron-light interactions beyond the adiabatic approximation: recoil engineering and spectral interferometry. Adv. Phys. X **3**(1), 1499438 (2018). https://doi.org/10.1080/23746149.2018.1499438
180. T.H. Taminiau, F.D. Stefani, N.F. van Hulst, Optical nanorod antennas modeled as cavities for dipolar emitters: Evolution of sub- and super-radiant modes. Nano Lett. **11**, 1020–1024 (2011). https://doi.org/10.1021/nl103828n
181. T. Taubner, F. Eilmann, R. Hillenbrand, Nanoscale-resolved subsurface imaging by scattering-type near-field optical microscopy. Opt. Express **13**, 8893 (2005)
182. A. Teulle, A. Sanchot, E. Ishow, J. Sharma, E. Dujardin, Photochemical mapping of the multimodal plasmonic response of 2D gold crystals. J. Phys. Chem. C **121**(29), 15908–15914 (2017)
183. I. Tijunelyte, I. Kherbouche, S. Gam-Derouich, M. Nguyen, C.N. Lidgi-Guigui, M.L. de la Chapelle, A. Lamouri, G. Lévi, J. Aubard, A. Chevillot-Biraud, C. Mangeney, N. Felidj, Multi-functionalization of lithographically designed gold nanodisks by plasmon-mediated reduction of aryl diazonium salts. Nanosc Horiz **3**, 53 (2018)
184. L. Tong, Q.S. Wei, A. Wei, J.X. Cheng, Gold nanorods as contrast agents for biological imaging: Optical properties, surface conjugation and photothermal effects. Photochem. Photobiol. **85**, 21–32 (2009). https://doi.org/10.1111/j.1751-1097.2008.00507.x
185. Z. Thollar, C. Wadell, T. Matsukata, N. Yamamoto, T. Sannomiya, Three-dimensional multipole rotation in spherical silver nanoparticles observed by cathodoluminescence. ACS Photonics **5**, 2555–2560 (2018). https://doi.org/10.1021/acsphotonics.7b01293
186. K. Ueno, S. Takabatake, Y. Nishijima, V. Mizeikis, Y. Yokota, H. Misawa, Nanogap-assisted surface plasmon nanolithography. J. Phys. Chem. Lett. **1**(3), 657–662 (2010)
187. A.V. Uskov, I.E. Protsenko, R.S. Ikhsanov, V.E. Babicheva, S.V. Zhukovsky, A.V. Lavrinenko, E.P. O'Reilly, X. Hongxing, Internal photoemission from plasmonic nanoparticles: comparison between surface and volume photoelectric effects. Nanoscale **6**, 4716 (2014)
188. R. Vogelgesang, A. Dmitriev, Real-space imaging of nanoplasmonic resonances. Analyst **135**, 1175–1181 (2010). https://doi.org/10.1039/C000887G
189. B. Voigtländer, *Atomic force microscopy*, 2nd edn. (Springer, Cham, 2019)
190. J.A. Veerman, A.M. Otter, L. Kuipers, N.F. van Hulst, High definition aperture probes for near-field optical microscopy fabricated by focused ion beam milling. Appl. Phys. Lett. **72**, 3115–3117 (1998). https://doi.org/10.1063/1.121564
191. E.J.R. Vesseur, R.D. Waele, M. Kuttge, A. Polman, Direct observation of plasmonic modes in au nanowires using high-resolution cathodo luminescence spectroscopy. Nano Lett. **7**, 2843–2846 (2007)
192. B. Vohnsen, S. Bozhevolnyi, R. Olesen, Study of shear force technique for near-field microscopy with an uncoated fiber tip. Ultramicroscopy **61**, 207–213 (1995)

193. G. Volpe, M. Noack, S.S. Aćimović, C. Reinhardt, R. Quidant, Near-field mapping of plasmonic antennas by multiphoton absorption in poly(methyl methacrylate). Nano Lett. **12**(9), 4864–4868 (2012)
194. Z. Vörös, R. Johnsen, A simple demonstration of frustrated total internal reflection. Am. J. Phys. **76**, 246 (2008)
195. F. Wang, Y. Ron Shen, General properties of local plasmons in metal nanostructures. Phys. Rev. Lett. **97**, 206806 (2006)
196. Y. Wang, S. Wang, S. Zhang, O.A. Scherman, J.J. Baumberg, T. Ding, X. Hongxing, Plasmon-directed polymerization: Regulating polymer growth with light. Nano Res. **11**, 6384–6390 (2018)
197. K. Wang, R. Dahan, M. Shentcis, Y. Kauffmann, A.B. Hayun, O. Reinhardt, S. Tsesses, I. Kaminer, Coherent interaction between free electrons and a photonic cavity. Nature **582**, 50 (2020)
198. R.W. Wood, On a remarkable case of uneven distribution of light in a diffraction grating spectrum. Philos. Mag. **4**, 396 (1902)
199. R. C. Word, T. Dornan, R. Könenkamp. Photoemission from localized surface plasmons in fractal metal nanostructures. Appl. Phys. Lett. 96, 251110 (2010); 10.1063/1.3457921
200. Y. Wu, G. Li, J.P. Camden, Probing nanoparticle plasmons with electron energy loss spectroscopy. Chem. Rev. **118**, 2994–3031 (2018). https://doi.org/10.1021/acs.chemrev.7b00354
201. Y. Wu, H. Zhongwei, X.-T. Kong, J.C. Idrobo, A.G. Nixon, P.D. Rack, D.J. Masiello, J.P. Camden, Infrared plasmonics: STEM-EELS characterization of Fabry-Pérot resonance damping in gold nanowires. Phys. Rev. B **101**, 085409 (2020)
202. A. Yurtsever, A. Zewail, Direct visualization of near-fields in nanoplasmonics and nanophotonics. Nano Lett. **12**, 3334 (2012)
203. G. Wurtz, R.A. Bachelot, P. Royer, A reflection-mode apertureless scanning near-field optical microscope developed from a commercial scanning probe microscope. Rev. Sci. Instrum. **69**, 1735–1743 (1998)
204. G. Wurtz, D. Burget, C. Carre, in *Photopolymerization-Induced Materialization of the Dipolar Response from Isolated Metallic Nanoparticles*. Proceedings of the SPIE 5458, Optical Micro- and Nanometrology in Manufacturing Technology (17 August 2004); https://doi.org/10.1117/12.545655
205. A.V. Zayats, I.I. Smolyaninov, Near-field photonics: Surface plasmon polaritons and localized surface plasmons. J. Optics A Pure Appl. Optics **5**, S16–S50 (2003)
206. C. Zhan, M. Moskovits, Z.-Q. Tian, Recent progress and prospects in plasmon-mediated chemical reaction. Matter **3**, 42–56 (2020)
207. Y. Zhang, G. Demesy, M. Haggui, D. Gerard, J. Béal, S. Dodson, Q. Xiong, J. Plain, N. Bonod, R. Bachelot, Nanoscale switching of near-infrared hot spots in plasmonic oligomers probed by two-photon absorption in photopolymers. ACS Photon. **5**(3), 918–928 (2017)
208. X. Zhou, C. Deeb, S. Kostcheev, G.P. Wiederrecht, P.-M. Adam, J. Béal, J. Plain, D.J. Gosztola, J. Grand, N. Félidj, H. Wang, A. Vial, R. Bachelot, Selective functionalization of the nanogap of a plasmonic dimer. ACS Photonics **2**(1), 121–129 (2014)
209. S. Zu, T. Han, M. Jiang, F. Lin, X. Zhu, Z. Fang, Deep-subwavelength resolving and manipulating of hidden chirality in achiral nanostructures. ACS Nano **12**, 3908–3916 (2018). https://doi.org/10.1021/acsnano.8b01380
210. S. Zu, T. Han, M. Jiang, Z. Liu, Q. Jiang, F. Lin, X. Zhu, Z. Fang, Imaging of plasmonic chiral radiative local density of states with cathodoluminescence nanoscopy. Nano Lett. **19**, 775–780 (2019). https://doi.org/10.1021/acs.nanolett.8b03850

Chapter 5
On-Chip Nanoscale Light Sources

Jacob Kokinda, Xi Li, and Qing Gu

Abstract As technology continues to require faster and more energy-efficient chip-scale communication, the carrier of information is evolving from electrons to photons. Following a similar downscaling trend in electronic integrated circuits, photonic devices are under a revolution of miniaturization. However, it has proven to be difficult to reduce the size of photonics without compromising effectiveness and speed. In this chapter, we review recent progress in nanophotonic devices, specifically, light sources. While there are many different materials and strategies considered in this research field, the advancements in photonics made possible by plasmonic, photonic crystal, and metamaterial-based devices will be highlighted in this review with a focus on three figures-of-merit: energy thresholds, scalability, and modulation speed.

5.1 Introduction

In 1965, Gordon Moore predicted that the number of transistors in an integrated circuit (IC) would double every 2 years [1]. This empirical law has been used to guide research and development of the semiconductor industry and has become one of the major driving forces of the advancements in electronics ever since. Photonics is on a similar path following this downscaling trend to improve energy efficiency and integration density of photonic components for integrated photonic-electronic ICs [2]. Thylén et al. and Smit et al. claim that in the early days of

J. Kokinda · X. Li
Department of Electrical and Computer Engineering, North Carolina State University, Raleigh, NC, USA

Q. Gu (✉)
Department of Electrical and Computer Engineering, North Carolina State University, Raleigh, NC, USA

Department of Physics, North Carolina State University, Raleigh, NC, USA
e-mail: qgu3@ncsu.edu

R. Gordon (ed.), *Advances in Near-Field Optics*, Springer Series in Optical Sciences 244, https://doi.org/10.1007/978-3-031-34742-9_5

photonic integrated circuits (PICs), the integration density followed an improvement rate comparable to Moore's Law and has the potential to carry on as long as the development of novel materials and fabrication techniques continue to improve [3, 4]. Waveguides, photodetectors, and light sources are examples of PIC building blocks that require continued miniaturization. This chapter focuses on the current designs and experimental strategies for semiconductor nanoscale light sources.

The most prevalent kinds of light sources are the light emitting diode (LED) and the light amplification by stimulated emission of radiation (laser). Both kinds of devices are used in areas ranging from advanced technological systems to daily consumer products; however, there exists a key difference in their operation principles [5]. LEDs generate photons from spontaneous emission, which occurs when an excited electron decays from the conduction band to the valence band, emitting a photon with a random phase and spatial direction. On the contrary, lasers generate photons from stimulated emission and typically require a confined cavity to provide compensation for the cavity loss. Stimulated emission occurs when an electron in the excited state is stimulated by an incoming photon and releases its energy as a new photon that has the same frequency, phase, and spatial direction as the incoming photon. This recurring process leads to an amplification of the incident radiation. A necessary condition for lasing emission is population inversion, which is defined as the state of a medium where the electron population is higher at the conduction band than at the valance band. Since population inversion would never occur naturally, an external pumping source is used to provide excitation to fulfill this condition. The pumping source can be either another light source or an electric current, the choice of which determines the applications and practicality of the device. When specific designs are discussed later in this chapter, this concept will be revisited.

One requirement for on-chip light sources is low energy cost. A figure-of-merit that quantifies the energy efficiency of an optical cavity is its quality factor (Q-factor). Q-factor is defined as the ratio of energy stored in a specific electromagnetic mode to energy lost during an oscillation cycle. Cavity loss can be introduced by parasitic effects, such as nonradiative recombination, as well as cavity leakage and output radiation from the light source. Optimizing the Q-factor is done by improving optical confinement and reducing parasitic losses. Thus, a high cavity Q-factor leads to a more efficient light source and is an important consideration for the devices studied in this chapter.

With the aim of downscaling, researchers have designed different types of cavities to reduce the footprint of the light sources. Nanoscale cavities can be categorized into two types by their material composition – purely dielectric cavities and metal cavities. Researchers first used metal as cavity cladding, which greatly improves the confinement of the photonic modes and the cavity Q-factor. After this, devices such as plasmonic lasers and SPASERs were developed to support plasmonic modes, i.e., confined modes at the metal-dielectric interface. In order to support plasmonic modes, the metals are specifically chosen based on their low dampening rate and high intraband transition energies. In addition, the negative electric permittivity of metal is exploited in designing metamaterials.

In modern times, semiconductor devices dominate the field of light emitters as they can be scaled down to the micro/nano scale and have properties that allow for efficient light-matter interaction in addition to electrical operation. This technology is critical for integration with other solid-state systems like semiconductor electronics. The journey to achieve room temperature semiconductor light sources started all the way back in 1960 when the first ever laser was demonstrated. Theodore Maiman created a solid-state pink ruby light source with coherent emission and single-frequency operation at Hughes Research Laboratory [6]. This was done by using a flash lamp to excite ruby, which produced a high-power short-pulsed laser output. The first continuous wave beam was demonstrated shortly after, in the same year, by Ali Javan, William Bennett Jr., and Donald Herriott at Bell Labs using an electrically pumped helium-neon laser [7]. Javan predicted this design with his theory on gas lasers, which was published the year before. There were many incremental advancements from the 1960s through the 1980s that helped build up to the designs of modern lasers and LEDs. These included the first semiconductor laser (GaAs) that used electric current to directly create infrared light and the first semiconductor laser diode (GaAsP) that produced visible red light, both being realized in 1962. The problem with these semiconductor laser designs was that they required cryogenic temperatures for operation. However, in 1970, Bell Labs and Loffe Physcio-Technical Institute simultaneously demonstrated a room-temperature continuous-wave semiconductor laser [8]. Jumping forward to 1982, Peter Moulton at Massachusetts Institute of Technology produced a titanium-sapphire laser that was used to create ultrafast pulses in the picosecond and femtosecond ranges [9]. From these advancements, quantum wells [10] and quantum dot gain media [11] were developed in the 1990s along with the metal-clad cavity design being proposed in 2007 [12, 13]. Novel laser designs today are built upon this accumulated knowledge to create highly efficient devices that we will review in the coming sections. LEDs have been in continuous development and improvement since 1907; nonetheless, they were not considered for communication purposes until recently [14].

At the nanoscale, cavity quantum electrodynamic (QED) effects greatly impact the operation and performance of light sources. With the device dimensions approaching half its operating wavelength, the diffraction limit, which sets a lower limit to the device size, is often reached [15]. The diffraction limit declares that the cavity size, in all three dimensions, cannot be smaller than half the wavelength of the propagating light to sustain the energy oscillation between electric energy and magnetic energy every half period. Otherwise, the energy will be lost through destabilization. In addition, there is also a significant increase in parasitic losses at the nanoscale. There are several processes that contribute to parasitic loss, including Auger recombination [16], surface recombination [17], and self-heating, where both Auger and surface recombination are non-radiative processes. Auger recombination is the transfer of energy from the recombination of an electron-hole pair to another electron or hole instead of the energy being emitted in the form of a photon. The Auger recombination rate is proportional to the third power of the carrier density and it is therefore more prominent at high pumping power. Surface recombination is a result of defects or impurities at the surface of a semiconductor (surface traps) and is proportional to carrier density. While there are other factors that influence the surface recombination rate, it is directly proportional to the surface-to-volume ratio. Hence, this type of loss becomes more prominent in nanoscale devices. Self-heating also becomes a critical issue where the dissipation of thermal energy is limited to smaller volumes. While there are many negative effects at this scale, there is also a vastly important effect that enables these devices to operate with improved characteristics, which is known as the Purcell effect [18, 19]. The Purcell effect is a phenomenon quantified by Edward Mills Purcell in the 1940s that claims the spontaneous emission rate of a radiating dipole can be enhanced inside a cavity. The associated spontaneous emission enhancement factor is termed the Purcell Factor. The Purcell Factor is inversely proportional to the modal volume of the cavity and is a result of the increase in the optical density of states. Various nanoscale lasers and LEDs designs have been demonstrated based on this effect. It is shown that nanoLEDs can be a competitive candidate for on-chip high-speed communications as they can have extremely fast spontaneous emission rates and no pumping thresholds which leads to high modulation speeds and low current requirements, respectively [14]. Since nanoLEDs have a niche role in chip-scale optical communication systems, more details of this kind of nanoscale light source will be covered in the first section of the chapter.

Using different materials and technologies, many types of lasers have been created with their own advantages in terms of Q-factor, emission rate, emission intensity, and device size. Depending on the application, certain advantages are desired more than others. In this chapter, we will discuss three types of lasers: plasmonic lasers, photonic crystal lasers, and metamaterial lasers. The second section focuses on plasmonic lasers that are characterized by their metal-dielectric interfaces that support plasmonic modes [20]. These devices exploit small device dimensions due to their ability to overcome the diffraction limit through the storage of optical energy in surface plasmon polaritons. Important applications of plasmonic lasers include sensing and biological probing, which are made possible by their extremely high sensitivity of resonant wavelength to local environment changes and small size. In the third section, we will introduce photonic crystal lasers that utilize periodic variations in the refractive indices, on the scale of or larger than the wavelength of light, to create high Q-factor cavities [21]. In this type of cavity, a photonic bandgap is formed where only the energy inside the bandgap can exist in the cavity. It is possible for the energy inside a cavity to be effectively forced into a single electromagnetic mode. These devices find their applications as thresholdless lasers and high-power single-mode lasers. In addition, photonic crystal-based devices can also be used on-chip in photonic circuits to employ wavelength-division multiplexing (WDM). Relying on the ability to manipulate the crystal lattice size and shape, the cavity resonant wavelength can be tuned with advanced nanofabrication techniques. In the last section, we will introduce another type of laser based on periodic structures, known as metamaterials, in which the lattice period is much smaller than the wavelength of light [22, 23]. The property of negative refraction can occur with this type of material, leading to prominent applications in subwavelength imaging due to its inherent focusing abilities. In a highly anisotropic metamaterial, we can find hyperbolic dispersion characteristics that can be used to enhance the spontaneous emission rate and emission intensity of a cavity. While there are plenty of sub-wavelength imaging devices, designs integrating such metamaterials into light sources have not yet been realized. However, we expect to see efficient lasers and LEDs built on metamaterials in the near future.

In the proceeding sections, we will dive deeper into the physics of each type of light source, their advantages, the challenges they face, and examples of designs that have been theoretically proposed or experimentally realized by experts in the field.

There are a few points to keep in mind when evaluating the performance of different light source designs. First, many designs are optically pumped. While these simpler devices are good first-step "proof-of-concept" devices, their performance is often skewed towards the more ideal case. Therefore, the speed and power consumption will deviate from the technologically relevant electrically pumped devices. Second, for an incoherent light source that operates with optical injection and spontaneous emission, we avoid using the term "LED" as the diode structure does not exist. For this reason, such devices will be classified as nano-emitters. Third, due to the complicated nature of nanofabrication, some of the devices discussed in this chapter have only been demonstrated through simulation and theoretical calculations.

5.2 NanoLEDs

Long-haul communications have witnessed a significant paradigm shift from electrical communication via copper wires to fiber-optical communication over the last decades. The light sources used in an optical communication system are dominated by lasers due to their capability to generate light with high output power, coherent emission, small divergence angle, and fast modulation speed [14]. In short-distance communication, especially chip-to-chip and on-chip communication, the number of interconnects increases exponentially with the integration density. The biggest bottlenecks for chip-scale interconnects are data transmission speed and energy efficiency. The metal interconnects currently used on-chip have brought many issues, including limited bandwidth, high power dissipation, and parasitic capacitance [24]. In contrast, optical interconnects could potentially provide a solution to issues like capacitance coupling energy loss and limited bandwidth. Hence, building a fast and efficient short-range communication system requires the transition from electrical interconnects to optical interconnects. A crucial part of such a system is an efficient light source. At first glance, lasers can be the ideal candidate. However, as the laser cavity approaches sub-wavelength dimensions, it becomes increasingly inefficient due to poor optical confinement. Even though metal-clad lasers have helped resolve the confinement issue, metal introduces significant ohmic loss which in turn creates a large lasing threshold. Therefore, LEDs could be a better choice for sub-wavelength light sources as they can be operated at lower currents due to their lack of threshold.

In addition to the simpler driver circuitry and a smaller footprint than nanolasers, nano-scale LEDs can achieve high modulation speeds at low operation currents [2]. Modulation speed for LEDs rely on the majority of electrons returning to the ground state before the next emission cycle can begin; therefore, the rate of spontaneous emission is the main limiting factor. However, the slow intrinsic spontaneous emission rate can be enhanced by the Purcell effect. The Purcell Factor that quantifies this enhancement is shown in Eq. 5.1, with Q representing the cavity quality factor and V_{eff} representing the effective modal volume [18]. It can be observed that the spontaneous emission rate scales inversely with the modal volume; thus, the conditions for high Purcell Factor agree with the demand for a smaller device footprint. With the device's Q-Factor and effective volume being optimized, the spontaneous lifetime can be reduced to reach higher modulation speed. However, in open cavities the effective volume is ill-defined and needs a more rigorous definition, as described in Chap. 3:

$$F = \frac{3}{4\pi^2}\left(\frac{\lambda_c}{\eta_a}\right)^3 \frac{Q}{V_{\mathrm{eff}}} \tag{5.1}$$

Traditional LEDs avoid many of the parasitic effects that are present in lasers, such as spectral hole burning and gain saturation. NanoLEDs, however, experience these effects due to the fact that there is a specific mode profile in the cavity and high carrier density. The increased spontaneous emission rate (R_{sp}) and spontaneous emission coupling factor (β) cause an augmented photon density in both the spatial dimension and spectral dimension, leading to spatial hole burning and spectral hole burning, respectively [25, 26]. For spatial hole burning, carriers in the gain medium seek to replenish the drastically reduced exciton population for the specific area that is saturated by locally intense electromagnetic field distributions; however, the carrier diffusion speed is not high enough for new carriers to reach the area before the end of the spontaneous emission lifetime. Spectral hole burning is a similar process, referring to carriers replenishing excitons of a particular energy rather than a physical position. Although the faster spontaneous emission rate from the Purcell effect allows for high-speed operation of nanoLEDs, it also leads to negative effects such as spectral and spatial hole burning. However, with purely spontaneous emission, the inherent benefit is the low amount of energy needed for operation, as LEDs do not require gain to fully compensate for the cavity losses [14]. Therefore, the problem of hole burning does not have a significant impact compared to lasers.

To enhance the modulation speed based on the Purcell effect, Li et al. demonstrated a shifted-core nanocoaxial light source [27]. It consists of 170 nm-thick InGaAsP quantum-well gain material with a silver cladding and core (Fig. 5.1). The gain medium is sandwiched between SiO_2 dielectric layers in the vertical direction to form a Fabry-Perot cavity. The upper SiO_2 layer is 40 nm in height, and the bottom output layer is 10 nm in height. In the horizontal direction, the gain material has a uniform thickness of 140 nm. As the metal core is shifted to one side of the cavity, the mode volume decreases, the Purcell Factor increases, and modulation speed increases. This device has a footprint of merely one-third of its emission

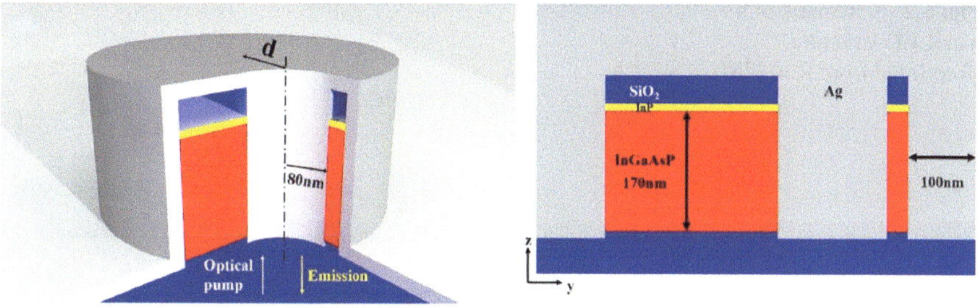

Fig. 5.1 3D and cross-sectional schematic of a shifted core coaxial nano-emitter. (Reprinted from Ref. [27])

wavelength of 1550 nm and a predicted modulation bandwidth of 62 GHz at a maximum core-shifting distance of 110 nm.

Considering the practical need for electrical pumping, Lau et al. demonstrated an LED that closely resembles a Fin-FET, which is intrinsically made for electrical pumping [28]. Ultra-small sub-wavelength dimensions were able to be achieved through the use of metal cladding to support plasmonic modes. A P-I-N junction is first formed with the intrinsic layer consisting of the optical gain material. Silver cladding covering the gain material enables the support of plasmonic modes at the interface between optical gain and cladding, which greatly improves radiation confinement (Fig. 5.2). Between the cladding and the intrinsic region, a 2 nm thick passivation layer is formed to prevent current leakage. Simulation shows that a mode volume of 0.01 $\left(\frac{2n_{\text{gain}}}{\lambda_0}\right)^3$ and Q-factor above 10 would be enough to achieve at least 100 GHz speed. A device with 20 nm m by 40 nm by 140 nm total volume was calculated to have a bandwidth of 660 GHz. Even though these results come from strictly simulations and calculations, future experimental implementation could potentially prove this design to be a great advancement in terms of both speed and device footprint.

Hoang et al. experimentally demonstrated an optically pumped device with ultra-fast spontaneous emission, using a plasmonic nano-patch antenna (NPA) integrated with colloidal semiconductor quantum dots [29]. Shown in Fig. 5.3a, the device consists of a silver nano-cube of 75 nm side length coupled to a 50 nm-thick gold film, separated by a 10 nm spacer consisting of quantum dots atop a thin dielectric layer of <1 nm. A Fabry-Perot fundamental plasmonic mode has resonance due to the reflections caused by the waveguide-like space under the nano-cube. Figure 5.3b shows the difference in decay time between the NPA-integrated quantum dots and quantum dots placed on a dielectric or metal layer without NPAs, clearly showing a faster spontaneous emission decay in NPA-integrated quantum dots. In order to fit the fluorescence decay curve, a bi-exponential function had to be used, resulting in two time constants. The fast time constant corresponds to the Purcell-enhanced spontaneous decay. The measured average spontaneous emission lifetime is 11 ps, corresponding to a 90 GHz modulation speed.

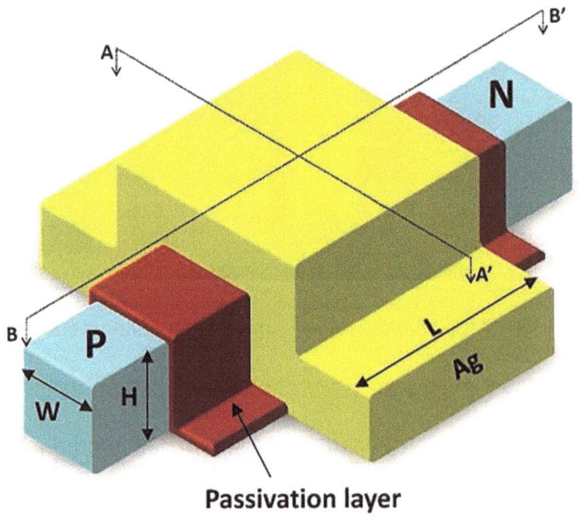

Fig. 5.2 Schematic of a nanoLED structure. (Reprinted from Ref. [28])

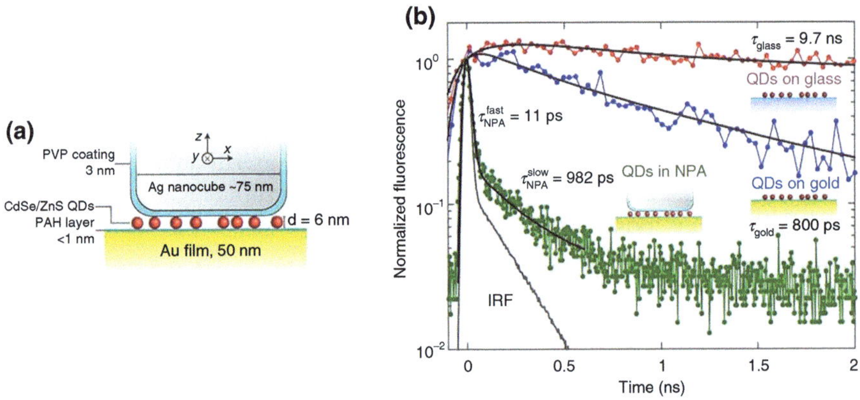

Fig. 5.3 (a) Cross-sectional schematic of a NPA. (b) Normalized time-resolved fluorescence of quantum dots on glass (red), on a gold film (blue) and in a NPA (green). (Reprinted from Ref. [29])

While modulation speed is the focus of many on-chip light source designs, another important aspect is often neglected. An on-chip light source must be integrable with other photonic devices, which involves coupling with waveguides that serve as optical interconnects. Due to small device aperture size and mode mismatch, coupling to waveguides commonly involves high loss. Therefore, increasing the coupling efficiency between nanoLEDs and waveguides is essential. This problem is largely dependent upon the percentage of energy in the cavity that can be coupled to the fundamental mode of the waveguide. When creating photonic integrated circuits, having a vast number of poor coupling LED-waveguide connections will lead to the detriment of the system's efficiency.

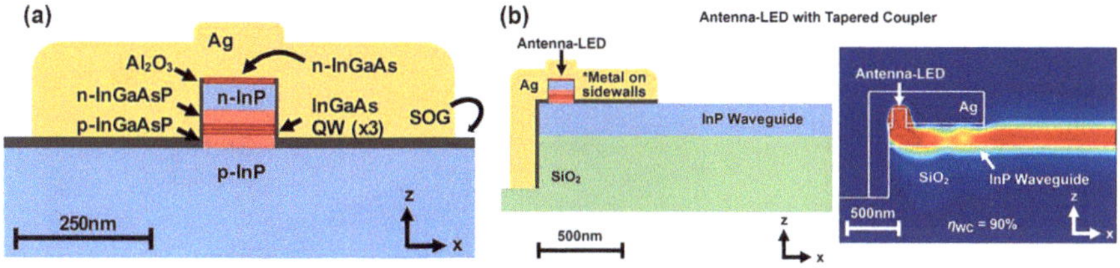

Fig. 5.4 (**a**) Vertical cross section schematic of an optical antenna-LED on a bulk InP substrate. (**b**) Vertical cross section and power flow of the fundamental waveguide mode for a InP tapered waveguide with metal wrapped around the waveguide facet and sidewalls. (Reprinted from Ref. [30])

To address the coupling efficiency issue, Andrade et al. proposed a cavity-backed slot antenna LED design, which can be coupled to a waveguide with extremely high coupling efficiencies and impressive modulation speeds [30]. In their structure, gold cladding is deposited on a multiple quantum well InGaAs ridge which sits on a InP substrate (Fig. 5.4a). The optimal location to couple the waveguide to this LED was found to be at the end of the waveguide, where the waveguide is tapered towards the emission point, with metal surrounding the tapering portion (Fig. 5.4b). This design was simulated to have a 90% coupling efficiency and a Purcell Factor of 162. After using machine learning–based optimizations, the efficiency and Purcell enhancement were slightly increased to 94% and 164, respectively. This Purcell Factor of 164 corresponds to a modulation speed of about 104 GHz. Even with purely theoretical results, this design can be considered very promising for further research in efficient waveguide-LED coupling.

An important milestone with light source – waveguide coupling was the first experimental demonstration of an electrically injected nanoLED integrated with a waveguide. Dolores-Calzadilla et al. designed a nano-pillar LED, made of bulk InGaAs gain layer sandwiched between a few doped InP carrier transport layers. Then, the device is bottom coupled to an InP waveguide [31]. The pillar, as shown in Fig. 5.5, is encapsulated by a SiO_2 electrical insulation layer followed by a silver cladding to create a metallo-dielectric cavity. After experimentation, analysis of the data suggested that the coupling efficiency was about 15%, based on the assumption that all the recombination was radiative and all the emitters were spectrally resonant with the desired mode. Since these assumptions are unrealistic, the results can be expected to be much lower. The demonstrated modulation speed was also low, around 5 GHz. Nonetheless, this first experimental realization of an electrically injected nanoLED – waveguide coupled device is a step towards the practical application of nanoLED.

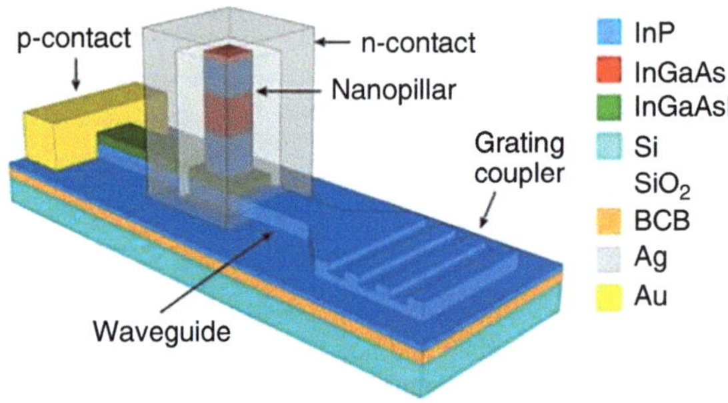

Fig. 5.5 Schematic representation of a nanopillar LED coupled to a waveguide. (Reprinted from Ref. [31])

5.3 Plasmonic Lasers

Plasmons are quasi-particles that describe a quanta of plasma oscillations (groups of free electrons) which are referred to as bulk plasmons [32]. Since metals have an abundance of free carriers, bulk plasmons can easily form with the influence of an electric field, such as light [33]. Surface plasmons are plasmons generated at the boundary between a metal and dielectric material and propagate along this interface [34]. Plasmonic lasers partially store energy, of an optical mode, in semiconductor-metal surface plasmon polariton (SPP) modes [20]. Incident photons create an exciton in the gain media, which can then nonradiatively recombine to excite a surface plasmon in the metal. The coupling with these plasmons and photons gives rise to SPPs. In addition to metal-dielectric interfaces, SPPs can also form at the interface of dielectric and other plasmonic materials, such as transparent metal oxide, as long as the permittivity of the plasmonic material is negative and has a greater magnitude than the dielectric material. The field is confined at the metal-dielectric interface, with the skin depth determining how far into the metal the field will penetrate. The tighter confinement can also be understood in terms of momentum conservation where higher k vectors correspond to higher momentum. Since electrons have mass (unlike photons), photons coupling with surface plasmons increases the momentum with respect to freely propagating photons in a dielectric, translating to larger k vectors. With larger k values, the effective wavelength is smaller, allowing for device dimensions below the diffraction limit of the lasing wavelength [15].

Another type of laser that is often confused with plasmonic laser is SPASER (surface plasmon amplification by stimulated emission of radiation). Figure 5.6 differentiates them by the amount of plasmonicity [35]. Plasmonicity is a measure of proximity to the SPP resonance. As the energy approaches resonance, more optical energy is stored in SPPs. The differentiation between a plasmonic laser and a SPASER does not have any physical significance; rather, it is a formality when characterizing the light source. The proceeding devices that we will look at are based on plasmonic modes and are classified as plasmonic lasers.

The nanowire is a popular cavity choice of plasmonic lasers. A nanowire is a quasi-1D structure that can serve as an efficient waveguide or cavity due to the

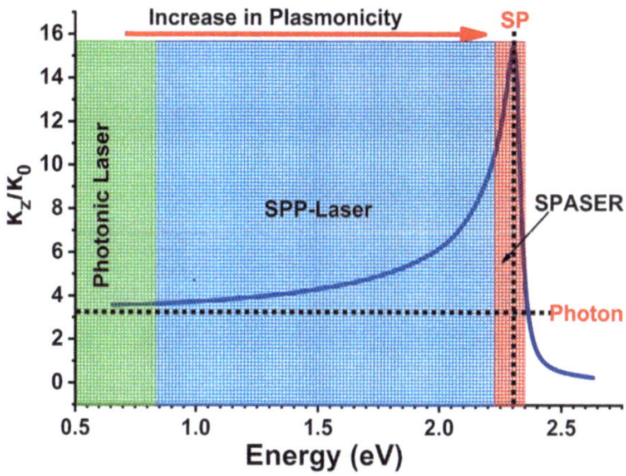

Fig. 5.6 The normalized wave vector versus energy at a semiconductor-silver interface. The three ranges of energy are divided into classifications depending on the proximity to the SPP resonance energy. The photonic laser represents the traditional laser; the SPP-Laser represents the "plasmonic laser"; and the SPASER represents a laser that largely operates at or right around the resonance energy. (Reprinted from Ref. [35])

index of refraction contrast between the wire and the surrounding material. The general approach to the plasmonic nanowire cavity design is a semiconductor wire placed on a low-index passivation dielectric material that separates the nanowire from the metal substrate. This cavity supports both photonic and plasmonic modes, which hybridize in the transverse direction and resonate between the end facets of the nanowire. The main issues of this design have to do with parasitic loss. These losses include ohmic damping, SPP scattering, and radiative emission at the ends of the wire. Scattering can be a result of the metal surface roughness or the interfaces of different materials [36]. Ohmic damping is intrinsically high in plasmonic lasers, which increases the required pumping energy. All these loss terms contribute to a high lasing threshold, such that cryogenic temperatures are required for the device to lase.

One approach to reduce the lasing threshold is to reduce the ohmic damping loss. Ohmic damping is drastically increased when the energy of light is near the interband transition energy of the metal, corresponding to violet or ultraviolet radiation (referred to as the bending-back effect) [33]. Despite the high metal loss, an ultraviolet plasmonic nanowire laser was demonstrated experimentally by using a closed-contact planar interface to reduce scattering and increase exciton-surface plasmon energy transfer efficiency. This in turn reduces the ohmic damping loss and the lasing threshold [36]. The reduced plasmon damping and scattering can also be attributed to the long length of the nanowire and surface smoothness resulting from improvements in fabrication technology and strategies, respectively (Fig. 5.7b). In this design, demonstrated by Zhang et al., a triangular cross-sectional nanowire is used to achieve the planar interface (Fig. 5.7a). Adjusting the thickness of the

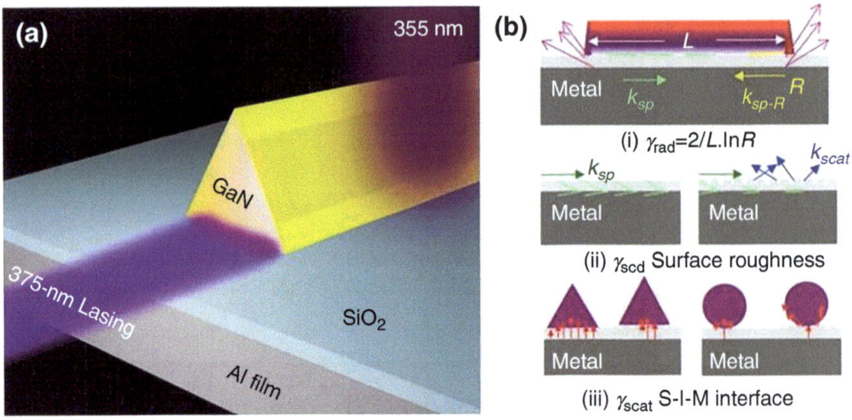

Fig. 5.7 (**a**) Schematic of a triangular cross-sectioned GaN nanowire on an Al film substrate with a SiO_2 passivation layer. (**b**) Schematic of surface plasmon damping via (i) radiative emission, (ii) scattering on a rough surface, and (iii) scattering on a semiconductor-insulator-metal interface. (Reprinted from Ref. [36])

insulator layer between the nanowire and metal is important for optimizing the distance between the generated excitons and surface plasmons. If the distance is too large, the efficiency of surface plasmon population inversion will be insufficient. On the other hand, if the distance is too small, there can be nonradiative effects such as electron transfer which will lower the gain of the active material. For this design, the optimal thickness was found to be 5–10 nm. It was reported that the laser was able to lase at 370 nm with an optical pumping threshold of 3.5 MW/cm^2 at room temperature. This threshold is a drastic improvement from other room-temperature visible wavelength plasmonic lasers.

Besides reducing ohmic loss, improving the gain material quality, such as increasing the gain coefficient and reducing the non-radiative recombination rates, is another approach to reduce the lasing threshold. An important attempt at creating a low threshold plasmonic nanowire laser was shown by Yu et al. [37]. Their design used a hybrid organic-inorganic perovskite gain material that requires less costly fabrication processes and was claimed to reduce the threshold of the device through the benefits of bandgap tunability and low non-radiative recombination rates. With a similar structure as the device from Zhang et al. [36], a 300 nm long and 150 nm wide nanowire is placed on a silver film with a 10 nm dielectric spacing material separating the two. The gain material was chosen to be $CH_3NH_3PbI_3$ and the dielectric material MgF_2. Large surface contact between the wire and the substrate was ensured to facilitate efficient exciton-plasmon coupling and reduced scattering. After experimentally testing many different devices of this same design, the lasing thresholds were consistently low with a range of 13.5–56.3 $\mu J/cm^2$.

In addition to nanowire plasmonic lasers, there are many other types of devices that make use of the plasmonic nature of metal clad cavities including nano-patch, nano-pan, and nanopillar lasers. As the line between a plasmonic laser and SPASER can sometimes be vague, the designs mentioned above can also be considered SPASERs. In addition to being the light source in chip-scale optical communication systems, plasmonic lasers can serve as sensors. Plasmonic laser sensor designs based on wavelength or intensity modulation have been realized in recent years, and a few examples of these devices are explored below.

Laser sensors based on wavelength sensing relies on the change of resonant wavelength in different environments. The resonant wavelength in plasmonic lasers is fixed with a given cavity dimension and solid gain material. However, if different gain material is dissolved in a fluid that is to be sensed, the change in the refractive indices of the fluids will lead to different lasing wavelengths. This is because a change in the refractive index corresponds to an altered dielectric screening effect. With this augmented screening effect, the restoring force is reduced causing the surface plasmon resonance frequency to shift. Yang et al. demonstrated successful static and dynamic wavelength sensing experimentally [38]. Their initial design was an array of gold nanoparticles placed on a polyurethane substrate (Fig. 5.8a–b). Next, the polyurethane was replaced with dielectric liquid solvents for static and dynamic wavelength tuning experiments (shown in Fig. 5.8c–d). The resonant wavelength's dependence on the dielectric liquid solvents suggests that this plasmonic laser can be used as a sensor.

In addition to changes in the emission wavelength due to local environment changes, sensing can also be accomplished through changes in the emission intensity. Ma et al. claim that intensity modulation is a more reliable measure for high sensitivity systems [39]. Their study was on explosives detection with plasmonic lasers and the device they proposed was a plasmonic cavity with a metal/dielectric/gain material structure (Fig. 5.9a). The semiconductor nanocrystal gain material was made to have a small thickness and an exaggerated length with the purpose of favoring the absorption of the analyte molecules. The small physical volume of the gain medium limits the number of carriers, which increases the sensitivity of lasing intensity to surface events. This structure was placed in a sealed chamber where there are two ports for gas exchange. Nitrogen (N_2) was used as a background reference to dilute the target chemical Dinitrotoluene (DNT). As the gas flows through the chamber, the target chemical was absorbed by the semiconductor surface which altered the lasing peak intensity, but with no observable change in the emission wavelength. As the gas flow alternates between pure nitrogen and diluted DNT, the emission intensity showed a clear evolution that demonstrates the reversibility of the process. Fig. 5.9b, c shows the design concept and the experimental results with varying concentrations of DNT.

Plasmonic devices are effective in biological sensing because of their ability to defeat the diffraction limit by supporting plasmonic modes at the deeply subwavelength scale. This ability is especially imperative for intracellular level sensing [40]. Generally, SPASERS are proposed for biological laser sensing due to their extremely confined optical mode, but the downside is their extremely high thresholds. With ongoing efforts dedicated to lowering the lasing threshold, the application of SPASERs will hopefully become more prominent.

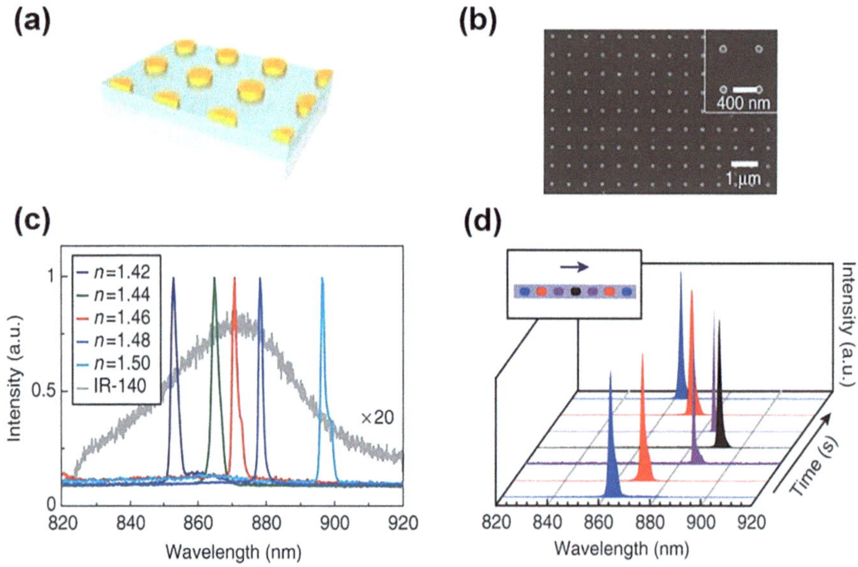

Fig. 5.8 (**a**) Gold nanoparticles placed on a substrate of polyurethane (in blue) supported by glass (in white). (**b**) Scanning electron microscope (SEM) image of Au nanoparticles. (**c**) Static tuning experimental results of emission intensity with different refractive index environments based on the replacement of dielectric solvent. (**d**) Dynamic tuning experimental results of emission intensity and lasing wavelength shifts. This is done by placing the nanoparticle array in a microfluidic device where the solvents can be introduced in real time, one after another. (Reprinted from Ref. [38])

To satisfy the requirements of a biological probe, the sensor must be biologically safe (non-toxic) to human cells and water-soluble. In a design used by Galanzha et al., a nanoparticle structure with a gold core surrounded by a silica shell is doped with disodium fluorescein (uranine) gain [41]. A certain combination of these materials allows the nanoparticle to be soluble in water and have low toxicity. With the nanoparticle diameter being 22 nm and 60 nm, the threshold was measured to be 20–30 mJ/cm^2 and 1–3 mJ/cm^2, respectively. An interesting phenomenon was found when pumping the SPASER above its lasing threshold. Due to the excessive heat generated during pumping, the liquid around the nanoparticle will evaporate and form a bubble enclosing the particle. This bubble leads to a drastic change in the dielectric polarizability of the SPASER. As the liquid is replaced by a pocket of air in the bubble, it results in an increased frequency of the plasmonic mode in the cavity. If the cavity is designed such that the increased frequency of the plasmonic mode is aligned with the peak of the material gain spectrum, an increased number of plasmons will result (shown in Fig. 5.10). This particular design was proposed not only as a biosensor, but also as a device that can destroy cancer cells. Accompanied by other devices previously mentioned, the developmental direction of plasmonic nanolasers and their advantages over traditional lasers are clearly seen.

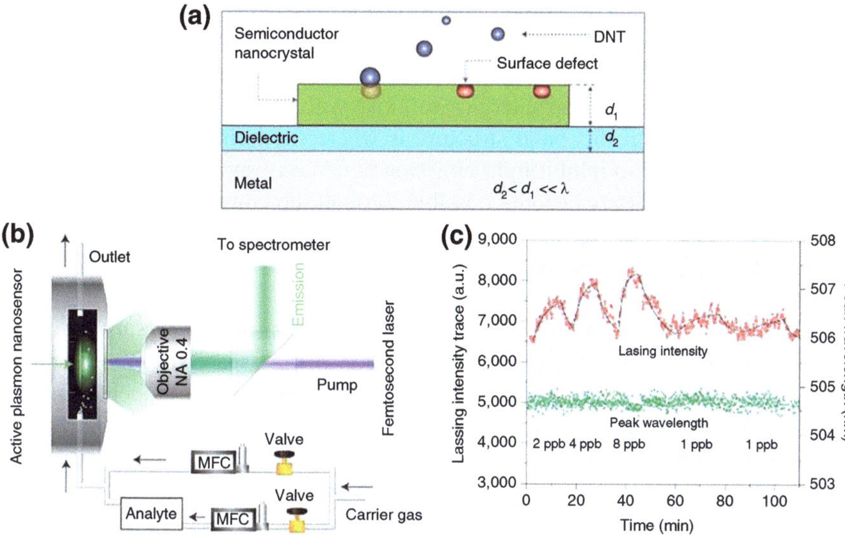

Fig. 5.9 (**a**) Metal/dielectric/gain structure demonstrating the semiconductor crystal acting as both the gain medium and sensing material. (**b**) Experimental setup of gas exchange, optical pumping, and measurement. (**c**) Experimental data showing the varying emission intensities of the sensor as a response to the different DNT concentrations in parts per billion (ppb). (Reprinted from Ref. [39])

Fig. 5.10 Light-in vs. light-out curve (red) with the "kink" signifying threshold and emission linewidth evolution (blue) in which a drastic decrease in linewidth occurs around the threshold. (Reprinted from Ref. [41])

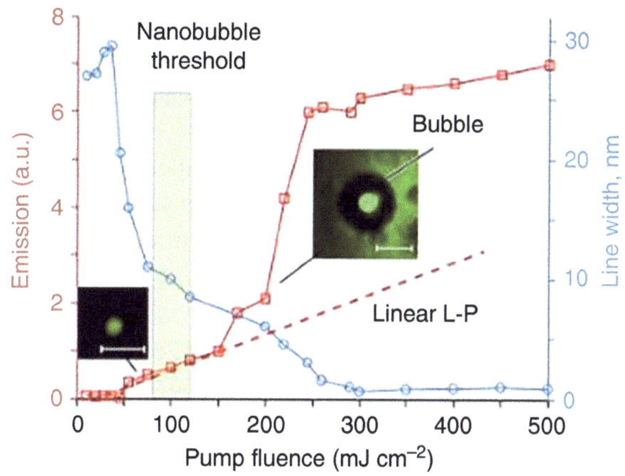

5.4 Photonic Crystal Lasers

Photonic crystal lasers exhibit unique characteristics that are unable to be achieved through traditional means. Much like how atoms form lattice structures with varying periodic formations, these crystals exhibit periodic changes in the refractive index by varying the media through which light travels. While most of these nanostructures are artificial in the field of photonic devices, photonic crystals can also be found in nature like in feathers on colorful birds. In photonic crystals, a photonic bandgap is formed similar to electronic bandgaps in solid state materials, allowing only certain photon energies to exist within the crystal. The photonic energy at the edges of this band structure is where the lasing mode occurs. This property can be used to inhibit light emission at certain wavelengths [21]. The so-called thresholdless lasers become possible through the combined use of quantum dots with photonic crystals that support only a few cavity modes [42]. Broad-area coherent lasers can also be realized by modifying the lattice structure of these crystals.

Thresholdless lasers take advantage of the prohibition of unnecessary modes of light and the restriction of spontaneous emission in free-space modes. When the laser supports only one mode, all spontaneous emission will be funneled into the cavity mode that would eventually become the lasing mode, making the typical transition between spontaneous and stimulated emission (threshold) nonexistent. Introducing a defect in the crystal and forcing emission directly into the defect resonance mode effectively creates a subwavelength mode volume that supports a single photonic mode. The choice of gain material is quantum dots because of their narrow gain spectra; therefore, it is easier for the gain spectrum to overlap with only one cavity resonance mode [43]. Reduction of nonradiative emission channels forces the energy into radiative modes, which in this case is only the defect resonance mode. Additionally, because of the low level of nonradiative recombination, quantum dots also have the advantage of becoming transparent with a small amount of excitation energy, which is important for keeping a high Q-factor as nonradiative recombination is a major loss mechanism. Practically, one needs to make sure the quantum dot bandgap energy is precisely matched with the cavity resonance energy to ensure the effects explained above. Interestingly and counter-intuitively, a phenomenon called the quantum anti-Zeno effect has been discovered that effectively relaxes this strict requirement [44]. In this case, despite a large mismatch between the quantum dot bandgap energy and cavity resonance, emission from the fundamental cavity mode is enhanced. The cause of the quantum anti-Zeno effect can be explained through energy broadening of the cavity due to surrounding electron systems and phonons. Energy transfer occurs through the tail-end of this cavity energy distribution and is then emitted into free-space. With this phenomenon, the realization of a thresholdless laser becomes increasingly practical due to the more tolerable energy mismatch.

Broad-area coherent lasers serve a different purpose than other lasers that have been discussed in previous sections. The goal, as the name suggests, is to produce coherent emission covering an extremely large area. Even with this unique purpose, the laser is still desired to have the properties of single frequency emission and have the smallest beam divergence possible. Other properties pursued by this type of laser include beam pattern control and polarization control. The operating principle is to construct standing waves resulting from Bragg reflection. As the forward and backward propagating waves travel through the varying refractive indices within the crystal, the two waves interfere and form a standing wave, thus creating a cavity.

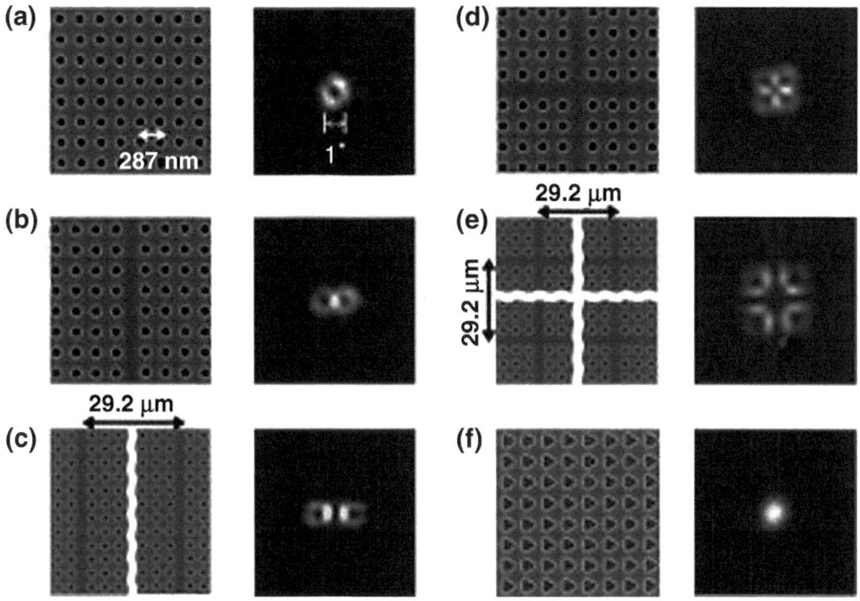

Fig. 5.11 SEM images of the fabricated photonic crystal lattices (left) and their corresponding beam patterns (right). (**a**) Lattice period with no shift, (**b**) One lattice shift in one direction. (**c**) Two lattice shifts in one direction. (**d**) one lattice shift in both directions. (**e**) Two lattice shifts in both directions. (**f**) Triangular lattice points with no phase shift. (Reprinted from Ref. [42])

The dimensions of a photonic crystal lattice correspond to how many directions are periodic with respect to the refractive index. In a 1D photonic crystal, only one direction has a periodic structure, allowing light to freely propagate in the other two spatial dimensions. In a 2D crystal, light can only travel in one direction without boundary conditions. For these 2D crystals, the standing waves can cover the entire surface of the plane [45]. Because the third dimension is not periodically nanostructured, light can be diffracted perpendicular to the plane of the crystal, allowing for a surface-emitting characteristic, they are, therefore, termed photonic crystal surface-emitting lasers (PCSELs). This is important for enabling the control of the beam pattern and polarization, as varying the shape and spacing of the lattice structure alters the field distribution, which results in a different beam shape and/or polarization. Some examples of this effect are shown in Fig. 5.11 [42]. In this square lattice nanohole design, periodically patterned circular lattice points produce a doughnut-shaped beam (Fig. 5.11a). Introducing shifts in the lattice produces higher orders of doughnut-shaped beams as a result of interference among different photonic crystal sections (Fig. 5.11b–e). Lastly, changing the circular lattice points to triangular lattice points changes the doughnut-shaped beam profile to a solid circular one (Fig. 5.11f).

A typical PCSEL structure is shown in Fig. 5.12 [46]. A photonic crystal is positioned in between cladding layers, with electrodes placed on the top and bottom.

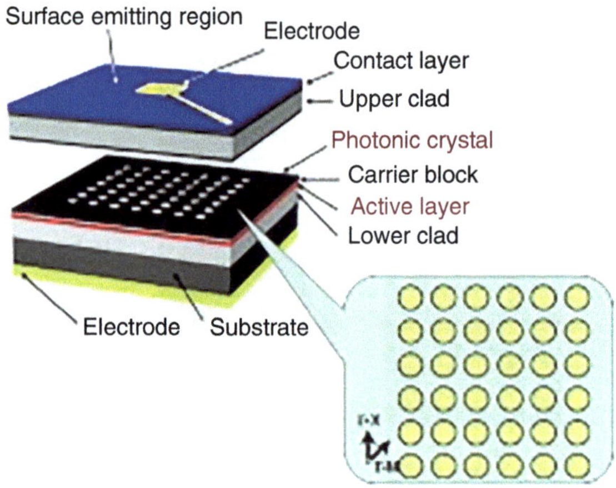

Fig. 5.12 Schematic of a photonic crystal surface emitting laser (PCSEL). The photonic crystal is embedded near the active layer with an electron blocking medium in between. The light emission is through the plane at the top of the device. (Reprinted from Ref. [46])

The gain medium is placed under the crystal, allowing for the band-edge modes to evanescently couple to the active layer. Lasing occurs and emits radiation vertically from the top of the structure, illuminating a very large output area. While a square lattice is used in this device, other lattice structures such as triangular and octagonal lattices are also often used.

Kunishi et al. showed that an advantage of using triangular lattice points, such as those in Fig. 5.11f, is an increase in output power [47]. The emission pattern from symmetric lattice points will result in a cancellation of fields in the center leaving a doughnut shape. In order to get rid of the destructive interference and increase the Q-factor, an asymmetric lattice must be implemented which is exemplified with the triangular pattern. A visual demonstration can be found in Fig. 5.13a. In their design, triangular air rods were formed through electron beam lithography and reactive ion etching. Continuous wave lasing at room temperature was achieved, and the output power of 46 mW was the highest reported value for a single-mode surface-emitting laser. The output power curve with respect to injection current is shown in Fig. 5.13b.

A few years later, Sakaguchi et al. found a different way to manipulate the Q-factor and output power of a photonic crystal laser without forming triangular lattice points [48]. The principle behind this design was to control the phase difference between the upward and downward emitted light by implementing a distributed Bragg reflector at the bottom of the device. This can be tuned for constructive interference which maximizes the output power, thus minimizing the Q factor. In this chapter, the results showed a peak output power of, a very impressive, >700 mW can be achieved (shown in Fig. 5.14).

Another important area of application for photonic crystals is large-scale integration for optical chip-scale communication and memory. With the growing demand to reduce energy consumption and heat generation, photonics can offer a solution through nano-cavities and waveguides, which can exploit the WDM capabilities of light-based computing. It has been demonstrated that a serial integrated optical

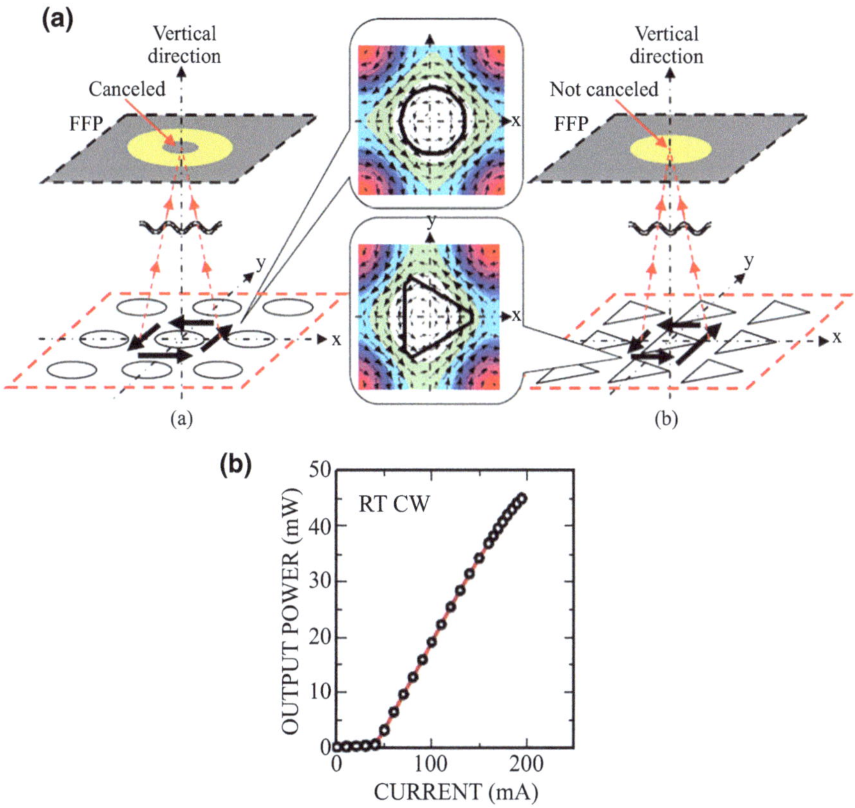

Fig. 5.13 (**a**) Schematic of the expected electromagnetic field distribution and far-field pattern with circular and triangular lattice points. (**b**) Experimental output power versus current of the device. (Reprinted from Ref. [47])

Fig. 5.14 Experimental output power versus current of the distributed Bragg reflector laser. (Reprinted from Ref. [48])

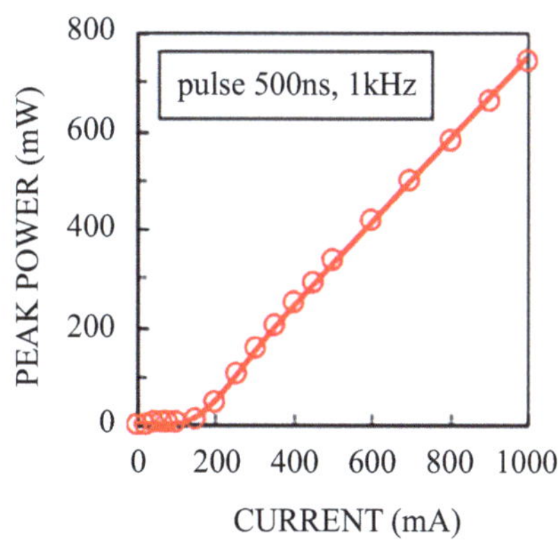

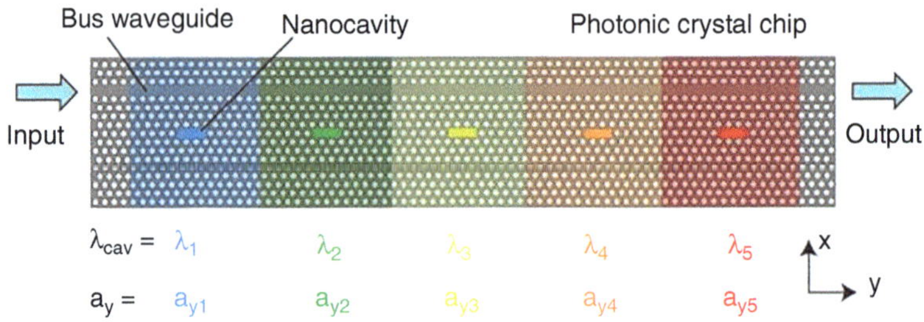

Fig. 5.15 Proposed design of a nanocavity-based optical RAM system on a photonic crystal chip. (Reprinted from Ref. [49])

random access memory system can be realized efficiently on a photonic crystal chip [49]. Many previous large-scale optical integration ideas have been proposed but have not been compatible with on-chip configuration; thus, it is significant and unique that these chips can host high-Q nano-cavity and efficient waveguide designs based on their strong optical confinement and tunability. High contrast switching for memory operations requires sufficiently high energy to be stored in the cavity to properly distinguish between memory addresses and large enough free spectral range so that there is no spectral overlap between electromagnetic modes. However, there is a consequence of the increased free spectral range: it results in limited bandwidth of the system. Optimization of these aspects makes the design a complicated process and therefore needs further investigation and improvements beyond the results shown in the following paper.

Kuramochi et al. demonstrate the serial integration using a common bus waveguide coupled to separate nano-cavities supporting different resonant wavelengths (a visual is shown in Fig. 5.15) [49]. By varying the lattice constant, the resonant wavelength can be tuned for each cavity. They demonstrated a working 28-bit and 105-bit memory system that can operate as RAM. Two photonic crystals, one on the Si platform and one on the InP platform, were measured and compared. In particular, the photonic crystal on the InP platform is a buried-heterostructure crystal in which InGaAsP is placed inside the cavities to act as an optically nonlinear medium. The advantage of this design is its ability for efficient carrier confinement and enhanced heat escape. Through the experiments, the InGaAsP/InP photonic crystal was found to require less power for operation. However, it requires a much more complex fabrication process. As such, the Si photonic crystal was seen as the better choice for the present technology, as the resonant wavelengths can be more precisely tuned and the resonant mode had a narrower spectral width. As fabrication technology continues to improve, the buried-heterostructure photonic crystal will eventually offer a more power efficient system.

Addressing the need for improvements in fabrication technology, a paper by Asano et al. describes an important discovery that led to a record setting Q-factor for photonic crystal cavities [50]. For Si, it was found that there are some discrepancies with the natural oxidation processes when creating the interface between Si and silicon dioxide. These discrepancies lead to significant fluctuations in the Q-factors for seemingly identical cavities. The so-called *cleaning* of the bottom side of the Si nanostructures was a solution to creating a controlled, predictable, and record high Q-factor. The *cleaning* refers to the process of controlled surface oxidation and removal that is repeated several times to ensure low fluctuations in radii and positions of air holes. Removal of the oxide was done through dipping the cavity in an HF solution. It was claimed that after four repetitions, the Q-factor was consistent and has a record high value of eleven million. This strategy opens the door for advancements in fabrication and thus the ability to realize more efficient devices and systems, especially in the realm of photonic crystals.

5.5 Metamaterials

Photonic metamaterials are defined as an artificial media made up of two or more materials in a periodically varying structure, with a period (lattice constant) smaller than the light's wavelength. At first, this may be mistaken for a photonic crystal, but the key difference between them is the lattice constant. The lattice constant must be much smaller than the wavelength of light propagating through the structure for it to behave as a metamaterial [51]. With this property, light experiences the effective parameters of the media including electric permittivity and magnetic permeability resulting in a single homogeneous refractive index. Namely, the periodic nanostructure can be considered a homogeneous media (a.k.a. a material) by averaging over the whole atomic scale [48]. With this ability, metamaterials can act as a "material," enabling the realization of properties that are not found in nature and were previously thought of as only theoretical phenomena – hence the name "meta" material. This is in stark contrast to photonic crystals in which the propagating light experiences varying refractive indices throughout the media.

Metamaterials originated from research in negative-index media. Negative refractive indices arise from the modification of both the electric permittivity and magnetic permeability parameters when they are less than zero for a certain frequency band of light and a certain propagation direction [52]. With a negative index, the wave refracts off the surface normal to the material on the same side as the incident light, as shown in Fig. 5.16a. An important application of this property is subwavelength imaging. With the biggest limitation for imaging being the diffraction limit, dealing with an object that is smaller than the wavelength of light being used makes it near-impossible to resolve the object's features. When using a metamaterial with a negative index of refraction, the waves can be focused back to the central axis, allowing for a perfect image of the subject to be copied from one side of the material to the other (Fig. 5.16b). By using a host of plasmon-like modes that occur at the surfaces of these materials with negative effective parameters, the higher order rapidly decaying parts of the image can be amplified and focused to the center axis conclusively allowing subwavelength imaging [53].

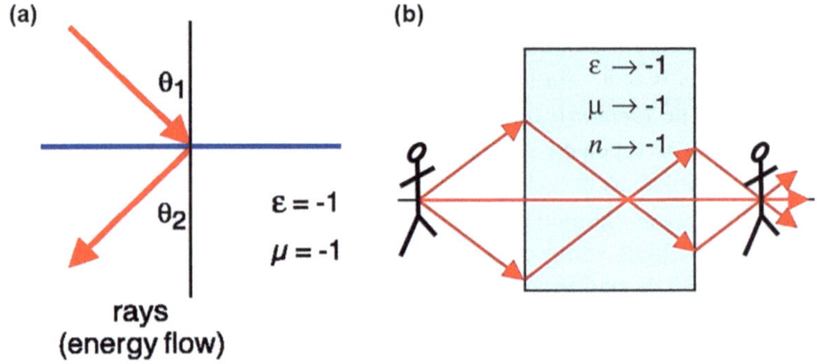

Fig. 5.16 (**a**) Demonstration of how a light ray, incident on a negative index material, is refracted in the opposite way relative to the surface normal. (**b**) Far field translation of an object into a perfect image due to a negative index. (Reprinted from Ref. [53])

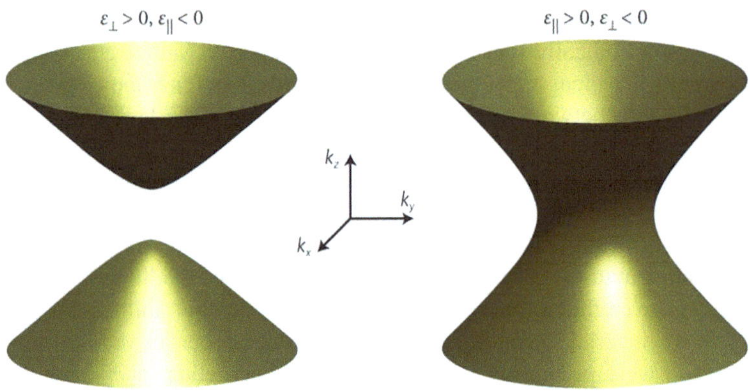

Fig. 5.17 Iso-frequency surfaces of extraordinary waves in hyperbolic metamaterials. (Reprinted from Ref. [54])

These super lenses and devices based on this principle were heavily focused on in the early 2000s.

Another emerging research area in recent years has been hyperbolic metamaterials (HMMs), which is a special class of a highly anisotropic media in terms of the effective parameters, permittivity and permeability. The hyperbolic nature refers to the shape of the iso-frequency contour (Fig. 5.17) [54]. The iso-frequency contour is a way of determining the allowed wave vectors in the material. Based on the contour demonstrating an indefinite dispersion (hyperbolic dispersion), it can be concluded that there is an essentially infinite optical density of states [55]. This arises from opposite signs of permittivity for one or two of the three principal axes (representing x, y, and z directions). The free carrier motion below the plasma frequency is what gives metals its negative permittivity; therefore, restricting this flow would change the sign from negative to positive [54]. HMMs are constructed through a

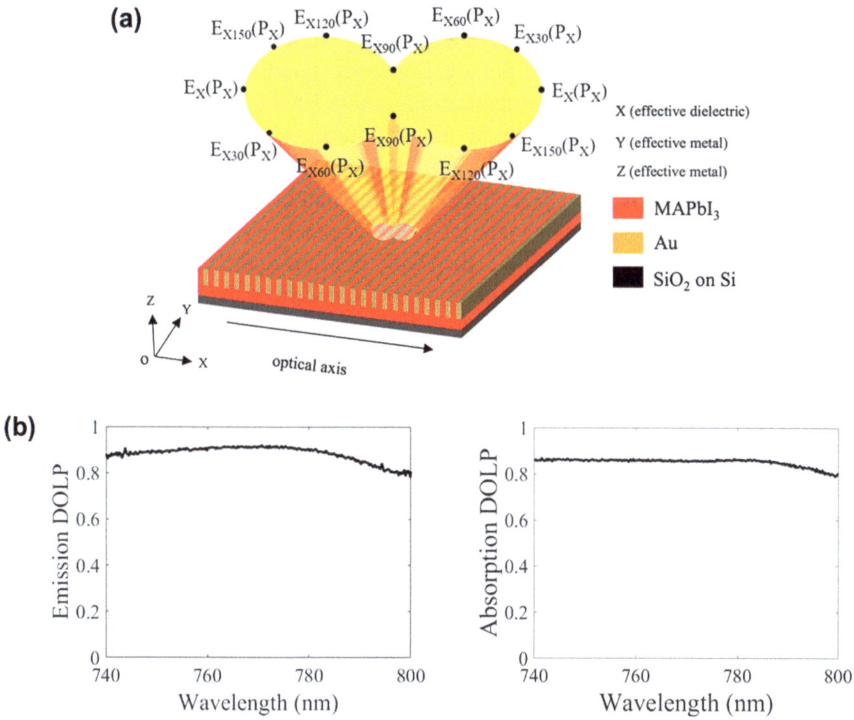

Fig. 5.18 (**a**) Schematic of an active HMM with the emission beam pattern that shows the angle-dependent emission characteristics. (**b**) Spectrum of the emission and absorption degree of linear polarization (DOLP). (Reprinted from Ref. [56])

metal-dielectric layered structure. The ability to manipulate the free carrier motion in certain directions for frequencies below the plasma frequency enables negative permittivity values in only specific directions. This results in the opposite signs in the effective parameter tensors. Not only can this be used for negative refractive index materials and subwavelength imaging, but enhanced spontaneous emission rates can also be realized through HMMs.

An inherent problem of HMMs is the ohmic loss due to the metal constituent. This motivated the idea to add a gain medium in HMM to compensate for such losses, which is essential for achieving the desired large wave vector values as well as integration with light sources. These HMMs are referred to as Active HMMs. Li et al. demonstrated the first active HMM with periodicity below 1 μm with a metal-dielectric multilayer on the Si platform, in which the dielectric constituent is composed entirely of $MAPbI_3$ perovskite gain material (Fig. 5.18a) [56]. Silver was chosen as the metal constituent due to its low loss in the visible spectrum compared with other metals. To confirm hyperbolic behavior from the HMM, polarization anisotropy is measured. Based on the polarization angle, the light will experience light-matter interactions with a certain metal/dielectric ratio. Quantifying the strength of anisotropy, the magnitude of hyperbolicity was measured through the absorption and emission degree of linear polarization (DOLP), as shown in Fig. 5.18b, where a higher degree of polarization corresponds to a stronger anisotropy. Note that in addition to loss compensation, active HMMs can be used in on-chip light source designs due to their highly linear polarized luminescent properties.

In order to find application in light emitting devices, HMMs must exhibit benefits in emission characteristics. Through the enhancement of the Purcell Factor, HMMs demonstrate such an effect. In a material, the optical density of states is proportional to area of the iso-frequency surface, therefore alluding to the fact that there can be an infinite number of states due to the hyperbolic nature of the dispersion. Relating the density of states to the Purcell effect, it would seem like it is possible to have an infinite Purcell Factor. This is obviously not realistic. The reason for this lies in the fact that the larger wave vectors allowed in the material would approach a limit to where the wave vector no longer fits within the Brillouin Zone and the optical density of states is nonexistent [57]. The Purcell Factor is also limited by the coupling problems between the photon source and the HMM where large wave vectors are unable to propagate and are classified as evanescent fields, scaling inversely with the distance between the source and HMM [58]. Even with these limitations, the Purcell Factor can be greatly increased.

Lu et al. claims that the Purcell enhancement is a result of the readily available high density of plasmonic states that are introduced through the metal layers in the HMM [59]. Using a polymethyl methacrylate (PMMA) material with fluorescent molecules randomly distributed, an HMM composed of silver and silicon is placed on the PMMA to improve the emission characteristics. Through simulation, they show that there is a large Purcell enhancement over a much wider band of wavelengths than that from traditional plasmonic nanocavities. This is attributed to the narrow band nature of plasmonic resonance frequencies. In contrast, HMMs provide the ability for enhancement of many different wave vectors due to the large plasmonic density of states. Another important finding was that different metal filling ratios of silver were determined to change the wavelengths at which the Purcell Factor is maximized, showing the capability for HMMs to have tunability with respect to the emission spectrum. An important aspect of light emission in addition to the Purcell Factor is the radiative enhancement. An inherent problem of HMM devices is outcoupling large wave vector components which is a significant portion of the energy emitted by the source. Lu et al. implemented a grating coupler atop the HMM to enhance the outcoupling of the high wave vector components of energy in contrast to the uniform HMM structure (the comparison is shown in Fig. 5.19). Through experimental testing, the grating period of the nanopatterned HMM was shown to vary the magnitude of which the radiation is enhanced. A few years later, the same group replicated this type of experiment by using nanopatterned HMMs with InGaN quantum wells [60]. This shows again that the smaller the pattern the better the outcoupling of the high wave vector modes, resulting in enhanced emission intensity and Purcell Factor. As these results suggest, HMMs show the potential to have a significant impact on light-emitting devices.

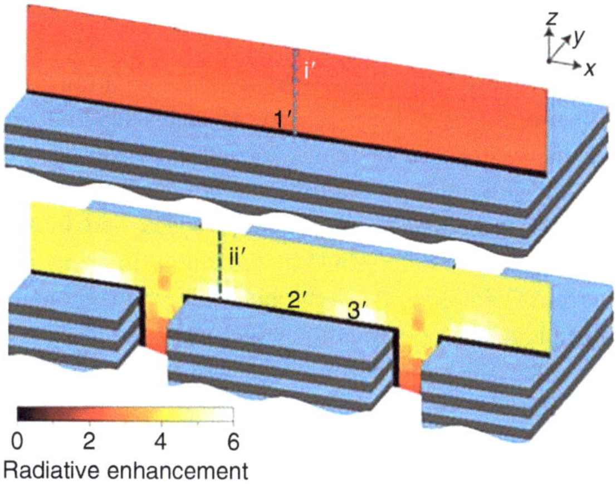

Fig. 5.19 Cross-sectional mapping of the radiative enhancement for uniform and nanopatterned multilayer HMMs at 600 nm. (Reprinted from Ref. [59])

Looking into the future of HMM and its integration into novel light emitting devices, it does not seem practical to fabricate nano-gratings with such small dimensions to improve high wave vector outcoupling. The inherent issue of outcoupling stems from the momentum mismatch between the HMM and surrounding media. In wave vector space, the grating provides a way of bridging the iso-frequency curves of air and the HMM by adding an extra vector with small enough periodicity to intersect from the air to the HMM. An alternate method is shown by Ferrari et al. to alter the HMM iso-frequency contour in such a way that it overlaps with that of free space, thus sharing allowed wave vectors [61]. This was proposed through dispersion engineering by modifying the metal filling ratio to change the dispersion characteristics. With proper outcoupling, HMMs have a promising future for integration with photonic devices through their ability to enhance the spontaneous emission rate and increase emission intensities.

The future of nanoscale light sources depends not only on the individual growth of the areas covered in this chapter but also how they can be integrated together with existing technology. In this chapter we focused on three main factors for light sources: (1) thresholds, (2) scalability, and (3) modulation speeds. Lasers and LEDs show promising results through the use of plasmonics, photonic crystals and metamaterials. Each exploiting characteristics that cannot be attained with traditional materials and structures. With plasmonics, the electromagnetic modes can be confined below the diffraction limit allowing for extreme scalability. The small size of such lasers was shown to be particularly beneficial for plasmonic sensing. In addition, the small size increases the practicality of integration with the current electronic components needed for a system level design. The current work in this field aims to reduce the lasing threshold and manage the inevitable ohmic loss. Photonic crystals utilize photonic bandgaps to restrict unnecessary modes of light, effectively eliminating the threshold conditions for lasers. This also allows for the manipulation of photonic lattices to drastically enhance the emission rate, thus the modulation speed of the device. As a bonus, these crystals open up the possibilities for efficient mode selection for on-chip optical systems such as wavelength-division-multiplexing. With the strict requirement of wavelength comparable lattice sizes, scalability remains an area of investigation going forward. Metamaterials have a theoretically infinite density of states leading to ultra-high modulation speeds for the realization of femto-second emission devices. The main area of focus for future integration is the difficulty in fabrication in integrated systems when large quantities are need. In the next decade and beyond, the direction of nanoscale devices will be determined by the ability to bring these different areas together with current technology through ohmic loss management, scalability improvements, and advancements in fabrication technology. As photonics and material science evolves, the integration of such devices will become more practical and compatible with mature industrial technology and processes for the ultimate goal of commercializing photonic integrated circuits and optoelectronic systems.

References

1. G.E. Moore, Cramming more components onto integrated circuits, Reprinted from Electronics, volume 38, number 8, April 19, 1965, pp.114 ff. IEEE Solid-State Circuits Soc. Newsl. **11**(3), 33–35 (2009). https://doi.org/10.1109/N-SSC.2006.4785860
2. B. Romeira, A. Fiore, Physical limits of nanoLEDs and nanolasers for optical communications. Proc. IEEE **108**(5), 735–748 (2020). https://doi.org/10.1109/JPROC.2019.2912293
3. L. Thylén, S. He, L. Wosinski, D. Dai, The Moore's Law for photonic integrated circuits. J. Zhejiang Univ. Sci. A **7**(12), 1961–1967 (2006). https://doi.org/10.1631/JZUS.2006.A1961
4. M. Smit, J. van der Tol, M. Hill, Moore's law in photonics. Laser Photonics Rev. **6**(1), 1–13 (2012). https://doi.org/10.1002/LPOR.201100001
5. L.A. Coldren, S.W. Corzine, M.L. Mašanović, *Diode Lasers and Photonic Integrated Circuits* (Wiley, Hoboken, 2012). https://doi.org/10.1002/9781118148167
6. (16) (PDF) RUBY LASER. https://www.researchgate.net/publication/360611681_RUBY_LASER. Accessed 6 Sept 2022
7. A. Javan, W.R. Bennett, D.R. Herriott, Population inversion and continuous optical maser oscillation in a gas discharge containing a He-Ne mixture. Phys. Rev. Lett. **6**(3), 106–110 (1961). https://doi.org/10.1103/PHYSREVLETT.6.106/FIGURE/1/THUMB
8. A History of the Laser: 1960–2019 | Features | June 2019 | Photonics Spectra. https://www.photonics.com/Articles/A_History_of_the_Laser_1960_-_2019/a42279. Accessed 11 Oct 2022
9. P. Moulton, Ti-doped sapphire: tunable solid-state laser. Optics News **8**(6), 9–9 (1982). https://doi.org/10.1364/ON.8.6.000009
10. Y. Arakawa, A. Yariv, Quantum well lasers – Gain, spectra, dynamics. IEEE J. Quantum Electron. **22**(9), 1887–1899 (1986). https://doi.org/10.1109/JQE.1986.1073185
11. D. Vandyshev, D.K. Okorokov, V. Yu, Lasing on a transition between quantum-well levels in a quantum dot. JETP Lett. **54**(8), 441–444 (1991) [Online]. Available: https://www.researchgate.net/publication/253590034
12. A. Maslov, C.A. Ning, Size reduction of a semiconductor nanowire laser by using metal coating. Phys. Simul. Optoelectron. Devices XV **6468**, 138–144 (2007). https://doi.org/10.1117/12.723786
13. A.V. Maslov, C.Z. Ning, Metal-encased semiconductor nanowires as waveguides for ultrasmall lasers. Conf. Quant. Electr. Laser Sci. (QELS) Tech. Dig. Ser. (2007). https://doi.org/10.1109/QELS.2007.4431377

14. B.L. Murillo-Borjas, X. Li, Q. Gu, High-speed nanoLEDs for chip-scale communication. Nano Commun. Netw. **30**, 100376 (2021). https://doi.org/10.1016/J.NANCOM.2021.100376
15. J.B. Khurgin, How to deal with the loss in plasmonics and metamaterials. Nat. Nanotechnol. **10**(1), 2–6 (2015). https://doi.org/10.1038/nnano.2014.310
16. E. Gaubas, E. Simoen, A. Haugi, Auger recombination in direct-gap semiconductors: Band-structure effects. J. Phys. C Solid State Phys. **16**(21), 4159 (1983). https://doi.org/10.1088/0022-3719/16/21/017
17. D.J. Fitzgerald, A.S. Grove, Surface recombination in semiconductors. Surf. Sci. **9**(2), 347–369 (1968). https://doi.org/10.1016/0039-6028(68)90182-9
18. J.-M. Gerard, B. Gayral, Strong purcell effect for InAs quantum boxes in three-dimensional solid-state microcavities. J. Lightwave Technol. **17**(11) (2089, 1999) [Online]. Available: http://opg.optica.org/jlt/abstract.cfm?URI=jlt-17-11-2089
19. E.M. Purcell, Spontaneous emission probabilities at radio frequencies, in *Confined Electrons and Photons: New Physics and Applications*, ed. by E. Burstein, C. Weisbuch, (Springer, Boston, 1995), p. 839. https://doi.org/10.1007/978-1-4615-1963-8_40
20. S.I. Azzam et al., Ten years of spasers and plasmonic nanolasers. Light Sci. Appl. **9**(1), 1–21 (2020). https://doi.org/10.1038/s41377-020-0319-7
21. J.D. Joannopoulos, S.G. Johnson, J.N. Winn, R.D. Meade, *Photonic Crystals: Molding the Flow of Light* (Princeton University Press, Princeton, 2011). https://doi.org/10.2307/J.CTVCM4GZ9
22. P. Shekhar, J. Atkinson, Z. Jacob, Hyperbolic metamaterials: Fundamentals and applications. Nano Converg. **1**(1), 1–17 (2014). https://doi.org/10.1186/S40580-014-0014-6/FIGURES/7
23. I.A. Kolmychek et al., Optical properties of hyperbolic metamaterials (brief review). JETP Lett. **114**(11), 653–664 (2021). https://doi.org/10.1134/S0021364021230089/FIGURES/9
24. D.A.B. Miller, Rationale and challenges for optical interconnects to electronic chips. Proc. IEEE **88**(6), 728–749 (2000). https://doi.org/10.1109/5.867687
25. J.J. Zayhowski, Limits imposed by spatial hole burning on the single-mode operation of standing-wave laser cavities. Opt. Lett. **15**(8), 431–433 (1990). https://doi.org/10.1364/OL.15.000431
26. C.H. Brito Cruz, J.P. Gordon, P.C. Becker, R.L. Fork, C.V. Shank, Dynamics of spectral hole burning. IEEE J. Quantum Electron. **24**(2), 261–269 (1988). https://doi.org/10.1109/3.122
27. X. Li, Q. Gu, Ultrafast shifted-core coaxial nano-emitter. Opt. Express **26**(12), 15177–15185 (2018). https://doi.org/10.1364/OE.26.015177
28. E.K. Lau, A. Lakhani, R.S. Tucker, M.C. Wu, Enhanced modulation bandwidth of nanocavity light emitting devices. Opt. Express **17**(10), 7790–7799 (2009). https://doi.org/10.1364/OE.17.007790
29. T.B. Hoang, G.M. Akselrod, C. Argyropoulos, J. Huang, D.R. Smith, M.H. Mikkelsen, Ultrafast spontaneous emission source using plasmonic nanoantennas. Nat. Commun. **6**(1), 7788 (2015). https://doi.org/10.1038/ncomms8788
30. N.M. Andrade, S. Hooten, S.A. Fortuna, K. Han, E. Yablonovitch, M.C. Wu, Inverse design optimization for efficient coupling of an electrically injected optical antenna-LED to a single-mode waveguide. Opt. Express **27**(14), 19802–19814 (2019). https://doi.org/10.1364/OE.27.019802
31. V. Dolores-Calzadilla et al., Waveguide-coupled nanopillar metal-cavity light-emitting diodes on silicon. Nat. Commun. **8**(1), 14323 (2017). https://doi.org/10.1038/ncomms14323
32. A. Kumar, P. Tyagi, R. Srivastava, D.S. Mehta, M.N. Kamalasanan, Energy transfer process between exciton and surface plasmon: Complete transition from Forster to surface energy transfer. Appl. Phys. Lett. **102**(20), 203304 (2013). https://doi.org/10.1063/1.4806979
33. K. Kolwas, A. Derkachova, Impact of the interband transitions in gold and silver on the dynamics of propagating and localized surface plasmons. Nanomaterials **10**(7), 1411 (2020). https://doi.org/10.3390/NANO10071411
34. H. Raether, *Surface Plasmons on Smooth Surfaces* (Springer, Berlin, 1988), pp. 4–39. https://doi.org/10.1007/BFB0048319

35. C.Z. Ning, Spaser or plasmonic nanolaser?-Reminiscences of discussions and arguments with Mark Stockman. Nanophotonics **10**(14), 3619–3622 (2021). https://doi.org/10.1515/NANOPH-2021-0291/MACHINEREADABLECITATION/RIS
36. Q. Zhang et al., A room temperature low-threshold ultraviolet plasmonic nanolaser. Nat. Commun. **5**(1), 1–9 (2014). https://doi.org/10.1038/ncomms5953
37. H. Yu et al., Organic–inorganic perovskite plasmonic nanowire lasers with a low threshold and a good thermal stability. Nanoscale **8**(47), 19536–19540 (2016). https://doi.org/10.1039/C6NR06891J
38. A. Yang et al., Real-time tunable lasing from plasmonic nanocavity arrays. Nat. Commun. **6**(1), 1–7 (2015). https://doi.org/10.1038/ncomms7939
39. R.M. Ma, S. Ota, Y. Li, S. Yang, X. Zhang, Explosives detection in a lasing plasmon nanocavity. Nat. Nanotechnol. **9**(8), 600–604 (2014). https://doi.org/10.1038/NNANO.2014.135
40. Y.C. Chen, X. Fan, Biological lasers for biomedical applications. Adv. Optic. Mater. **7**(17), 1900377 (2019). https://doi.org/10.1002/ADOM.201900377
41. E.I. Galanzha et al., Spaser as a biological probe. Nat. Commun. **8**(1), 1–7 (2017). https://doi.org/10.1038/ncomms15528
42. S. Noda, Photonic crystal lasers – Ultimate nanolasers and broad-area coherent lasers [Invited]. JOSA B **27**(11), B1–B8 (2010). https://doi.org/10.1364/JOSAB.27.0000B1
43. K. Kounoike, M. Yamaguchi, M. Fujita, T. Asano, J. Nakanishi, S. Noda, Investigation of spontaneous emission from quantum dots embedded in two-dimensional photonic-crystal slab. Electron. Lett. **41**(25), 57–58 (2005). https://doi.org/10.1049/EL:20053011
44. M. Yamaguchi et al., Photon emission by nanocavity-enhanced quantum anti-Zeno effect in solid-state cavity quantum-electrodynamics. *Opt. Express*__16__(22), 18067–18081 (2008). https:/ /doi.org/10.1364/OE.16.018067
45. M. Imada, A. Chutinan, S. Noda, M. Mochizuki, Multidirectionally distributed feedback photonic crystal lasers. Phys. Rev. B **65**(19), 195306 (2002). https://doi.org/10.1103/PhysRevB.65.195306
46. S. Noda, K. Kitamura, T. Okino, D. Yasuda, Y. Tanaka, Photonic-crystal surface-emitting lasers: Review and introduction of modulated-photonic crystals. IEEE J. Select. Topics Quant. Electr. **23**(6) (2017). https://doi.org/10.1109/JSTQE.2017.2696883
47. T. Sakaguchi, et al., *Surface-Emitting Photonic-Crystal Laser with 35W Peak Power*. Conference on Lasers and Electro-Optics/International Quantum Electronics Conference (2009), paper CTuH1, p. CTuH1 (2009). https://doi.org/10.1364/CLEO.2009.CTUH1
48. High-Power Single-Lobed Surface-Emitting Photonic-Crystal Laser. https://opg.optica.org/abstract.cfm?uri=cleo-2006-CMKK1. Accessed 25 Aug 2022
49. E. Kuramochi et al., Large-scale integration of wavelength-addressable all-optical memories on a photonic crystal chip. Nat. Photonics **8**(6), 474–481 (2014). https://doi.org/10.1038/nphoton.2014.93
50. T. Asano et al., Photonic crystal nanocavity with a Q factor exceeding eleven million. Opt. Express ***25***(3), 1769–1777 (2017). https://doi.org/10.1364/OE.25.001769
51. (13) (PDF) Metamaterials and photonic crystals – potential applications for self-organized eutectic micro- and nanostructures. https://www.researchgate.net/publication/242726283_Metamaterials_and_photonic_crystals_-_potential_applications_for_self-organized_eutectic_micro-_and_nanostructures. Accessed 25 Aug 2022
52. Y. Liu, X. Zhang, Metamaterials: A new frontier of science and technology. Chem. Soc. Rev. **40**(5), 2494–2507 (2011). https://doi.org/10.1039/C0CS00184H
53. D.R. Smith, J.B. Pendry, M.C.K. Wiltshire, Metamaterials and negative refractive index. Science (1979) **305**(5685), 788–792 (2004). https://doi.org/10.1126/SCIENCE.1096796/ASSET/AE0DF28E-3FA8-42B0-9269-AFAD6619A0AD/ASSETS/GRAPHIC/ZSE0310427410004.JPEG
54. A. Poddubny, I. Iorsh, P. Belov, Y. Kivshar, Hyperbolic metamaterials. Nat. Photonics **7**(12), 948–957 (2013). https://doi.org/10.1038/nphoton.2013.243
55. G. Palermo, K.V. Sreekanth, G. Strangi, Hyperbolic dispersion metamaterials and metasurfaces. EPJ Appl. Metamater. **7**, 11 (2020). https://doi.org/10.1051/EPJAM/2020015

56. Z. Li et al., Active perovskite hyperbolic metasurface. ACS Photonics **7**(7), 1754–1761 (2020). https://doi.org/10.1021/ACSPHOTONICS.0C00391/ASSET/IMAGES/LARGE/PH0C00391_0004.JPEG
57. A.N. Poddubny, P.A. Belov, P. Ginzburg, A.V. Zayats, Y.S. Kivshar, Microscopic model of Purcell enhancement in hyperbolic metamaterials. Phys. Rev. B. Condens. Matter. Mater. Phys. **86**(3), 035148 (2012). https://doi.org/10.1103/PHYSREVB.86.035148/FIGURES/11/MEDIUM
58. O. Kidwai, S.V. Zhukovsky, J.E. Sipe, Dipole radiation near hyperbolic metamaterials: Applicability of effective-medium approximation. Opt. Lett. **36**(13), 2530–2532 (2011). https://doi.org/10.1364/OL.36.002530
59. D. Lu, J.J. Kan, E.E. Fullerton, Z. Liu, Enhancing spontaneous emission rates of molecules using nanopatterned multilayer hyperbolic metamaterials. Nat. Nanotechnol. **9**(1), 48–53 (2014). https://doi.org/10.1038/nnano.2013.276
60. D. Lu et al., Nanostructuring multilayer hyperbolic metamaterials for ultrafast and bright green InGaN quantum wells. Adv. Mater. **30**(15), 1706411 (2018). https://doi.org/10.1002/ADMA.201706411
61. L. Ferrari, J.S.T. Smalley, Y. Fainman, Z. Liu, Hyperbolic metamaterials for dispersion-assisted directional light emission. Nanoscale **9**(26), 9034–9048 (2017). https://doi.org/10.1039/C7NR00980A

Index

R. Gordon (ed.), *Advances in Near-Field Optics*, Springer Series in Optical Sciences 244, https://doi.org/10.1007/978-3-031-34742-9

MIX
Papier aus verantwortungsvollen Quellen
Paper from responsible sources
FSC® C105338

If you have any concerns about our products,
you can contact us on
ProductSafety@springernature.com

In case Publisher is established outside the EU,
the EU authorized representative is:
Springer Nature Customer Service Center GmbH
Europaplatz 3, 69115 Heidelberg, Germany

Printed by Libri Plureos GmbH
in Hamburg, Germany